Animal Learning
Survey and Analysis

NATO ADVANCED STUDY INSTITUTES SERIES

A series of edited volumes comprising multifaceted studies of contemporary scientific issues by some of the best scientific minds in the world, assembled in cooperation with NATO Scientific Affairs Division.

Series A: Life Sciences

Recent Volumes in this Series

The series is published by an international board of publishers in conjunction with NATO Scientific Affairs Division

A	Life Sciences	Plenum Publishing Corporation
B	Physics	New York and London
C	Mathematical and Physical Sciences	D. Reidel Publishing Company Dordrecht and Boston
D	Behavioral and Social Sciences	Sijthoff International Publishing Company Leiden
E	Applied Sciences	Noordhoff International Publishing Leiden

Animal Learning
Survey and Analysis

M. E. Bitterman
University of Hawaii
Honolulu, Hawaii

V. M. LoLordo
Dalhousie University
Halifax, Nova Scotia, Canada

J. Bruce Overmier
University of Minnesota
Minneapolis, Minnesota

and

Michael E. Rashotte
Florida State University
Tallahassee, Florida

With a Foreword by
Gabriel Horn

PLENUM PRESS ● NEW YORK AND LONDON
Published in cooperation with NATO Scientific Affairs Division

Library of Congress Cataloging in Publication Data

Nato Advanced Study Institute on Animal Learning, Reisensburg, Ger., 1976.
    Animal learning.

    (NATO advanced study institutes series: Series A, Life sciences; v. 19)
    Bibliography: p.
    Includes index.
    1. Animal intelligence. 2. Learning, Psychology of. I. Bitterman, M.E. II. North
Atlantic Treaty Organization. III. Title. IV. Series.
QL785.N37 1976   156'.3'1   78-9894
ISBN 0-306-40061-8

Proceedings of the NATO Advanced Study Institute on Animal Learning
held at Schloss Reisensburg, Federal Republic of Germany, November 29—
December 11, 1976

© 1979 Plenum Press, New York
A Division of Plenum Publishing Corporation
227 West 17th Street, New York, N.Y. 10011

Advanced Study Institute on Animal Learning, Reisensburg, Ger., 1976.

Foreword

For ten days, a number of neuroscientists met at Reisensburg to attend a series of lectures and discussions, an Institute, on animal learning. The students were drawn from a wide variety of disciplines, including anatomy, biochemistry, pharmacology, physiology and zoology. It is probably true to say that many of them had at best a sketchy knowledge about the learning behavior of animals, about the conditions which are necessary for learning to take place and about the theories that psychologists have constructed about the learning processes. Was the Institute of any benefit to those neuroscientists whose interests lay in studying the functioning of the nervous system by manipulating it or probing it in some direct way?

Some twenty years ago the answer to this question would probably have been "No"; and there is a very good reason why this view might have been held, especially by students of the mammalian nervous system. At that time most investigators used anaesthetised animals, or animals immobilized in some other way such as by surgically isolating the brain from the spinal cord, by dividing the brain at various levels or through the use of paralyzing agents. These conditions achieved two things. On the one hand, they allowed substantial advances to be made, particularly in the analysis of sensory processing and in the analysis of the neuronal mechanisms of relatively simple reflex action. On the other hand, the experimental conditions virtually eliminated complex behavior. To parody an attitude that these conditions generated in some investigators:

(1) We study the nervous system of immobilized animals,
(2) Immobilized animals do not behave,
(3) *Ergo* behavior does not exist as a problem for us.

Clearly, individuals who held such views had no interest in listening to what students of animal behavior had to say.

When, for a variety of reasons, experimenters began to record from the brains of unanaesthetized animals, a dramatic change in attitude took place. It was found that unanaesthetized animals were not still, they explored the environment, they seemed to attend to

one thing and then to another, they gave orientation reactions to
novel stimuli, they discriminated between objects and they seemed
to learn. In other words, the animals <u>behaved</u>. In the face of
the disconcerting ability of animals to <u>avoid</u> doing what the
experimenters wished them to do, many experimenters returned to work
with the anaesthetized "preparation." Others, however, persisted,
new techniques were developed and the stage was set for studying
the neural basis of a wide range of behaviors, for example, thirst,
hunger, sexual activity, locomotion, attention and learning.

In considering ways of studying the neural bases of learning it
may be valuable to divide the field up into two broad groups,
naturally recognizing that there is considerable overlap between
them. In the first group, which would include classical conditioning and habituation, the relationship between the stimulus
and response is fairly well specified. Perhaps because of this
specificity it did not prove too difficult to find neurones whose
activity changed in parallel with some of the behavioral changes.
In the case of habituation, for example, soon after neurones with
the appropriate response properties had been detected, neural networks controlling simple behavioral responses were proposed, and
habituation of the behavioral response was visualized as an
expression of the changes in synaptic transmission in the network.
These ideas, which were generated in the '60s, were closely tied
to experimental evidence and, no doubt, because of this the field
has proved to be a fruitful and expanding one.

In the second group of learning situations, the experimental
analysis at the neural level has proved to be more difficult.
These are the situations where the relationship between stimulus,
response and reward are more loosely drawn, where the response is
complex and involves movement of the animal rather than the
entrainment of such built-in responses as, for example, salivation
and where the past history of the animal may be of prime importance
in determining the response. When an animal is trained to perform
a task many things happen. In addition to learning something about
its environment, the animal may move about more, receive more
sensory stimulation, be more attentive and more stressed than the
untrained control. A given change in the state of the brain may
be brought about by any or all of these things. How are we to
tease them apart, to get at those processes of acquisition and
storage where a particular experience exerts a lasting and
specific effect on behavior -- those processes which we call
"learning"? The experimenter must distinguish clearly between
brain changes that are necessary for and specifically related to a
particular behavior experience and the brain changes that are
merely <u>associated</u> with the experience. Recognition of this point
is crucial for a successful attack to be made on the neural bases
of learning. Furthermore, neurobiologists must not expect that a
single, crucial experiment will allow them to identify brain changes

that directly reflect these processes. When the neural bases of
learning are ultimately understood it will be the result of a whole
series of experiments. In order to conduct such experiments it is
essential that investigators understand the rules which govern and
the constraints which operate on the behavior. And in this enter-
prise the brain manipulators -- the anatomists, biochemists,
pharmacologists, physiologists, and others -- need to collaborate
closely with students of animal behavior.

If things go well, it is probable that in the next few years
neurobiologists will have a fairly clear understanding of the neural
bases of certain aspects of highly complex forms of behavior, including the mechanisms of information storage. Such understanding
will not, however, tell us how the nervous system, with its
immensely complex feed-back and feed-forward loops and its intricate
arrangements of interconnections "works" in the control of the
behavior. Micro-sampling methods, including such techniques as
electron microscopy and microelectrode recording, may be of limited
value here and we are bound to turn to the techniques of systems
analysis to attempt to understand how these networks function in
the control of behaviors such as those that were discussed at the
meeting.

Most, if not all of those, who attended the Institute would
agree that they have profited from it no matter how skeptical they
may have been at the outset about what they had to learn from
students of animal behavior. This success was achieved through the
almost herculean efforts of the lecturers, the authors of this
volume, both at the formal sessions during the day and at the many
informal discussions that took place at night, often extending into
the early hours of the morning. As a result, the answer to the
question which I posed at the beginning of this Preface -- "Was the
Institute of any benefit to those neuroscientists whose interests
lay in studying the function of the nervous system by manipulating
it or probing it in some direct way?" -- is, in my view, unequivo-
cably "Yes." I hope that the readers of this book, though denied
direct contact with its authors, will reach the same conclusion.

Gabriel Horn
Department of Anatomy
University of Bristol

28 November 1977

Preface

This volume is based on a series of lectures given by us at a North Atlantic Treaty Organization Advanced Study Institute on Animal Learning held at Schloss Reisensburg, Federal Republic of Germany, from November 29 through December 11, 1976. The lectures were intended to communicate the core methods and issues in the psychology of animal learning to colleagues in neighboring disciplines. In no sense were the lectures, nor are these chapters, considered by the authors as exhaustive treatments of the vast learning literature. Professor M. Blancheteau (Montpellier), G. Horn (then at Bristol, now at Cambridge), R. Menzel (Berlin), M. Poli (Milano), H. Ursin (Bergen), and J. M. H. Vossen (Nijmegen) served as discussants throughout and skillfully fostered interdisciplinary communication. To them, to Professor M. Lindauer (Wurzburg) who contributed to the planning of the Institute, and to the 40 other participants from 16 countries, we are grateful for many fresh insights gained in the course of long and interesting discussions.

The principal costs of the Institute were met by the NATO Office of Scientific Affairs. Two of the participants from the United States were supported in part by the National Science Foundation and three participants from non-NATO countries (Mexico, Spain, and Switzerland) by Hoffman-LaRoche, Inc. We wish to thank the staff of the University of Minnesota's Center for Research in Human Learning for their conscientious preparation and redaction of the copy for this volume.

All of the figures in this volume citing a reference (e.g., After Azrin & Holz, 1961) were taken from the cited copyrighted works. The copyright to these figures is held by the publisher of those works, and the figures are used here with the permissions of the authors and copyright holders.

These chapters should prove valuable to persons who desire a mature introduction to and critical overview of the phenomena, problems, and theories of the psychology of animal learning.

                                        M.E.B.
                                        V.M.L.
                                        J.B.O.
                                        M.E.R.

July, 1978

Contents

1. HISTORICAL INTRODUCTION

M. E. Bitterman

University of Hawaii
Honolulu, Hawaii, USA

The study of animal intelligence was brought into the laboratory at the end of the nineteenth century, with Darwin's theory
of human evolution providing the principal impetus. The anatomical evidence alone was not convincing, Darwin (1871) conceded;
evidence of mental continuity between men and animals also was
required, which he found in the casual observations of hunters and
naturalists, zoo-keepers and pet-lovers. Were animals capable of
reasoning? A hunter named Colquhoun once "winged two wild ducks,
which fell on the opposite side of a stream; his retriever tried
to bring over both at once but could not succeed; she then, though
never before known to ruffle a feather, deliberately killed one,
brought the other over, and returned for the dead bird" (p. 46).
A captive female baboon, scratched one day by a kitten which she
had "adopted," examined the kitten's feet and promptly bit off the
claws. It was anecdotal evidence such as this on which Darwin
based his conclusion that there is nothing new in the mind of man
-- no characteristic or capability which is not shown also, at
least in incipient form, by some lower animal.

With Darwin's encouragement, Romanes (1881) prepared a com-
pilation of the anecdotal materials which was published under the
title of _Animal Intelligence_. Although Romanes screened the materials carefully, relying only on observations which were made by
reputable persons under favorable conditions and "about which
there scarcely could have been a mistake" (p. 91), it must be
evident to the contemporary reader that a great many mistakes, both
of fact and of interpretation, were made. Consider, for example,

a report by a Mrs. Hutton, who, having killed some soldier ants,
returned after half an hour to find the dead bodies surrounded by
a large number of ants and "determined to watch the proceedings
closely.... In a few minutes, two of the ants advanced and took
up the dead body of one of their comrades; then two others, and so
on, until all were ready to march. First walked two ants bearing
a body, then two without a burden; then two others with another
dead ant, and so on, until the line was extended to about forty
pairs, and the procession now moved slowly onwards, followed by an
irregular body of about two hundred ants. Occasionally the two
laden ants stopped, and laying down the dead ant, it was taken up
by the two walking unburdened behind them, and thus, by occasionally relieving each other, they arrived at a sandy spot near the
sea. The body of ants now commenced digging with their jaws a
number of holes in the ground, into each of which a dead ant was
laid, where they now laboured on until they had filled up the ants'
graves. This did not quite finish the remarkable circumstances
attending this funeral of the ants. Some six or seven of the ants
had attempted to run off without performing their share of the task
of digging; these were caught and brought back, when they were at
once attacked by the body of ants and killed upon the spot. A
single grave was quickly dug, and they were all dropped into it"
(pp. 91-92).

The conclusions reached by Romanes in a chapter on the intelligence of cats are especially interesting because they differ so
strikingly from those reached in the earliest experiments. Consider
a case reported in Nature by a Dr. Frost: "Our servants have become
accustomed during the late frost to throw the crumbs remaining from
the breakfast-table to the birds, and I have several times noticed
that our cat used to wait there in ambush in the expectation of
obtaining a hearty meal from one or two of the assembled birds.
Now, so far, this circumstance in itself is not an 'example of
abstract reasoning.' But to continue. For the last few days this
practice of feeding the birds has been left off. The cat, however,
with an almost incredible amount of forethought, was observed by
myself, together with two other members of the household, to scatter
crumbs on the grass with the obvious intention of enticing the birds"
(Romanes, 1881, p. 418). Even for Romanes, this account created some
suspicion, but he felt obliged to give it weight because similar
observations were made by his friend Dr. Klein (F.R.S.) and others,
and because scattering crumbs to attract birds did not seem to
involve ideas very much more complex than other, more widely corrobo-
rated instances of feline intelligence, particularly those which
displayed understanding of mechanical appliances.

There were numerous reports in the literature of cats which
were able to open doors, an achievement which seemed to Romanes to

imply the operation of "complex" processes. "First the animal
must have observed that the door is opened by the hand grasping
the handle and moving the latch. Next she must reason, by 'the
logic of feelings'--If a hand can do it, why not a paw? Then,
strongly moved by this idea, she makes the first trial" (pp. 421-
422). There were many reports, too, of cats that used door-
knockers, but Romanes was more impressed with the data on doorbells.
"For here", he wrote, "it is not merely that cats perfectly well
understand the use of bells as calls, but I have one or two cases
of cats jumping at bell-_wires_ passing from outside into houses the
doors of which the cats desired to be opened. My informants tell
me that they do not know how these cats, from any process of
observation, can have surmised that pulling the wire in an exposed
part of its length would have the effect of ringing the bell; for
they can never have observed any one pulling the wires. I can only
suggest that in these cases the animals must have observed that
when the bells were rung the wires moved, and that the doors were
afterwards opened; then a process of inference must have led them
to try whether jumping at the wires would produce the same effects"
(pp. 423-424).

ORIGINS OF THE EXPERIMENTAL MOVEMENT

A central figure in the transition from anecdotalism to the
experimental study of animal intelligence was an English psychologist named C. Lloyd Morgan, whose Introduction to Comparative
Psychology appeared in 1894. Morgan emphasized the importance of
systematic and sustained investigation, urging those who observed
what appeared to be an intelligent bit of behavior not to be content merely to record it but to try to discover its true nature.
Morgan had a dog named Tony that could open his garden gate. To
anyone who saw only the smooth, finished performance, it might
have seemed that Tony understood the workings of the latch, but
Morgan, who had observed the development of the behavior, could
offer a simpler interpretation. The response occurred first quite
by chance in the course of a great deal of excited running back and
forth along the fence. The dog put its head out through the vertical railings at various points, happening after a while to put its
head out beneath the latch, which lifted as the head was withdrawn,
and the gate swung open. Thereafter, the dog was required always
to open the gate for itself. Performance improved slowly, with
useless movements (such as putting the head out at wrong places)
gradually being eliminated, and only after about three weeks could
the animal go directly to the latch and lift it without any in-
effectual fumbling. Morgan made similar observations in his work
with young chicks. The animals were kept in a corner of his study

which was partitioned off with old newspapers, and one of the chicks learned to escape from the enclosure in the course of pecking at the number on one of the pages.

Although Morgan was an evolutionist and certainly not pre-disposed to find an impassable gap between the minds of men and animals, he rejected the undisciplined anthropomorphism of the anecdotalists, advocating instead a principle of parsimony which became known as Morgan's canon. "In no case," he ruled, "may we interpret an action as the outcome of the exercise of a higher psychical faculty if it can be interpreted as the outcome of the exercise of one which stands lower in the psychological scale" (p. 53). The achievements of Morgan's own animals seemed to be based on trial and error with accidental success. An action per-formed at first without view to any particular outcome was dis-covered to have pleasurable consequences, and its repetition could be traced to associations gradually formed between the situation, the action, and the consequences. No assumption of logical infer-ence or conceptual thought was required.

It was the work of Thorndike perhaps more than any other which ushered in the experimental era. The work was reported in a series of papers, beginning in 1898 with his doctoral dissertation, which was later repreinted in a volume entitled Animal Intelligence: Experimental Studies (Thorndike, 1911). Like Morgan, Thorndike was interested in the sources of human intelligence, was critical of the excesses of the anecdotalists, and saw the need for careful, analytical study. The experimental situations used by Thorndike resembled those with which Morgan had worked more informally. His principal method was to put a hungry animal into a small enclosure called a puzzle box or problem box, which it could get out of by performing some simple action or series of actions such as pulling a loop of wire, manipulating a lever, or stepping on a treadle. One of these boxes is shown in Figure 1. Outside the enclosure was some food which the animal could see from inside, and a record was kept of the time required to obtain the food on each of a series of trials. The learning curves of a cat in two of the boxes--plots of time (ordinate) against trials--are shown in Figure 2. A number of advantages for this method were claimed by Thorndike: it was objective (in the sense that the influence of the observer was minimal), quantitative, reproducible, flexi-ble (capable of wide variation in complexity), natural (at least in the sense that it presented problems like those dealt with in the anecdotal literature), efficient (permitting large samples of animals to be studied), and adaptable to the characteristics of a variety of different species.

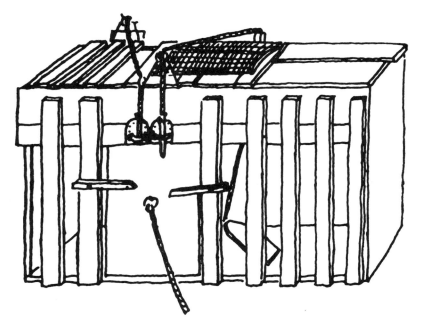

Figure 1.1   One of the problem boxes used by Thorndike (1911).

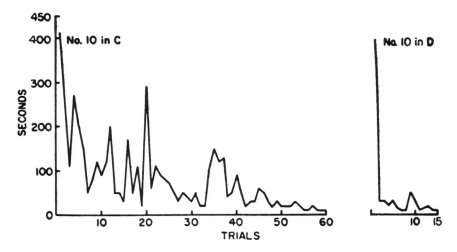

Figure 1.2   The performance of one of Thorndike's cats in two of
             his problem boxes.

Typically, there was nothing about the behavior of cats and dogs in these situations to suggest understanding, but only a scramble to get out, which often seemed motivated as much by the aversiveness of the enclosure as by the food, and in the course of which the correct response seemed to occur quite by accident. In subsequent trials, unsuccessful responses gradually dropped out of the behavioral repertoire until in the end the successful response was made almost as soon as the animal was placed in the situation. The speed of learning varied considerably from subject to subject as well as from one problem to another, and occasionally improvement was quite precipitous -- especially in amimals that had mastered several previous problems and in which, therefore, the tendency to make certain responses that were irrelevant in all of the problems had been greatly weakened. Monkeys seemed on the whole to learn more quickly than cats and dogs, but there was no reason to think that the basis of solution was qualitatively different. Thorndike speculated that the behavior of all animals, including man, might be governed by what he called the laws of Exercise and Effect. According to the former, each occurrence of a response in the presence of a stimulus tends to increase the strength of the tendency for the stimulus to evoke the response. According to the latter, the strength of the tendency for a stimulus to evoke a response is increased by pleasant consequences and decreased by unpleasant consequences. Of the two principles, Effect seemed to be the more powerful; in the course of its training in a problem box, an animal might make some incorrect response much more frequently than the correct (rewarded) response, yet the correct response would prevail.

In the first decades of this century, many new animals -- vertebrates mostly, but invertebrates as well -- were studied in Thorndikian situations. Problem boxes of a variety of designs were constructed. Thorndike had begun to analyze discriminative learning (rewarding the subject for responding to one of two stimuli but not for responding to the other), and this line of inquiry, too, was continued out of interest both in the sensory capacities of the animals studied and in their learning. Thorndike based his denial of understanding in part on the pattern of improvement shown by his animals in discriminative training; even with readily distinguishable stimuli, there was a gradual increase in the probability of response to the positive and a gradual decrease in the probability of response to the negative that looked more like the operation of an associative process than of a conceptual one. Thorndike's work with mazes also was continued. Shown in Figure 3 is the popular Hampton Court pattern introduced by Small (1901), along with some relative learning curves plotted in terms of numbers of errors (ordinate) per trial expressed as a proportion of the number of

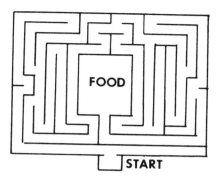

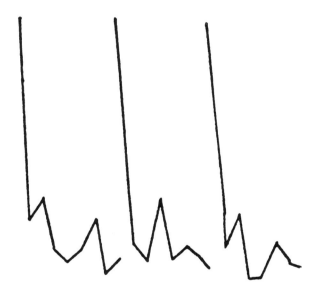

Figure 1.3   At the top is the Hampton Court maze developed by Small (1901).   At the bottom are relative error curves for three different animals trained in the maze -- rat (Small, 1901), sparrow (Porter, 1904), and monkey (Kinnaman, 1902).

errors that were made on the first trial. One of the three
curves -- which are strikingly similar -- is for a rat trained by
Small, one for a monkey trained by Kinnaman (1902), and one for a
sparrow trained by Porter (1904). The results of these experiments
seemed to make it perfectly clear that problem solving in animals
is in large measure associative or reproductive in character, as
Morgan and Thorndike had proposed, rather than insightful or crea-
tive, but could there be anything more?

Hobhouse (1901) concluded from his own work that dogs and cats
are capable only of reproductive solutions, but in primates he found
something akin to originality. For his experiments, he developed a
series of ingenious tasks which were suggested in part by the anecdotal literature. One of their virtues was simplicity -- an animal
incapable of understanding the pulley system in a Thorndikian problem
box was not necessarily incapable of understanding anything at all.
Another virtue of Hobhouse's tasks was that they were "open" in the
sense that all critical features of any situation were in plain view.
An animal might be required to reach a piece of food with the aid of
a stick; to reach a more distant piece of food with the aid of a
longer stick which was itself reached with a shorter stick; or --
as diagrammed in Figure 4 -- to reach food on top of a table with
the aid of a box brought near the table. A problem which seemed to
be solved in a meaningful way by a chimpanzee but was failed by a
monkey required a bit of food to be pushed out of a heavy pipe with
a long stick. Understanding might develop rather gradually, Hobhouse
found -- for example, monkeys seemed to be able to use sticks properly only after considerable practice -- but in the end specific
experience was transcended, the products of previous practice
selected and recombined to meet the demands of a new problem. Influential arguments for understanding in animals were offered also
by Köhler (1925) on the basis of his experiments with chimpanzees
during the first World War, and by Yerkes (1916), whose extensive
work with primates began at about the same time. Like Hobhouse,
Yerkes observed the emergence of understanding in the course of
experience and emphasized the importance of ontogenetic study, himself providing a model account of the way a young gorilla learned
to use sticks as tools (Yerkes, 1927).

It should not be imagined, however, that all studies of primates
in simple, open problems produced such results. Analyzing the
ability of monkeys to reach food with a stick or a length of cloth,
Watson (1908) found no sign of comprehension even after a considerable amount of training. He wondered whether Hobhouse had worked
with animals of greater native ability or had read into their
behavior more than actually was there. While Köhler and Yerkes
believed that insight is a perfectly objective property of behavior,

it seemed evident to others that a good deal of subjective judgment
is involved. How easy it is to be mistaken about "instant understanding" was demonstrated by Holmes (1916), who tested a monkey
with a length of board set near its cage. When a piece of apple
was placed on the far end of the board, which was out of reach,
the animal would draw in the board by the nearer end and take the
apple. The same thing happened repeatedly, however, when the apple
was placed, not on the board, but on the floor, six inches away
from the far end. Because of contradictions in the data and because
of the subjectivity of the concept of understanding, this line of
investigation soon was largely abandoned in favor of the study of
learning by trial and error, for which objective and quantitative
techniques were available and many more were being developed.
Since the results of the first experiments with a variety of
species were qualitatively identical, it seemed reasonable to suppose that the underlying associative processes might be the same
in all animals and that further efforts might therefore safely be
concentrated on one. Before long, the small, docile, readily available albino rat became the principal subject.

Almost at the same time as the Thorndikian tradition was being
born in western Europe and America, the Pavlovian tradition was born

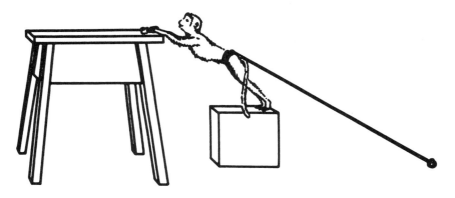

Figure 1.4 A problem studied by Hobhouse (1901). To be able to
           reach the food on the top of the table, the tethered
           animal had to bring the box into the appropriate
           position.

in Russia. There were striking differences between Pavlov and
Thorndike. Pavlov was an established scientist who already had won
the Nobel prize for his work on the digestive glands when he began
to study conditioning, while Thorndike was only twenty-four when
his dissertation was published. Pavlov was a physiologist rather
than a psychologist, interested in brain rather than in mind. The
methods of the two men also were different, Pavlov's designed to
study the way in which stimuli come to function as signals, and
Thorndike's to understand adaptive behavior. That the two tradi-
tions had much in common, however, was evident as soon as contact
between them was made. Watson (1916) in America was strongly
impressed with the psychological significance of Pavlov's work and
helped to publicize it. Another important source of information
was Anrep's translation into English of Conditioned Reflexes
(Pavlov, 1927).

CLASSICAL AND INSTRUMENTAL CONDITIONING

Of the various terms which have been used to distinguish the
methods of Pavlov and Thorndike, the most widely used and the most
satisfactory perhaps are classical and instrumental conditioning
(Hilgard and Marquis, 1940). In classical conditioning, an animal
is exposed to sequences or conjunctions of stimuli, and the effects
of the experience on its behavior are examined. In the simplest
case, changes in the functional properties of one stimulus (S_1)
that result from its repeated pairing with a second stimulus (S_2)
are studied. A tone paired with food comes, as Pavlov found, to
elicit salivation in dogs; a light paired with shock comes to
elicit agitated behavior in goldfish. Motivational as well as
response-eliciting properties can be conditioned. A hungry rat
placed repeatedly in a box (S_1) containing food (S_2) will subsequently learn a complex maze leading to the empty box, which has
become a conditioned or secondary reward (Williams, 1929). In
most cases, S_2 has certain well-defined motivational or response-
eliciting properties which can be shown to transfer to S_1, but the
pairing of relatively neutral stimuli also has some interesting
effects that will be considered later. If the observed change in
the properties of S_1 are to be attributed to the sequencing of the
two stimuli, the effects of experience with the stimuli independently of sequence must, of course, be controlled. A good procedure is
to use a different sequence for another group of animals -- say, to
present the two stimuli in random order. It will not do simply to
omit S_2, although that mistake often has been made.

In instrumental conditioning, the experimenter sets out to
manipulate behavior by manipulating its consequences. It may be

helpful to think of an instrumental conditioning experiment as a classical conditioning experiment in which the sequence of stimuli is contingent upon the behavior of the animal. Instead, for example, of putting an animal in a box (S_1) containing food (S_2), the animal is put in an empty box and food is given only after it has made some defined response (R_1), either manipulative (such as pressing a lever) or simply locomotor (going to a particular place). The response-contingency may be either positive or negative, which is to say that R_1 may either produce S_2 or prevent it (as, for example, when the animal is permitted to avoid shock by jumping out of the box). In the positive case, the sequence of events is $S_1R_1S_2$ or S_1--; in the negative case, S_1-S_2 or S_1R_1-. Where the contingency is positive, instrumental training cannot proceed unless the probability that R_1 will be elicited by S_1 is greater than zero, since S_2 is not presented unless R_1 occurs. Instrumental conditioning with a negative contingency has a special relation to classical conditioning, since S_1 and S_2 are paired when R_1 does not occur; where the initial probability of R_1 to S_1 is zero, the classical and instrumental procedures are identical until R_1 does occur and we depend, in fact, on the pairing of S_1 and S_2 to produce it.

Since instrumental conditioning experiments vary with respect to the desirability of S_2 as well as the sign of the R_1S_2 contingency, they fall into four major categories as shown in Table I.

TABLE I

Kinds of Instrumental Conditioning Experiments

	Desirable S_2	Undesirable S_2
Positive R_1S_2 Contingency	Reward, Escape	Punishment
Negative R_1S_2 Contingency	Omission	Avoidance

With a positive contingency and a desirable S_2 (e.g. pressing a lever produces food), we have reward training. Escape training is a special case of reward training in which the desirable consequence of response is the termination of an undesirable stimulus (e.g. pressing a lever terminates shock). With a positive contingency and an undesirable S_2 (pressing a lever produces shock), we have punishment training. With a negative contingency and an undesirable S_2 (pressing a lever prevents shock), we have avoidance training, and with a negative contingency and a desirable S_2 (pressing a lever prevents food), omission training. The 2 X 2 table is a considerable simplification, of course, since it dichotomizes two continuous variables. The correlation between R_1 and S_2 may be less than 1.00, and it may even be zero -- which reduces instrumental conditioning to classical. Furthermore, S_2 may vary widely in desirability or aversiveness. There has, in fact, been substantial interest in the case in which S_2 is neutral.

The term reward often is used to describe the desirable consequences of response in reward training and in this sense of the term we can say that response in omission training prevents reward. Similarly, undesirable consequences of response or of failure to respond may be called punishment; in avoidance training, response prevents punishment, and punishment training itself may be referred to as passive avoidance training in the sense that the animal can avoid punishment by doing nothing. The presentation of reward or punishment in an instrumental experiment is called reinforcement, and the same term is used for the presentation of S_2 in classical conditioning where, of course, reinforcement is not response-contingent. It is the response-contingency which permits us to distinguish between rewards and punishments: S_2 is a reward if it tends to increase the probability of R_1 when the contingency is positive and to decrease the probability when the contingency is negative; S_2 is a punishment if it tends to have the opposite effects. Rewards and punishments cannot be distinguished in terms of the direction of the change in responding which they produce in classical experiments.

The problem box is classified as a unitary situation because the experimenter is concerned with a single response, such as pressing a lever, and measures the readiness with which it is made. Another unitary situation which soon came into wide use is the runway -- a straight alley along which the animal runs from a starting box to a goal box, the measure of performance being the time taken to go from one point to the other; a runway is a maze without choice-points or blind alleys. The simplest true maze has a single choice-point, often in the shape of a T- or Y-junction, with each arm leading to a separate goal box. In such a maze or in a problem box with two

(or more) manipulanda, the experimenter is concerned with which of
the alternative responses will occur; we say then that the animal
is in a choice situation. Both unitary and choice situations may
be chained. The simplest example is a maze of many choice-points --
such as the Hampton Court maze or the multiple-T maze (consisting
of a succession of T-shaped junctions) -- in which correct response
at the first choice-point takes the animal to the second, and so
forth, until the goal is reached. In a unitary situation, the
animal may be required to operate a series of devices presented in
a fixed sequence before the reward is given.

Another important distinction is between generalized and
discriminative situations. The problem box, runway, and maze as
thus far considered fall into the first category because the training conditions are constant from trial to trial. In a discriminative
task, some distinguishable property of the situation is varied
systematically and with it the consequences of response. For example,
food may be found in the goal box when the runway is white but not
when it is black; or food may be found in the right-hand goal box
of a simple T-maze when the stem (the approach to the choice-point)
is white but in the left-hand goal box when the stem is black.
Shown in Figure 5 is the jumping apparatus developed by Lashley
(1938) for the study of visual discrimination in rats. It displays
two cards, one of which, when struck, gives way easily and admits
the animal to a feeding platform in the rear; if the animal jumps
at the other card, which is locked in place, it falls into a net
below. Another discriminative situation which has been widely used
is the Wisconsin apparatus, originally developed for primates (Harlow,
1949) but then adapted for other animals, such as cats, rats, and
birds. The response required is to displace one or another of a set
of stimulus-objects, each of which covers a food-well that may or may
not contain food. Discrimination is, of course, studied in classical
as well as in instrumental situations; for example, one tone is paired
with food but another is not.

In the earliest instrumental conditioning experiments, training
was given in discrete trials. The animal was introduced into the
apparatus at the start of a trial, removed from it at the end, and
then returned after an intertrial interval which could be quite short
(massed trials) or very long (spaced trials). Then Skinner (1938)
automated the problem box and began to develop another kind of training described as free operant. Since the reward in Skinner's apparatus is delivered to the animal by a feeder, the animal is free to
respond again as soon as it has taken the food -- the intertrial
interval is in effect zero -- and a cumulative graphic record of
responding is made. The preferred measure of performance in free
operant experiments is not latency of responding (time per response)

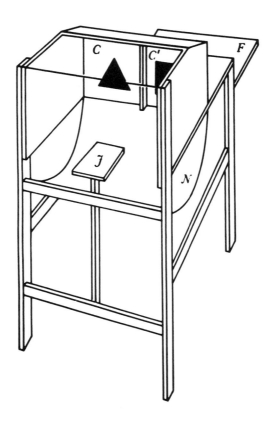

Figure 1.5 The jumping apparatus developed by Lashley (1938).
J, jumping platform; C and C', stimulus cards; F,
food platform; N, net.

but rate (number of responses per unit time), which can be read from the slope of the cumulative record. The acquisition curve for a rat trained to press a lever under these conditions is shown in Figure 6; a quite steady rate of responding may be seen to have developed after only a few reinforcements.

The main interest in free operant experiments has not been, however, in the acquisition of the instrumental response, but in the conditions that govern its rate of occurrence after it has become part of the animal's behavioral repertoire. Skinner discovered early, for example, that a high rate of responding can be maintained even with low frequencies of reward, and he began to study the effects on performance of various schedules of reinforcement. In the schedule called continuous reinforcement or CRF, each response is rewarded. In a fixed ratio or FR schedule, every nth response is rewarded (e.g., every tenth response in FR-10). In a variable ratio or VR schedule, e.g., VR-10, the required number of responses varies in a quasi-random fashion but averages 10. The so-called interval schedules are time-based rather than frequency-based. In a fixed interval or FI schedule, e.g., FI-1 min, response is rewarded only after 1 min has elapsed since the previous reinforcement. Variable interval or VI schedules bear the same relation to FI schedules as VR schedules to FR schedules. While all the examples given thus far are of reward training, free operant procedures are suitable also for the study of punishment, avoidance, and omission. A widely used free operant method for the study of avoidance was developed by Sidman (1953). In Sidman avoidance training, a clock which schedules shock is reset by each response. If the clock is set at 30 sec, the animal is shocked every 30 sec when it fails to respond; each response postpones shock for the same period.

Free operant experiments typically are fully automated, with programming equipment presenting stimuli and keeping track of responses, which makes for efficiency (a single experimenter can take data simultaneously from a number of subjects) as well as for accuracy and objectivity. It does not follow, however, that all automated experiments are free operant experiments (a common misconception). With a motor-driven lever which is introduced to start each trial and retracted at the end, and with a feeder which brings food to the animal instead of an exit door through which the animal goes to food, discrete-trials experiments quite like those done in the old Thorndikian problem boxes are possible. It should be noted that the training situations developed by Skinner are quite suitable also for the study of classical conditioning. Consider, for example, the finding of Brown and Jenkins (1968) that the key-pecking response long used for the study of instrumental conditioning

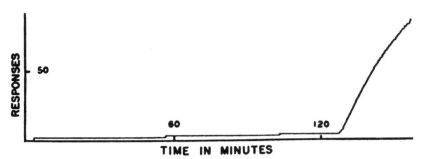

Figure 1.6   Cumulative record of the acquisition of lever pressing
             in a rat.   The ordinate shows the cumulative number
             of responses made since the start of the session.
             From Skinner (1938).

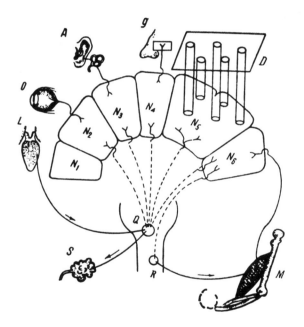

Figure 1.7   Pavlov's diagram of the neural mechanism of conditioning.
             L, taste receptors; O, visual receptors; A, auditory
             receptors; G, olfactory receptors; D, pressure and
             temperature receptors; N_1-N_6, cortical sensory areas;
             Q, salivary motor center; S, salivary gland; R, skeletal
             motor center; M, skeletal muscle and movement receptors.
             From Frolov (1937).

in pigeons can be classically conditioned as well. If the illumination of the key -- a small plastic disk on one wall of the box --
is followed repeatedly by the presentation of food, the pigeon will
soon begin to peck the key when it is illuminated even though the
response is not required to produce the food. Comparable results
have been obtained with monkeys (Sidman and Fletcher, 1968), rats
(Petersen, Ackil, Fromme, and Hearst, 1972), and goldfish (Woodard
and Bitterman, 1974).

EARLY THEORIES OF CONDITIONING

     Classical and instrumental conditioning are powerful procedures
for modifying behavior, and it is not surprising, therefore, that as
soon as they were discovered there should have been a great deal of
speculation as to how they work. How does it happen, for example,
that a light followed repeatedly by food soon begins to evoke salivation? Pavlov's conception of the process is illustrated in a
diagram preserved by Frolov (1937), which is shown in Figure 7. A
new neural connection was assumed to develop between a cortical
sensory center activated by the light and a subcortical motor center
for salivation as a result of the contiguous activation of the two
centers. A more general (more functional) statement of the same
idea, not tied to specific anatomical structures, is to be found in
the so-called S-R contiguity principle, which says simply that an
association or connection between a stimulus and a response is
established and strengthened by their contiguous occurrence. In
this view, the contiguity of light and food serves only to produce
contiguity between light and salivation, the response evoked by the
food. An obvious alternative to the S-R contiguity principle is the
S-S contiguity principle, according to which contiguous stimuli are
associated. In this view, the light becomes at least to some extent
the functional equivalent of the food (in neural terms, the two
sensory centers are connected) and evokes salivation because the
food does so. A way of deciding between the two principles is provided by the sensory preconditioning experiment, conceived early by
Thorndike (1911) but not carried out properly until later. In the
first stage of the experiment, two neutral stimuli, S_1 and S_2, are
paired repeatedly. In the second stage, the animal is trained to
make some new response, R, to S_2, either by a classical or by an
instrumental procedure. Will S_1 then also elicit R? According to
the S-S contiguity principle it should, but according to the S-R
contiguity principle it should not.

     To account for instrumental conditioning seems somewhat more
difficult. Suppose that in the presence of S_1 the animal makes a
variety of responses, R_a, R_b, ..., R_n, one of them, R_a, always is

followed by food (S_2), which in turn elicits a series of consummatory responses (R_2) including but not limited to salivation. By either of the contiguity principles, we should expect some tendency to develop for S_1 to evoke R_2 (by direct connection to R_2 or by way of a connection with S_2), but why R_a? S_{Ra}, the feedback from R_a, also is contiguous with S_2 and R_2, but how can that help? It may be interesting to consider an early S-S contiguity interpretation offered by Morgan (1894). Whenever the animal makes R_a in the presence of S_1, an S_1-S_{Ra} connection develops; when R_b is made, an S_1-S_{Rb} connection develops; and so forth. That is, reintroduced to S_1 on a subsequent occasion, the animal has a basis for "remembering" its previous responses, and a response made more frequently in the past will be remembered best. Whenever R_a is made, furthermore, an S_{Ra}-S_2 connection develops, which means that whenever the correct response is remembered the reward also is remembered: S_1-S_{Ra}-S_2. Now assume that there is an innate connection between the sensory feedback center for any response and the motor center for that response (e.g., S_{Ra}-R_a), and assume, too, that each sensory center has an innate connection to a hedonic "control" center which reads its pleasantness and tends to facilitate ongoing response-tendencies when the reading is pleasant but to suppress them when the reading is unpleasant. The animal will then make R_a when it is remembered because S_{Ra} is connected both to R_a and to S_2, which produces facilitation by the control center. If S_2 were unpleasant, of course, the tendency to make R_a would be suppressed. Morgan's belongs to a class of theories which came to be called <u>cognitive</u> theories because learning about and remembering responses and their consequences are assumed to play a central role. One of the most important cognitive theorists of later years was Tolman (1932).

Thorndike (1911) had a different theory, the first of another class called <u>reinforcement</u> theories. His assumption was that rewarding a response made in the presence of a stimulus tends to establish and strengthen a connection between the stimulus and the response. It may be well to distinguish two statements of Thorndike's Law of Effect, one empirical and the other theoretical. According to the empirical law, rewarding a response in the presence of a stimulus tends to increase the probability that the stimulus will evoke the response. The theoretical law, which came to be called the <u>S-R reinforcement principle</u>, purports to tell us why that should be so. It says that the probability increases because rewards strengthen S-R connections. Clearly, one can accept the empirical law (as did Morgan, who formulated it even before Thorndike) but reject the theoretical law, and a great many interesting experiments have been performed to decide between the cognitive and reinforcement theories. Consider, for example, an experiment on <u>latent learning</u>. In the first stage, R_a to S_1 is followed by an S_2 which is not rewarding, either because it is intrinsically neutral (such as a light) or because

it is pleasant only under other than the prevailing motivational
conditions (such as food for a completely satiated animal). Then
S_2 is given some reward value, either by pairing it with an effective reward or by changing the motivational state (depriving the
animal of food). Will S_1 now show an increased tendency to evoke
R_a? According to the cognitive interpretation it should, but
according to the reinforcement interpretation it should not.

If instrumental conditioning is to be accounted for in cognitive terms, some supplementary assumptions are necessary to trans-
late the new, purely sensory linkages into appropriate action. The
problem has been dealt with in different ways. Morgan, as noted
already, postulated an innate $S_{Ra}-R_a$ connection and a hedonic control
center to activate it selectively. An alternative to $S_{Ra}-R_a$ proposed
later by Mowrer (1960) is to assume that the animal always is rapidly
scanning its repertoire of responses; when R_a is scanned, S_{Ra} (the
feedback presumably is central rather than proprioceptive) evokes S_2,
which then facilitates R_a much in the manner of Morgan. An advantage
of the S-R reinforcement principle once seemed to be that it requires
no such supplementary activation assumptions, but that idea soon was
dispelled. Experiments reported by Tolman (1932) showed, for example,
that a rat which has learned a maze for food when hungry does not perform very well when it is less hungry; that is, the mere existence of
an S-R connection does not mean that S will always evoke R. Hull
(1943), a leading systematizer of reinforcement theory, proposed that
S-R connections are activated by deprivation (or <u>drive</u>). Later
(Hull, 1952), he added activation by anticipated <u>reward</u>, the anticipation based not on an S_1-S_2 connection but on a connection between
S_1 and a conditionable component of the consummatory response to S_2 --
what he called a <u>fractional anticipatory goal response</u> and symbolized
as r_G.

Just as the cognitive theorists proposed to account for instrumental conditioning in terms of S-S contiguity, a principle suggested
by classical conditioning, so Hull (1943) proposed to account for
classical conditioning in terms of the S-R reinforcement principle,
which, of course, was suggested by instrumental conditioning. The
S-R reinforcement principle makes no reference, Hull noted, to the
contingency of reward on the occurrence of response, but only to the
contiguity of response and reward. The function of the contingency
in instrumental conditioning is simply to ensure response-reward contiguity, which is present without the contingency in classical conditioning. In a Pavlovian experiment with food as S_2, the food may be
assumed to play two roles, one of which is to elicit salivation in
the presence of S_1 and the other to reward it. Pavlov found good
salivary conditioning also with weak acid as S_2; here salivation in
the presence of S_1 may be assumed to be rewarded by dilution of the

acid just as the termination of an aversive stimulus rewards the instrumental response in escape training. Hull's interpretation suggested a number of tests, among them experiments on sensory preconditioning, but quite apart from the correctness of the interpretation an important point had been made. However reward works -- whether it is learned about, or whether it strengthens S-R connections -- we must be alert always to the possibility of adventitious reward (adventitious in the sense that it is unplanned by the experimenter and not under his control) as well as to the possibility of adventitious response-reward contiguity. The dilution of acid by saliva is a good example of unplanned reward. Unplanned response-reward contiguity is to be found in any classical conditioning experiment with a rewarding S_2; response to S_1 always is followed by S_2 despite the fact that the response does not produce it.

Various theorists decided that two principles might be necessary to deal with all of the data, although they differed considerably as to what the principles might be. Morgan (1894) relied on S-R contiguity to explain the low-latency performance which he found in highly trained animals and which did not seem to provide much opportunity for remembering previous responses and their consequences: these cognitive processes might be short-circuited, he speculated, by direct S-R connections. Maier and Schneirla (1942) combined the S-S contiguity and S-R reinforcement principles, while Mowrer (1947) suggested at one point that autonomic responses (such as salivation) are connected to stimuli by contiguity but skeletal responses (such as pressing a lever) by reinforcement. Skinner (1938) had earlier proposed another classification of conditionable responses -- one class, called respondents, said to be reflexly "elicited" by stimuli, and a second class, called operants, said simply to be "emitted" by the animal; respondents were assumed to be conditioned by contiguity and operants by reinforcement. A good review of two-process theories has been provided by Rescorla and Solomon (1967).

A phenomenon closely related to conditioning which posed some closely related explanatory problems is extinction. It was soon discovered that a response conditioned to S_1, either classically or instrumentally, gradually disappears (extinguishes) in the course of repeated presentations of S_1 when S_2 is entirely withheld. Why does that happen? An immediate implication of the phenomenon, of course, was that something must be wrong with the S-R contiguity principle, according to which the mere contiguity of S and R is sufficient to strengthen the connection between them. Two contrasting interpretations of extinction (by no means mutually exclusive) were considered. One was that the connections established and strengthened in conditioning simply weaken and disappear in extinction. The second was that some competing process develops in extinction. In his work on classical conditioning, Pavlov (1927) found reason to believe that the competing process is an inhibitory one, different in kind from the excitatory process which he assumed to underlie conditioning. Another view was that of Guthrie (1930), who was more inclined to think in instrumental terms and who held the two processes to be the same in kind; in extinction, he proposed, the animal simply learns to do something else. How well these early conceptions of conditioning and extinction fared, we shall have an opportunity to see in the following chapters dealing with the research which was stimulated by them.

REFERENCES

Brown, P. L., & Jenkins, H. M. Auto-shaping of the pigeon's key-peck. JOURNAL OF THE EXPERIMENTAL ANALYSIS OF BEHAVIOR, 1968, 11, 1-8.

Darwin, C. THE DESCENT OF MAN AND SELECTION IN RELATION TO SEX. Vol. 1. New York: D. Appleton, 1871.

Frolov, Y. P. PAVLOV AND HIS SCHOOL: THE THEORY OF CONDITIONED REFLEXES. New York: Oxford University Press, 1937.

Guthrie, E. R. Conditioning as a principle of learning. PSYCHOLOGICAL REVIEW, 1930, 37, 412-428.

Guthrie, E. R. THE PSYCHOLOGY OF LEARNING. New York: Harper & Brothers, 1935.

Harlow, H. F. The formation of learning sets. PSYCHOLOGICAL REVIEW, 1949, 56, 55-61.

Hilgard, E. R., & Marquis, D. G. CONDITIONING AND LEARNING. New York: Appleton-Century, 1940.

Hobhouse, L. T. MIND IN EVOLUTION. London: Macmillan, 1901.

Holmes, S. J. STUDIES IN ANIMAL BEHAVIOR. Boston: Richard C. Badger, 1916.

Hull, C. L. PRINCIPLES OF BEHAVIOR. New York: Appleton-Century, 1943.

Hull, C. L. A BEHAVIOR SYSTEM. New Haven: Yale University Press, 1952.

Kinnaman, A. J. Mental life of two Macacus rhesus monkeys in captivity. II. AMERICAN JOURNAL OF PSYCHOLOGY, 1902, 13, 173-218.

Köhler, W. THE MENTALITY OF APES. New York: Harcourt Brace, 1925.

Lashley, K. S. The mechanism of vision: XV. Preliminary studies
    of the rat's capacity for detail vision. JOURNAL OF GENERAL
    PSYCHOLOGY, 1938, 18, 123-193.

Maier, N. R. F., & Schneirla, T. C. Mechanisms in conditioning.
    PSYCHOLOGICAL REVIEW, 1942, 49, 117-134.

Morgan, C. L. AN INTRODUCTION TO COMPARATIVE PSYCHOLOGY. London:
    Walter Scott, 1894.

Mowrer, O. H. On the dual nature of learning: A reinterpretation
    of "conditioning" and "problem-solving." HARVARD EDUCATIONAL
    REVIEW, 1947, 17, 102-148.

Mowrer, O. H. LEARNING THEORY AND BEHAVIOR. New York: John
    Wiley & Sons, 1960.

Pavlov, I. P. CONDITIONED REFLEXES. Oxford: Oxford University
    Press, 1927.

Peterson, G. B., Ackil, J., Fromme, G. P., & Hearst, E. Condi-
    tioned approach and contact behavior toward signals for food
    or brain-stimulation reinforcement. SCIENCE, 1972, 172, 1009-
    1011.

Porter, J. P. A preliminary study of the psychology of the English
    sparrow. AMERICAN JOURNAL OF PSYCHOLOGY, 1904, 15, 313-346.

Rescorla, R. A., & Solomon, R. S. Two-process learning theory:
    Relationships between Pavlovian conditioning and instrumental
    learning. PSYCHOLOGICAL REVIEW, 1967, 74, 151-182.

Romanes, G. J. ANIMAL INTELLIGENCE. New York: D. Appleton, 1881.

Sidman, M. Avoidance conditioning with brief shock and no extero-
    ceptive warning signal. SCIENCE, 1953, 118, 157-158.

Sidman, M., & Fletcher, F. G. A demonstration of auto-shaping with
    monkeys. JOURNAL OF THE EXPERIMENTAL ANALYSIS OF BEHAVIOR,
    1968, 11, 307-309.

Skinner, B. F. THE BEHAVIOR OF ORGANISMS. New York: Appleton-
    Century-Crofts, 1938.

Small, W. S. Experimental study of the mental processes of the rat.
    AMERICAN JOURNAL OF PSYCHOLOGY, 1901, 12, 206-239.

Thorndike, E. L. ANIMAL INTELLIGENCE: EXPERIMENTAL STUDIES. New
    York: Macmillan, 1911.

Tolman, E. C. PURPOSIVE BEHAVIOR IN ANIMALS AND MEN. New York:
    Century, 1932.

Watson, J. B. Imitation in monkeys. PSYCHOLOGICAL BULLETIN, 1908,
    5, 169-178.

Watson, J. B. The place of the conditioned reflex in psychology.
    PSYCHOLOGICAL REVIEW, 1916, 23, 89-116.

Williams, K. A. The reward value of a conditioned stimulus. UNIVER-
    SITY OF CALIFORNIA PUBLICATIONS IN PSYCHOLOGY, 1929, 4, 31-55.

Woodard, W. T., & Bitterman, M. E. Autoshaping in the goldfish.
    BEHAVIOR RESEARCH METHODS & INSTRUMENTATION, 1974, 6, 409-410.

Yerkes, R. M. The mental life of monkeys and apes: A study of
    ideational behavior. BEHAVIORAL MONOGRAPHS, 1916, 3, 1-145.

Yerkes, R. M. The mind of a gorilla. GENETIC PSYCHOLOGICAL MONO-
    GRAPHS, 1927, 2, 1-193.

2. CLASSICAL CONDITIONING: THE PAVLOVIAN PERSPECTIVE

Vincent M. LoLordo

Dalhousie University
Halifax, Nova Scotia, Canada

Three treatments of an area of research called classical or
Pavlovian conditioning will be presented in the next three chapters.
The three treatments reflect three conceptual frameworks for organ-
izing the experimental literature on classical conditioning. Each
treatment reflects an increase in theoretical sophistication,
though none replaces all the key concepts of earlier treatments
with novel ones. Taken together, the three chapters should provide
an historical account of our ways of thinking about classical conditioning, and of the interplay of method and theory in this area.
In the first chapter, which outlines the Pavlovian perspective, a
large number of terms will be introduced. It should be some con-
solation that, once these terms are understood, they will be very
useful throughout these chapters.

Definition of a Classical Conditioning Experiment

Generally, in a classical conditioning experiment an organism
is presented with a stimulus which elicits, or is reliably followed
by, a response. The stimulus is called the unconditional or unconditioned stimulus (UCS), and the response, the unconditional or
unconditioned response (UCR). The designation "unconditional"
simply means that the response to the UCS does not depend on any
manipulation that occurs during the conditioning experiment, and
implies no assumptions about the ontogenetic status of the correlation between UCS and UCR. For example, it is not being assumed
that the UCR was present in the neonatal organism, or that it was

learned. However, it remains an empirical question whether such
factors affect the capacity of the UCS and the UCR to promote conditioning, or even affect the nature of the laws which describe
that conditioning.

The other discrete stimulus which is presented in a classical
conditioning experiment is the conditioned (conditional) stimulus
or CS. This is the stimulus to which a response is to be conditioned. Such a response is thus called a conditioned response or
CR. CSs are often called neutral stimuli, which means that they
do not initially elicit the to-be-conditioned response.

The procedural hallmark of classical conditioning, and the
characteristic which best distinguishes classical conditioning
from instrumental training, is the use of response-independent
stimulation. The UCS, whether it is electric shock which elicits
leg flexion, or food which elicits salivation, is presented without regard for the animal's prior or ongoing behavior.

The important features of a classical conditioning experiment
are best illustrated by an experiment from Pavlov's laboratory
(Anrep, 1920). The experiment was conducted in a new laboratory
especially designed for the study of classical conditioning. In
this laboratory the dog subjects were effectively isolated from
extraneous stimuli, including those emanating from the experimenter.
A periscope permitted observation of the dog from another room,
and presentation of stimuli and recording of salivation from a
fistulated parotid duct were accomplished by remote control. In
order to further reduce extraneous stimuli produced by struggling
and attempts to escape, before any conditioning experiments were
conducted the dogs used in Pavlov's laboratory were first adapted
to standing on a table, loosely harnessed by the collar and limbs.
In Anrep's study all four dogs were veteran subjects, and would
generally

> "run as quickly as possible, and of their own
> accord, into the inner room, jump unaided onto
> the table and lift each paw in turn to be
> placed in the string loop" (Anrep, 1920, p. 372).

Anrep's dogs were food-deprived, and would salivate copiously in
response to biscuit powder, the UCS used in the experiment.

A tone CS occasionally sounded for 3-5 sec, and then a plate
containing a premeasured amount of biscuit powder (UCS) was moved
within the dog's reach by remote control. When the dog had eaten
the powder, the plate was withdrawn automatically. These CS-UCS
pairings or conditioning trials were separated by long intervals

during which no stimuli were presented, so that each dog received only three or four trials in a one- or two-hr daily session.

Figure 1 depicts changes in the amount of salivation and the latency of salivation, or interval between the beginning of the CS and the occurrence of salivation, as a function of conditioning trials. The data were collected on test trials on which the tone was presented for 30 sec; every tenth trial was such a test trial. Test trials are necessary in cases where the latency of the conditioned response is longer than the CS-UCS interval, or interval between the start of the CS and the start of the UCS (e.g., Gormezano, 1972), and conditioning would not otherwise be revealed, or when the effectiveness of various CS-UCS intervals is being assessed (Bitterman, 1965). More often, however, researchers measure the CRs occurring during the CS-UCS interval on conditioning trials.

Neither dog salivated between trials, or in response to the initial presentation of the CS. Thereafter the amplitude of the salivary CR to tone increased as trials progressed, reaching an asymptote after 30 conditioning trials for both dogs. Further, the latency of the CR decreased over trials, and also was stable after 30 trials. Thus salivation, which initially occurred only in response to the biscuit powder, came to occur in response to a conditioned stimulus which preceded, and was temporally contiguous with, the unconditioned stimulus. In this instance the amplitude and latency measures of the CR were highly correlated, painting the same picture of the increasing strength of the conditioned response. However, such high correlations among measures have not always been obtained, and care must be taken when comparing experiments which use different measures of the strength of conditioning.

Goals of Research on Classical Conditioning

In his early lectures on what we now call classical conditioning, Pavlov characterized the primary task of research:

> "to concentrate our whole attention upon the
> investigation of the correlation between the
> external phenomena [CSs and UCSs] and the
> reaction of the organism, which in our case
> is the salivary secretion..." (Pavlov, 1928;
> pp. 50-51; material in brackets mine).

Thus Pavlov's immediate goal was to understand how the various experimentally arranged temporal relationships between CSs

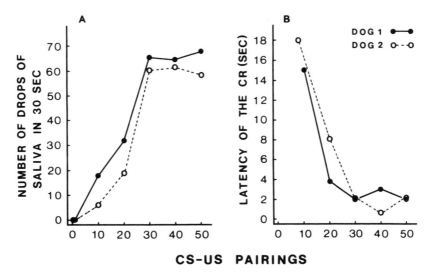

Figure 2.1   (A) The number of .01-ml drops of saliva secreted from
              the parotid glands of two dogs during 30-sec test pre-
              sentations of a tone CS which preceded food, as a func-
              tion of the number of CS-UCS pairings which preceded
              the test (Anrep, 1920); (B) The latency of the salivary
              response on the same test trials.

and UCSs determine changes in the probability of salivary responses
to the CSs.

     In pursuit of this goal, Pavlov decided to study the effects
of a great variety of environmental changes upon a single response
system, instead of exploring the effects of a limited range of
environmental changes upon many different responses. Pavlov's
decision was a wise one, for he was able to formulate fairly quickly
a surprisingly enduring set of statements describing the correlations between changes in the environment and changes in the probability of salivation.

     A research program which, like Pavlov's, studies only one
response system is bound to be limited to the extent that the
response is not representative of a class of "interesting" behaviors,
i.e., behaviors which seem important in the daily lives of animals.
In retrospect, salivation does seem to be a sufficiently representative response, since, as the next few chapters will illustrate, many
other responses are similarly affected by classical conditioning
operations.

Pavlov wanted to describe the correlations between changes in the environment and changes in behavior in order to understand the neurological mechanisms underlying conditioning. In this area Pavlov's contributions were less enduring, because he was working with a model of neural functioning that now seems inappropriate.

Contemporary researchers in classical conditioning generally share Pavlov's goals. The effect of various temporal relationships between CSs and UCSs on the formation of conditioned responses remains a major focus of research. The scope of this research has been expanded to include new experimental arrangements of conditioned and unconditioned stimuli. As methodological sophistication has increased, researchers have intensified their efforts to distinguish conditioned responses from behaviors which occur in conditioning experiments, but do not reflect associative processes. Competing theories of classical conditioning have been developed, and these have been related to accounts of other sorts of experiments on learning. A variety of different responses are being studied, in an effort to characterize differences in the responses of various systems to classical conditioning procedures. In recent years interest in the neural correlates of classical conditioning has grown.

This chapter will present a recent example of a classical conditioning experiment, and then use this example, along with Anrep's (1920) study of salivary conditioning in dogs, to introduce the Pavlovian account of classical conditioning.

Contiguity of CS and UCS

Presentation of a puff of air to the rabbit's cornea, or of an electric shock to the region around its eye, results very quickly in retraction of the eyeball, a sliding of the nictitating membrane across the nasal half of the cornea, and closure of the eyelids (Gormezano, 1966). The nictitating membrane rarely extends spontaneously, or as an initial response to an auditory CS, thus its use allows considerable scope for the demonstration of conditioning. When such putatively aversive UCSs as an airpuff or para-orbital shock are used, the procedure is often called defensive or aversive classical conditioning, as opposed to appetitive classical conditioning, which entails the use of putatively desirable events as UCSs.

In one of a series of parametric studies of the rabbit's classically conditioned nictitating membrane response, Smith, Coleman, and Gormezano (1969) assessed the effects of variations

in the CS-UCS interval. On conditioning trials, seven groups of
rabbits received a 50-msec auditory CS in various temporal relations to a 4-mA, 50-msec para-orbital shock UCS. For the present,
we will consider only four groups, for which the CS began 100, 200,
400, or 800 msec before the UCS. Such conditioning procedures,
in which the CS ends before the UCS begins, are called trace conditioning procedures, in contrast to delay conditioning procedures
in which the CS precedes but overlaps the UCS. The rabbits received
80 CS-UCS pairings at intertrial intervals of 50-70 sec on each of
eight sessions. Bitterman (1965) has pointed out that when all
measures of conditioning are made on trials on which CS and UCS
are paired, the CS-UCS interval is confounded with opportunity to
respond. This confound would favor longer CS-UCS intervals, and
would distort the function relating the growth of per cent CRs to
the CS-UCS interval. This problem can be circumvented by presentation of occasional CS-alone test trials on which all subjects have
somewhat more time than the longest CS-UCS interval in which to
respond. Every fifth trial Smith et al. presented a test trial on
which the CS was 1 sec long.

Figure 2 illustrates the percentage of test trials on which a
nictitating membrane CR occurred as a function of daily blocks of 21
test trials for the various groups. All groups of rabbits responded
on a large proportion of the test trials by the end of conditioning.
Groups 200 and 400 acquired the nictitating membrane response most
rapidly, reaching 100% CRs in a block of test trials within 300 conditioning trials. Group 800 reached 80% CRs within 600 trials, and
Groups 100, 60%, but performance was not asymptotic in these groups.
The latency of the conditioned response varied directly with the
CS-UCS interval. Parenthetically, since a 200-msec CS-UCS interval
resulted in a greater percent CRs, but a longer latency CR, than
a 100-msec interval, it is clear that response latency is not
always a simple measure of response strength.

On the basis of experiments on trace conditioning of the
salivary response which produced results formally similar to those
of Smith et al., Pavlov (1927) asserted that when the neural trace
of a conditioned stimulus was on repeated occasions present when
some unconditioned stimulus elicited a UCR, then the CS would
itself come to elicit a response (CR) like the UCR. To restate
this point, temporal contiguity of the neural trace of a CS with
the UCS and UCR was for Pavlov a necessary condition for classical
conditioning of a response to the CS.

Pavlov's stimulus trace notion has been developed by Hull
(1943, 1952), and more recently, modified by Gormezano (1972)
along the following lines. Suppose presentation of a peripheral

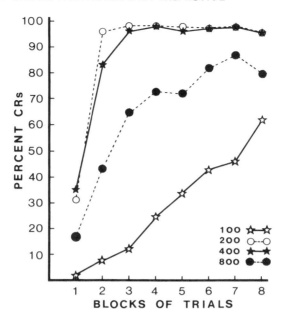

Figure 2.2   The mean percent nictitating membrane CRs as a function
of blocks of 21 test trials for groups of rabbits that
received trace conditioning with 100, 200, 400, and
800-msec CS-UCS intervals (Smith, Coleman, & Gormezano,
1969).

CS creates a neural trace in some critical "higher center" with a
latency of 30-40 msec, followed by rapid recruitment of neural
activity to some peak, and then a decline. Many variables, among
them CS modality, intensity, and rise time, should affect the slopes
of the functions describing the changes in the neural trace over
time. Nonetheless, the general inverted-U shape of the function
relating strength of the trace to time since CS onset should be
relatively constant. Thus, the CS-UCS interval should determine
the effective CS intensity at the time of UCS presentation. This
account makes several predictions. First, over some range of
CS-UCS intervals, the rate of conditioning should increase with
the CS-UCS interval, because the increasing interval should permit
the stimulus trace to grow to progressively larger values at the
time of UCS presentation. Second, the CR should be acquired most
rapidly at CS-UCS intervals that allow the trace to reach its peak
amplitude when the UCS is presented. Third, conditioning should

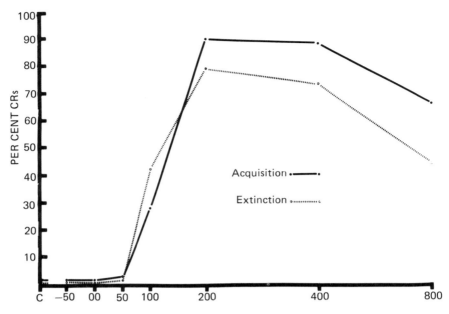

INTERSTIMULUS INTERVAL IN MSEC.

Figure 2.3   The mean percent nictitating membrane CRs for 168 test
             trials interspersed among conditioning trials as a
             function of CS-UCS interval for 7 groups of rabbits
             (Smith, Coleman, & Gormezano, 1969).

be progressively poorer at still longer CS-UCS intervals that
allow the trace of the CS to decay before the UCS is presented.
That is, the function relating CS-UCS interval and some measure of
the strength of conditioning, most commonly percent CRs, should
have an inverted-U shape.

     Figure 3 presents the overall percent CRs for 168 test trials
interspersed among conditioning trials for the four aforementioned
groups of rabbits in the study by Smith et al., as well as for
three additional groups that received the CS 50 msec before the
UCS, simultaneously with it, or 50 msec after the UCS (backward
conditioning).   The inverted-U shape of the obtained function
provides strong support for the modified Pavlovian account, as
does the absence of conditioning in the simultaneous conditioning,
backward conditioning, and 50-msec CS-UCS interval groups, for

which there should have been insufficient time for the growth of a strong stimulus trace. Parenthetically, when Patterson (1970) employed as the CS electrical stimulation of the inferior colliculus (an auditory relay nucleus), which should result in more rapid recruitment of a strong stimulus trace, a high level of nictitating membrane conditioning was obtained with a 50-msec CS-UCS interval.

For Pavlov, the development of a conditioned response reflected the growth of a connection between the neural representations of CS and UCS. In the studies by Anrep and Smith et al., the temporal contiguity between the neural trace of the CS or signal and the neural excitation produced by the UCS presumably led to a functional connection or association between the two, so that the CS came to evoke the neural excitation formerly evoked only by the UCS. Thus Pavlov called this procedure and its outcome "excitatory conditioning". Though stripped of its particular neurological referents by Western researchers, the term "excitatory conditioning" is still used as a label for cases in which a CS paired with some UCS comes to elicit a CR.

Most investigators have been interested in excitatory classical conditioning as a reflection of a learned association between CS and UCS. Given this shared associationist bias, researchers have tried to ascertain that their "excitatory CRs" were indeed conditioned, i.e., attributable to an associative process, and not to some non-associative factor. This is an extremely important point, and bears restatement: Simply setting up a relationship between a CS and a UCS that should produce an association between them does not ensure that a response which increases in probability during the CS reflects the formation of an association. It is possible that the response would have increased in probability during the CS even if the stimuli had been arranged so that there should have been no association between them. The student of association is thus confronted with the problem of devising experiments which allow him to state that his effects are associative.

Thus far the problem of controlling for non-associative effects in classical conditioning has been stated abstractly. It will now be concretized in two ways. First, the sorts of non-associative factors that have troubled researchers will be described. Then the proper control procedures, given a Pavlovian or contiguity-based account of conditioning, will be discussed.

Non-Associative Factors and Control Groups

Suppose that a CS is repeatedly paired with a UCS, and that some response increases in probability over successive presentations

of the CS.   Such a response might be a conditioned response,
reflecting the association of CS and UCS.   On the other hand, it
might be a pseudoconditioned or a sensitized response.

        Pseudoconditioning refers to the case in which a CS which
does not initially elicit a response like the unconditioned response
to the UCS comes to elicit such a response later, as a result of
intervening presentations of the UCS alone, even though the CS and
the UCS have never been paired.   In one widely-cited example of
pseudoconditioning, Grether (1938) presented four young rhesus
monkeys with the sound of an electric bell (CS), which elicited
at most a slight orientation response.   Later the monkeys received
10 presentations of the UCS, either the explosion of a small amount
of flash powder, or a "snake blowout" that uncurled toward the
monkey until the end was about 6 inches from the monkey's nose and
then rattled.   Each of these stimuli produced a short-lived fear
reaction, including struggling and vocalization.   When the bell
(CS) was again presented alone, the animals reacted as they did to
the UCSs.   This effect was retained for at least several days.
Similar results have been obtained in other studies, particularly
when noxious UCSs have been used (Harlow & Toltzien, 1940; Sears,
1934; Wickens & Wickens, 1942).

        "Sensitization" generally refers to cases in which the initial
response to the CS is facilitated by presentations of the UCS, even
though the CS and UCS have not been paired.   For example, if the
bell in Grether's experiment had originally elicited a fear reaction,
which would have disappeared if the UCS had never been presented,
then the outcome observed by Grether would have been called sensitization.   To the extent that the sensitized response is topographically similar to the CR observed in the experimental group which
receives CS-UCS pairings, then sensitization poses a problem for
the assessment of conditioning.   Sensitization has been observed
most often in studies of classical conditioning of the human eyeblink to CSs paired with an airpuff directed at the cornea (e.g.,
Grant & Adams, 1944).

        Other non-associative effects, both incremental and decremental,
will be discussed later.   However, armed with descriptions of sensitization and pseudoconditioning, we can now return to the problem
of how to assess the associative and non-associative contributions
to a given outcome, e.g., some relative frequency of CRs to a CS
which has been paired with some UCS.

        A researcher's approach to the assessment of non-associative
factors will be theory-bound, i.e., it will depend upon his notions
about the sorts of relationships between the CS and the UCS which
will promote the formation of an association between the two events.

Since the Pavlovian viewpoint maintains that the temporal contiguity between the trace of the CS and the occurrence of the UCS and UCR is the basis of the association formed during excitatory classical conditioning, the most straightforward way to assess the contribution of pseudoconditioning and sensitization to a putative example of excitatory conditioning would be to present a control group with the same number of CSs and UCSs per session as the experimental group, but with CS and UCS always separated by a long interval. Such a procedure would deprive the control group of the relationship between CS and UCS, namely temporal contiguity, which is essential for excitatory conditioning according to the Pavlovian view. Several specific procedures which fulfill this formal requirement have been used as controls for pseudoconditioning and sensitization. They include: (1) Backward conditioning, in which UCS onset and the occurrence of the UCR precede the start of the CS, so that the trace of the CS cannot be present when the UCS and the UCR occur, and (2) Unpaired presentations of CS and UCS, in which either the CS or the UCS occurs on a trial, with trials separated by relatively long intervals. Statistical comparisons between some measure of responding to the CS in these control groups, e.g., percent CRs over successive presentations of the CS, and the same measure in the experimental group (which received CS-UCS pairings) would allow a researcher to state that excitatory conditioning had occurred, i.e., that the growth of the response to the CS was an associative effect, beyond any non-associative effects that might have occurred.

In the nictitating membrane conditioning experiment of Smith et al. (1969), a group of rabbits that received backward conditioning trials in which onset of the 50-msec shock UCS preceded onset of the auditory CS by 50 msec produced virtually no nictitating membrane responses on test presentations of the CS. The same result was obtained in a control group that received only the CS on some trials, and only the UCS on others, with the intertrial interval halved in order to hold session length constant across groups. Thus there was no evidence for pseudoconditioning or sensitization of the rabbit's nictitating membrane response, and the growth of the response in the experimental groups can be called associative, given the Pavlovian viewpoint.

Though neither pseudoconditioning nor sensitization appears to play a major role in conditioning of either the salivary or nictitating membrane responses, the results of Grether's (1938) study, eyelid conditioning experiments with humans (e.g., Grant Adams, 1944) and other research using noxious UCSs (e.g., Harris, 1943), as well as recent research on classical conditioning of the rabbit's jaw movement response to a CS paired with water squirted into the mouth (Sheafor & Gormezano, 1972; Sheafor, 1975), indicate that the possibility of pseudoconditioning or sensitization should not be dismissed in the absence of appropriate control procedures.

Thus far pseudoconditioning and sensitization have been treated simply as factors to be assessed, so that the magnitude of more "interesting" conditioned, i.e., associative, effects can be determined. Although this has been the dominant point of view, there have always been investigators interested in the mechanism of pseudoconditioning, including some who, viewing the requirements for the formation of associations more liberally, have suggested that pseudoconditioned responses are learned, and depend upon the similarity of the CS and the UCS (cf. Rescorla & Furrow, 1977; Testa, 1975; Wickens & Wickens, 1942). Interest in this issue is growing rapidly, and the next few years should yield a marked increase in our understanding of sensitization and pseudoconditioning, and their relation to classical conditioning (Kandel, 1976).

Extinction and Inhibition

If excitatory conditioning were unopposed by another process, then once conditioned responses were established they might always be present. But the environment changes, and a conditioned stimulus that was formerly followed by a UCS may no longer be. Pavlov (1927) arranged such conditions in the laboratory. A CS which had been reliably followed by food, and which elicited conditioned salivation, subsequently was presented repeatedly without the UCS in a procedure called experimental extinction. The latency of the salivary CR gradually increased, and the amount of conditioned salivation gradually decreased, over extinction trials. Eventually the CS elicited no salivation.

According to Pavlov, strong conditioned responses recovered completely on a test trial, given a sufficient rest period following extinction. This phenomenon is called spontaneous recovery. Typically, however, in North American experiments spontaneous recovery has been incomplete (e.g., Wagner, Siegel, Thomas, & Ellison, 1964). Spontaneous recovery led Pavlov (1927) to reject the possibility that the excitatory connection formed during conditioning had decayed or disappeared during extinction. Instead he argued that inhibition of salivation was elicited by the unreinforced CS during extinction, and that the inhibition summed algebraically with the already conditioned excitation, resulting in the observed salivary response. On this argument, as extinction progressed, inhibition accumulated, and ultimately masked the excitatory effect. For Pavlov spontaneous recovery further implied that the inhibitory process faded more rapidly than excitation. Another phenomenon observed by Pavlov (1927) also suggested

that inhibition was relatively fragile. If an extinguished CS
which had elicited no salivation over the preceding few unreinforced
trials was suddenly accompanied by a novel, moderately intense
stimulus, salivation would occur, though the novel stimulus itself
would have elicited no salivation. Pavlov called this phenomenon
disinhibition, implying that an inhibitory process which was
masking conditioned excitation was somehow neutralized by presentation of the novel stimulus, thereby freeing the underlying excitation.

Pavlov maintained that extinction was only one of several
procedures which yield internal, or conditioned, inhibition.
Another such procedure he labelled "conditioned inhibition": one
CS (CS+) was reliably followed by the UCS, unless that CS+ was
preceded and overlapped by another CS (CS-), in which case the
compound was not reinforced. Initially the compound of CS- and CS+
would elicit some salivation, but after many sessions of the conditioned inhibition procedure neither the compound nor the CS-
presented alone resulted in salivation. If we identify conditioned
inhibition as an active tendency not to produce a particular CR,
then these data suggest that the CS- had become inhibitory, but
do not prove it. Alternatively, the CS-CS+ compound and the CS-
might have become neutral. This distinction can be restated as
follows: Following many sessions of the conditioned inhibition
procedure, did the CS- elicit a tendency not to salivate, or did it
fail to elicit any response tendency at all?

Pavlov answered this question to his satisfaction with a
further experiment. He reasoned that the demonstration that a
given CS is a conditioned inhibitor of some response requires that
the response would have otherwise occurred, if the inhibitor had
not been presented. Pavlov presented the CS- in compound with
another CS+, which also had been paired with the food UCS, and
found that the new compound CS elicited less salivation than did
the other CS+ alone. Pavlov interpreted the outcome of this procedure, which will be called a summation test (Rescorla, 1969),
as a reflection of the algebraic summation of excitatory and inhibitory tendencies which had been conditioned to CS+ and CS-,
respectively.

Pavlov (1927) also observed conditioned inhibition of salivation in a procedure related to, but somewhat simpler than, the
conditioned inhibition procedure. In this procedure, which has
been called discriminative or differential conditioning, one CS
(CS+) was reinforced, or followed by the UCS, whenever it occurred,
but a second CS (CS-) occurred on other trials and was never
reinforced. This procedure can be described as the conditioned
inhibition procedure with the CS+ omitted on non-reinforced trials.
Often the CS- elicited considerable salivation on its first few

presentations, particularly if it was physically similar to the
CS+. This phenomenon is known as stimulus generalization, referring
to the generalization or spread of the excitatory effects of one
CS to other stimuli. Stimulus generalization will be discussed
in detail in a later chapter; for the moment it is sufficient to
say that the amount of stimulus generalization tends to increase
with increasing physical similarity of the CS and the test stimulus.
In any case, with repeated application of the differential conditioning procedure during session after session, the amount of
salivation elicited by the CS- would gradually diminish, and would
often reach a zero or near-zero level.

Again, Pavlov asked whether the CS- which no longer elicited
salivation was an inhibitor of salivation, or simply a neutral
stimulus. Several sorts of evidence suggested that the CS- was
inhibitory. First, Pavlov observed disinhibition of salivation
when a novel stimulus, the odor of amyl acetate, was added to a
tone CS- which was perfectly differentiated. The tone elicited
four drops of saliva when accompanied by the amyl acetate, although
it had elicited no salivation when presented alone (Pavlov, 1927;
p. 128). Second, Pavlov observed a marked reduction in salivation
to the CS+ tone when it was presented only one min after a perfectly
differentiated CS- tone, instead of after the usual 20-30 min
intertrial interval. Pavlov (1927; p. 125) attributed this outcome
to the algebraic summation of the inhibitory after-effect of the
CS- with the excitatory effect of the CS+. Conceptually, this
procedure is similar to the summation test procedure which was used
to assess the status of the CS- following the conditioned inhibition
procedure.

One other procedure which, according to Pavlov, led to internal
or conditioned inhibition is called inhibition of temporal delay.
Pavlov (1927) identified as long-delay procedures those in which a
CS began a relatively long time, up to several min, before presentation of the UCS, and remained on until the UCS was presented. He
maintained that in long-delay procedures the asymptotic latency
of the salivary CR would be long, and proportional to the CS-
UCS interval. Further, he asserted that the early part of a long-
delay CS came to elicit inhibition of salivation. This assertion
was based on the disinhibitory effect upon salivation of moderately
intense, novel stimuli. Such stimuli, which would have elicited
no salivation when presented alone, evoked considerable salivation
when presented during the early part of a long-delay CS, as shown
in Table 1. In this study by Zavadsky in Pavlov's laboratory
(Pavlov, 1927; p. 93), tactile stimulation of the skin preceded a
squirt of weak citric acid solution into the dog's mouth by three
min and overlapped it. Typically the latency of the salivary CR
was one to two min, with the amount of saliva per 30-sec period

TABLE I

The number of drops of saliva elicited during successive
30-sec intervals during a three-min tactile CS that
preceded weak citric acid in a dog's mouth. On the
third of five trials the tactile CS was accompanied
by a novel stimulus (M), the ticking of a metronome
(Zavadsky, in Pavlov, 1927); the metronome alone
evoked no salivation.

Time	Stimulus	Drops of Saliva
9:50	tactile CS	0, 0, 3, 7, 11, 19
10:03	tactile CS	0, 0, 0, 5, 11, 13
10:15	tactile CS+M	4, 7, 7, 3, 5, 9
10:30	tactile CS	0, 0, 0, 3, 12, 14
10:45	tactile CS	0, 0, 5, 10, 17, 19

increasing progressively thereafter. However, when a novel
stimulus, the ticking of a metronome, was presented along with the
CS, the latency of salivation was markedly reduced, and the amount
of salivation during the first min of the CS increased considerably.
Pavlov asserted that this outcome revealed the presence of a
relatively fragile, inhibitory CR during the early part of the
long-delay CS. Parenthetically, presentation of the metronome also
somewhat suppressed salivation during the later, excitatory part of
the CS. In the Pavlovian literature this effect is called external
inhibition.

Although no one has used the outcome of a summation test to
demonstrate that the early part of a long-delay CS paired with food
is an inhibitor of salivation, several studies using other responses
(Rescorla, 1967, 1968; Rodnick, 1937) have obtained reduced conditioned responding to CS+s presented during the early part of long-
delay or long-trace CSs.

In summary, Pavlov maintained that an extinguished conditioned
stimulus, the non-reinforced CSs (CS-s) in the conditioned inhibition and differential inhibition procedures, and the early part of
a long-delay or long-trace conditioned stimulus asymptotically
elicited a conditioned inhibitory response. Conditioned inhibition
could be revealed by showing that responding to a CS+ which had been
paired with the same reinforcer was reduced when the CS+ was
compounded with the putative inhibitor, by demonstrating inhibitory
after-effects of the latter, or by demonstrating disinhibition. In
the next section a recent discussion of the experimental analysis
of inhibition will be considered.

Criteria for Inhibition

Thus far, the discussion of the criteria for inhibition has
not been very rigorous, although some of Pavlov's criteria have
been described. Rescorla (1969) has suggested that, if a given
change in the probability of a CR resulting from some relationship
between that CS and a UCS is called conditioned excitation, then
the term inhibition should be reserved for a change in the
probability of the CR in the opposite direction, as the result of
a different relationship between the CS and the UCS. For example,
if the probability of the rabbit's nictitating membrane response
during a CS increases over successive pairings of that CS with
para-orbital shock, and this phenomenon is called excitatory conditioning, then another stimulus (or the same stimulus in another
group of rabbits) should only be called an inhibitor if it comes to
elicit a decreased probability of the nictitating membrane response
as a result of being placed in some other relation to para-orbital

shock. Given this definition, to which we will return in the
next chapter, Rescorla (1969) proposed two experimental tests of a
putative inhibitor, summation tests and retardation of acquisition
tests.

Examples of the summation test, in which the suspected inhibitor is compounded with an excitatory CS based on the same UCS,
and responding to the CS- CS+ compound is compared to responding
to CS+ alone, have already been presented. If the CR to the
compound is reliably smaller or less probable than the CR to CS+
alone, then it is inferred that the CS+ and CS- elicit opposing
response tendencies, and the CS- is called an inhibitor. Sometimes
summation tests compound the suspected inhibitor with the UCS
itself (Wagner, Thomas, & Norton, 1967), or with some background
cues which are presumably excitatory)e.g., Rescorla & LoLordo,
1965), but the logic of the experiments remains the same.

The retardation of acquisition test is also based on the
definition of excitatory and inhibitory conditioned responses as
opposites. Given this definition, it should be more difficult to
transform an inhibitory CS into an excitatory CS than to transform
a neutral CS into an excitatory CS. A retardation of acquisition
test simply compares the rates at which an excitatory CR is
conditioned to a putatively inhibitory CS and to some control CS,
when each is repeatedly followed by the UCS. An excitatory CR
should develop more slowly in response to a formerly inhibitory
stimulus. Rescorla (1969) has suggested that evidence that a given
stimulus both reduces responding to a CS+ in a summation test and
is relatively slow to elicit an excitatory CR in a retardation
of acquisition test constitutes the strongest demonstration that
the stimulus is an inhibitor, because each test rules out a different sort of alternative account which does not require the
concept of inhibition. This point will be reconsidered in the
next chapter, after some alternatives to inhibition have been
discussed.

A methodological issue that arises from the proposed definition
of inhibition, and that parallels an issue discussed earlier with
regard to conditioned excitation, must be mentioned now. As in the
case of excitatory effects, the question "how can we ascertain that
an inhibitory effect of a CS is a conditioned effect?" arises
because of the associationist bias in our thinking about classical
conditioning. Suppose the CS- from differential conditioning comes
to elicit an inhibitory CR, as determined by the outcomes of
summation and retardation tests. What additional procedures, if
any, must be completed before we can be sure that the inhibitory
effect depended upon the critical relationship between stimuli, and
would not have occurred otherwise? Again paralleling the discussion

of conditioned excitation, the answer to this question depends on
one's particular specification of the "critical relationship,"
i.e., upon one's particular theory of inhibition. For this reason,
the answer to this question will be deferred until an account of
inhibition has been presented.

Pavlov's Account of Inhibition

In his 1927 book, <u>Conditioned Reflexes,</u> Pavlov maintained that
simply presenting a CS repeatedly was sufficient for the development
of inhibition in response to that CS. However, the inhibition would
be masked by the growth of a stronger, opposite effect, excitation,
if the CS was followed by the UCS. Thus the development of inhibition would be suspected fairly quickly only when the CS was
presented in the absence of the UCS, and there was no continued
growth of excitation, as in the case of extinction and the other
Pavlovian examples discussed thus far.

One aspect of Pavlov's account of inhibition is particularly
noteworthy. It seems to be a non-associative account in which the
development of inhibition is based upon sheer repetition of the
conditioned stimulus. This aspect of the account is difficult to
reconcile with Pavlov's characterization of the extinction, conditioned inhibition, differential conditioning, and long-delay
conditioning procedures as means of producing internal or conditioned inhibition.

Pavlov's claim that repetition of the CS is the basis of
inhibition has several implications. First, it suggests that the
paradigmatic case of inhibition should be repeated presentation
of a CS alone, prior to any other manipulation. Such a procedure,
which has been called the "latent inhibition" procedure in North
America, has been studied extensively in recent years. If a CS
is repeatedly presented, and no UCSs are presented, prior to the
arrangement of repeated pairings of that CS and some UCS, then
excitatory conditioning to the CS is retarded, relative to a
control group which received no preconditioning exposure to the
CS (Lubow, 1973). Considered in isolation, this "latent inhibition"
effect is consistent with Pavlov's account of inhibition. A
similar, though usually greater, retardation of excitatory classical conditioning has been observed when former CS-s from conditioned
inhibition (e.g., Marchant, Mis, & Moore, 1972; Marchant &
Moore, 1974; Rescorla & LoLordo, 1965) and differential conditioning (e.g., Hammond, 1968; Konorski & Szwejkowska, 1952; Wessells,
1973) procedures were repeatedly paired with the UCS. Thus the
data from retardation of acquisition tests are consistent with the
Pavlovian claim that simple repeated presentations of a CS render

that CS functionally equivalent to the conditioned (internal) inhibitory stimuli produced by conditioned and differential inhibition procedures. However, there are other data which demonstrate that this claim is incorrect, and that the so-called "latent inhibition" procedure does not yield an inhibitory stimulus (Rescorla, 1971; Riess & Wagner, 1972).

Rescorla's experiments are the more decisive, and they will be described in detail. They employed the Conditioned Emotional Response (CER) procedure (Estes and Skinner, 1941), which has become the most widely used procedure for the study of classical conditioning in North America. In the first phase of a CER experiment appropriately deprived animals are permitted to consume some positive reinforcer in the experimental chamber. Then they are given the reinforcer whenever they perform some instrumental response, e.g., pressing a lever. The most common schedule of positive reinforcement in CER studies is the variable-interval (VI) schedule, which reinforces the first instrumental response after a variable interval of time, averaging t, since the previous reinforcement. Variable interval schedules produce rather steady rates of responding throughout a session, such that the extent of variation from the baseline rate can be used to assess the effects of imposed stimuli.

The second phase of a CER experiment involves a classical conditioning procedure. Sometimes this procedure is superimposed upon the instrumental baseline; in other experiments it is conducted separately, with the response manipulandum unavailable to the animal. The second variation, which eliminates some confounding, will be described here. A typical procedure would include daily one- or two-hr sessions during which there would be four presentations of a three-min CS, each immediately followed by an electric shock UCS delivered to the feet of the animal through the grid floor.

In the final phase of the experiment the animal is again permitted to perform the instrumental response, the CS is presented several times, and its effect on the rate of instrumental responding is determined. It is generally assumed that the strength of classical conditioning that occurred in the preceding phase would be manifested in the strength of a peripheral CR that somehow interferes with performance of the instrumental response (e.g., Kamin, 1965), or of a conditioned motivational state that somehow cancels some of the appetitive motivation supporting the instrumental response (Millenson & de Villiers, 1972). In either case, the degree of suppression of the ongoing instrumental performance should directly reflect the strength of classical conditioning.

The most commonly used measure of the effects of a CS upon
some baseline instrumental response is the suppression ratio (Annau
and Kamin, 1961). This measure is expressed as the ratio A/(A+B),
where A is the number of responses during the CS, and B is the
number of responses in an equally long interval immediately prior
to the CS. The suppression ratio has a value of 0.5 when the CS
has no effect on instrumental behavior, a value greater than 0.5
when the CS has a facilitative effect, and a value approaching
zero as the CS progressively suppresses behavior. The suppression
ratio has been widely used on the assumption that it will be
constant despite variations in the baseline rate of responding
(Hoffman, 1969). However, recent data suggest that this is not the
case (Blackman, 1968; Millenson & de Villiers, 1972), and
Blackman (1977) has concluded that the possibility that the suppression ratio is not an uncontaminated measure of the CR should
lead researchers to present absolute response rates prior to and
during CSs, as well as suppression ratios (cf. Lea and Morgan,
1972). Further, comparisons between groups or conditions with
different baseline rates should be interpreted with caution.

Now we can return to Rescorla's experiments on the effects of
prior unreinforced presentations of the CS upon subsequent conditioning. First, confirming results already cited, he found that rats
which had received 120 unreinforced presentations of a 2-min auditory
CS subsequently acquired conditioned suppression of lever-pressing
during the CS more slowly than controls which had received no
pre-conditioning exposure to the CS. On the other hand, in a
second study the amounts of suppression to a visual CS+ and to a
simultaneous compound of the pre-exposed auditory CS and the CS+
were equivalent, suggesting that unreinforced presentations of a
CS do not render that CS inhibitory. Furthermore, since CS-s from
differential conditioning and conditioned inhibition procedures
do reduce responding to a CS+ in summation tests (e.g., Pavlov,
1927; Rescorla & Holland, 1977; Szwejkowska & Konorski, 1959), a
pre-exposed CS is not functionally equivalent to such stimuli.

Rescorla (1971) provided conclusive evidence against the claim
that unreinforced presentations of a CS render it inhibitory by
demonstrating that inhibitory classical conditioning is also
retarded if the CS has previously been presented in the absence of
reinforcement. One group of rats received 120 unreinforced
presentations of the 2-min auditory CS, while a control group
was simply placed in the chamber. Then both groups received eight
pairings of a visual CS and electric shock, to produce a CS+ which
could be used in a subsequent test. Finally, both groups were
administered a conditioned inhibition procedure. During each
2-hr session, one pairing of the visual CS+ and electric shock
and three unreinforced compounds of the visual and auditory CSs were

superimposed upon the instrumental baseline of food-reinforced lever-pressing. Both groups acquired virtually complete conditioned suppression to the visual CS+, reflecting strong excitatory conditioning. Figure 4 illustrates the course of the conditioning of inhibition to the auditory CS, as indicated by an increase in the difference between the suppression ratios to the compound and to the CS+ alone. An increasing difference score can be identified with an increasing tendency for the tone to reduce suppression conditioned to the light. Thus, these data reflect the acquisition by the tone of the capacity to inhibit the CER. Inhibitory conditioning to the tone was retarded in the group (Group T) which had received unreinforced presentations of the auditory CS as its first treatment. This result, which has also been obtained by Best (1975) in a study of conditioned aversions to gustatory cues in rats, indicates that the latent inhibition procedure does not yield conditioned, i.e., associative, inhibition. Since pre-exposure to a CS has equivalent, decremental effects upon subsequent excitatory and inhibitory classical conditioning to that CS, the effect of such pre-exposure is non-associative. Alternatively, perhaps an organism comes to ignore a CS which bears no relation to any significant event in its environment (e.g., Lubow, Alek, & Arzy, 1975; Mackintosh, 1975).

The Pavlovian conditioned inhibition and differential inhibition procedures do yield inhibitory stimuli, according to the outcomes of summation and retardation tests, however an extinguished CS apparently does not become an inhibitor. In a review of Pavlovian conditioned inhibition, Rescorla (1969) noted that there is no strong evidence that an extinguished CS becomes inhibitory. On the contrary, Reberg (1972) presented summation test data from a CER procedure with rats that suggest that an extinguished CS, even one extinguished far beyond the point at which it produces no response when presented alone, is still slightly excitatory.

Reberg (1972) trained rats to depress a lever for food reinforcement on a variable-interval schedule, and then superimposed pairings of CSs and shocks upon the appetitive baseline. In separate trials, auditory and visual CSs (CS1, CS2) each terminated with a brief electric shock. Excitatory conditioning was continued until lever-pressing was completely suppressed during each CS. CS1, which was to be used as a comparison stimulus in a subsequent summation test, was partially extinguished. Then extinction to CS2 began. One group received six extinction trials per day until CS2 no longer produced suppression (group .45). A second group (group .45+54) was given the same treatment, but received an additional 54 extinction trials after suppression had disappeared. The extra extinction trials were given because extinction is known to continue beyond the point at which the

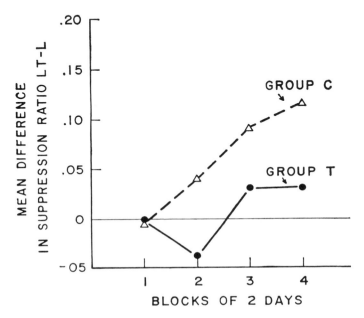

Figure 2.4   The mean difference in suppression ratios on light-tone
             and tone trials for groups of rats that received prior
             unreinforced exposures to the tone (Group T) and no
             such pre-conditioning exposure (Group C).   An increas-
             ing difference score reflects conditioning of inhibi-
             tion to the tone (Rescorla, 1971).

observable CR, in this case conditioned suppression, disappears. Reberg wanted to allow this "extinction below zero" (Pavlov, 1927) to occur, in order to maximize the possibility of producing an inhibitory CS by means of extinction.

The summation test began on the day after extinction was completed. Unreinforced presentations of CS1, CS2, and the compound of CS1 and CS2 were administered in various orders. The compound of CS1 and CS2 produced greater suppression of lever-pressing than did CS1 alone in both groups, suggesting that CS2 was still a weak excitatory CS even after 54 trials on which it produced no suppression. This conclusion is strengthened by other experiments from the same report which showed that: (1) the compound of a moderately suppressive CS1 and a CS2 which had been paired with shock only once produced greater suppression than did CS1 alone, and (2) the compound of a moderately suppressive CS1 and a CS which had never been paired with shock produced the same amount of suppression as CS1 alone.

Reberg's data are compatible with the common finding that reacquisition of an excitatory CR following extinction occurs more rapidly than original acquisition (e.g., Konorski & Szwejkowska, 1950), but contrary to the view that extinction renders a CS inhibitory. Despite this result, it is still plausible to assert that an associative process contributes to the decrement in an established CR over repeated presentations of the CS alone. For example, such a process could stop when the net associative strength of the extinguished CS reached zero, i.e., when the CS was neither excitatory nor inhibitory. This possibility will be considered in detail in a subsequent chapter.

In summary, although Pavlov's research on conditioned inhibition provided much of the empirical base of contemporary research on the topic, it is now apparent that procedures that Pavlov believed to have equivalent, inhibitory effects must be reclassified. Sheer repetition of an unreinforced CS, as occurs in the latent inhibition procedure, does not result in the same sort of conditioned, i.e., associative, inhibitory effect that is produced by the conditioned inhibition, differential inhibition, and very likely the inhibition of delay procedures. Nor does even protracted extinction yield the same sort of inhibitory stimulus, though associative effects may occur during extinction. The necessary and sufficient conditions for the establishment of conditioned inhibition, as well as its mechanism, remain lively issues today, and will form much of the basis for the next chapter.

Thus far the discussion has focussed on the question: "What environmental conditions are necessary and sufficient for the

development of excitatory and inhibitory stimulus control?" Most
of the experiments which have been considered here have used
salivary conditioning, nictitating membrane conditioning, or the
CER procedure, and little attention has been paid to the diversity
of behaviors which can be classically conditioned. The last section
of this chapter will focus on such diversity, by examining the
relation between the UCR and the CR in a variety of classical
conditioning procedures.

Morphology of CRs and UCRs

Pavlov's (1927) view of the morphological relationship between
the CR and UCR has been called stimulus substitution. Based
largely upon cases in which a CS repeatedly paired with the presentation of food to a restrained dog came to evoke a salivary
CR like the unconditioned salivary response to food, the view has
been interpreted to state that the CS simply comes to substitute
for the UCS in evoking the UCR. On this account the CR is the UCR,
moved forward in time. The stimulus substitution account would
seem to be most compatible with a theory of conditioning which
asserts that the CS is directly associated with the UCR, but it is
not incompatible with the claim that S-S (CS-UCS) associations are
the basis of classical conditioning.

Moore (1971, 1973) has proposed that a second variety of
stimulus-substitution hypothesis can also be found in Pavlov's
writings. This view, which will be called object-substitution
(Hearst & Jenkins, 1974), maintains that an animal will treat
the CS as if it were the UCS, so long as the stimulus properties
of the CS, e.g., its location, accessibility, three-dimensionality,
etc., permit such responses. The object-substitution view, which
can apply only to motor responses, asserts that animals will
approach and contact accessible, localized CSs which precede food.
Further, the form of the contact should be similar to the form of
the contact with the UCS.

When we consider studies in which the CS is localized, and
there is a motor UCR, the two stimulus-substitution views make
different predictions. The traditional account predicts that the
motor CRs will be directed towards the site of UCS delivery, but
the object-substitution account (Moore, 1971) predicts that the
CRs will be directed towards the site of the CS. Jenkins and Moore
(1973) have obtained data which support the object-substitution
account.

Pigeons were placed in a standard Skinner box which contained
a square aperture behind which was a grain magazine. Whenever grain

was brought within a pigeon's reach, this magazine was illuminated. Located on each side of the magazine, and at roughly the height of a standing pigeon's beak, were two translucent disks called keys. Various colors and patterns could be projected onto these keys from behind, and a microswitch recorded any contacts with the keys. In the experiment by Jenkins and Moore (1973) two sorts of conditioning trials were presented in a mixed order. On half the trials illumination of one key with red light was followed six sec later by the presentation of water; on the other half illumination of the other key with stripes was followed in 6 sec by grain. As Figure 5 illustrates, the birds learned to approach the CSs and contact them with their beaks. This classically conditioned key-pecking or autoshaping, first reported by Brown and Jenkins (1968), has been the focus of much recent research, because it points to the heretofore neglected possibility of the classical conditioning of directed responses. As a later chapter will demonstrate, classically conditioned, directed responses are playing an increasingly important role in the explanation of instrumental behavior.

The autoshaped pecks in the study by Jenkins and Moore (1973) were morphologically similar to contacts with the appropriate reinforcer. Pecks at the signal for grain were forceful, and were made with the beak open at the time of contact, much as a pigeon picks up grain. Pecks on the CS for water were less forceful, and the beak was closed or almost closed at the time of contact. The CR often included the slow, rhythmic, low-amplitude opening and closing of the beak which chracterizes drinking. Three of the six birds were even observed to lick towards the signal for water.

These data suggest that object substitution prevails over responses to the site of the UCS, but Moore (1971) has noted that pigeons sometimes peck at the food hopper during the CS, even when that CS is an illuminated key. Thus far there has been little research on the characteristics of stimuli that make them "attractive," or give them appropriate object-quality in a given situation, or on the determinants of response location when both the CS and UCS sites are attractive.

When the CS is diffuse, e.g., a change in the color of illumination in the experimental chamber, and the UCS is elevation of the grain magazine, then conditioned pecking still seems to occur. Shapiro, Jacobs, and LoLordo (1977) have reported that such pecking is directed towards the grain magazine, however Newlin (1974) observed only that it tended to occur in the third of the chamber that contained the food magazine.

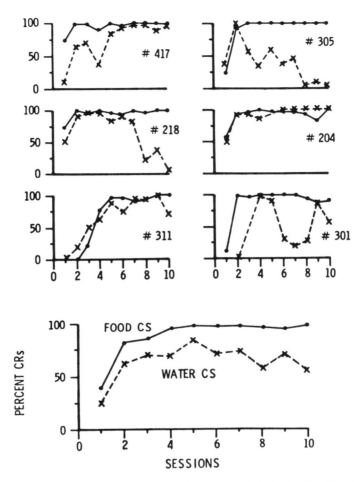

Figure 2.5   The proportions of trials on which individual pigeons
pecked at least once on keylight CSs which signalled
food (solid lines) and water (dashed lines), as well
as the group mean proportions, plotted as a function
of sessions.   The low water-key response rates shown
briefly by subjects 218 and 305 were due to reinforce-
ment-delivery or water-balance problems not experienced
by other subjects (Jenkins & Moore, 1973).

Thus far the discussion has been consistent with the claim that one or the other form of stimulus-substitution hypothesis predicts performance in classical conditioning experiments: object substitution in the case of free-moving animals responding to localized signals for appetitive UCSs, and the traditional view in the cases of non-localized CSs and in conditioning of non-directed skeletal responses and autonomic responses. However, several outcomes are difficult to reconcile with a stimulus-substitution account which asserts that the form of the CR must be like the form of the response to the UCS during conditioning.

There have been several recent demonstrations that animals will approach and contact a localized CS even when receipt of the UCS requires no approach response. For example, Wasserman (1973) placed three-day-old chicks in a cooled incubator and presented them with repeated classical conditioning trials in which eight-sec illumination of a green key on the wall immediately preceded four-sec activation of an overhead heat lamp. Wasserman describes the chicks' reactions to the heat lamp:

> "When first placed in the refrigerated chamber, the chicks showed highly agitated behavior, racing to and fro and cheeping loudly. The activation of the heat lamp produced a marked change in their behavior. The chicks stopped scurrying about, extended their wings, and often emitted twittering sounds. Sometimes the birds would stiffen into a "head up" posture; at other times they would lower their bodies and occasionally rub the floor with their chests." (Wasserman, 1973, p. 876).

The chicks began pecking the keylight CS after a median of eight CS-UCS pairings, and generally pecked on 70-80% of the trials thereafter. Sometimes they vocalized while pecking, but they did not extend their wings or lower their bodies. Data from a control group which received independent presentations of CS and UCS revealed that the pecking acquired by the chicks in the experimental group was indeed dependent upon the pairings of CS and UCS, i.e., it was an associative effect. Wasserman concluded that instrumental approach and contact of the UCS were unnecessary for the conditioning of approach and contact of the CS (cf. Woodruff & Williams, 1976).

Hogan (1974) has noted that chicks will approach a broody hen, peck at the feathers on the underpart of her body, and snuggle up to her--responses like those observed in response to the keylight by Wasserman--and that these behaviors often induce the hen to sit. This comment implies that the CRs observed by Wasserman are like the normal unconditioned responses to heat, and thus are compatible with a liberalized object-substitution account.

Timberlake and Grant (1975) have taken an ethologically oriented approach to the description of directed conditioned responses and their relation to the unconditioned response. They argued that when a localized CS preceded the presentation of some UCS, an entire system of species-typical behaviors commonly related to that UCS would be conditioned. Moreover, they asserted that the form of the CR depends on "which behaviors in the conditioned system are elicited and supported by the predictive stimulus" (Timberlake & Grant, 1975, p. 692). Thus when the insertion of a live rat strapped to a board signalled the presentation of food to a hungry rat, the conditioned contacts with the rat CS were very different from the contacts with food. Subjects approached the rat CS, contacted it with one or both forepaws, groomed it, and climbed over it, with at least three legs off the cage floor, i.e., they directed social behaviors towards it. They did not bite the CS rat. These outcomes cannot be easily explained by even a liberalized object-substitution account.

The stimulus substitution account also faces a problem when the same response system is affected by both CS and UCS, but the CR and UCR are in opposite directions. Some of the most interesting examples of this phenomenon come from studies in which pharmacologic agents are used as UCSs. Siegel (1975, 1976) has demonstrated such an effect in a provocative series of experiments on tolerance to the analgesic effects of morphine.

Morphine initially has potent analgesic effects. In one standard test, the latency with which a rat licks its paw after being placed on a hot surface is markedly increased if the test is administered 30 min after the injection of morphine, as compared to the latency for saline-injected controls. Unfortunately, from a medical point of view, the analgesic effect of morphine exhibits marked tolerance, i.e., the analgesic effect of a given dose of morphine decreases progressively with repeated administrations of the drug. Operationally, the rat's paw-lick latency decreases progressively on tests following repeated administrations of morphine.

Although there have been several physiological theories of narcotic analgesic tolerance (cf. Siegel, 1975), Siegel proposed that such tolerance could be explained on the basis of classical conditioning of "compensaotry CRs," i.e., CRs which would oppose, and thus compensate in advance for, the physiological responses to drugs. Siegel argued that with repeated administrations of morphine in a given environment, the cues associated with administration of the drug would come to evoke CRs which oppose the unconditioned responses responsible for analgesia. In short, conditioned hyperanalgesia would develop. The algebraic summation of the fixed

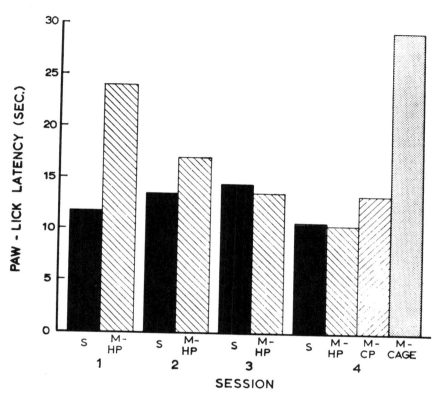

Figure 2.6   The mean paw-lick latency as a function of test sessions for four groups of rats (Siegel, 1975).

analgesic effect of the drug with a progressively stronger conditioned hyperalgesic effect would thus result in the observed
tolerance.

Siegel (1975) has supported his view of tolerance in an
impressive series of experiments. In one study he demonstrated
that rats became tolerant to the analgesic effect of a morphine
injection administered in a distinctive environment if they had
received three prior injections of morphine in the same environment,
but not if they had received the injections of morphine in the rat
colony. Figure 6 illustrates the data from this experiment, in
which the dependent variable was the mean paw-lick latency. Group
S, which received injections of physiological saline before each
hot plate test, exhibited relatively constant latencies. Group
M-HP, which was injected with a small amount of morphine (5 mg/kg)
30 min before each hot plate test, exhibited an initial long
latency, indicating an analgesic effect of morphine. However, the
analgesic effect declined over trials, and there was complete
tolerance by the fourth test, i.e., the latency of group M-HP
did not differ from that of group S. Tolerance was also shown by
rats in group M-CP, which was treated like group M-HP, except
that the hot plate was not turned on until the fourth test. Most
striking was the long paw-lick latency on the fourth test of
group M-Cage, which had received three prior morphine injections
in the colony, and which therefore should have exhibited no
conditioned hyperalgesic response in the distinctive test environment. In accord with Siegel's prediction, the paw-lick latency
after the fourth administration of morphine to group M-Cage was as
long as the latency for group M-HP after their first morphine
injection, i.e., there was no evidence of tolerance.

Siegel (1977) has noted that the variables which determine
whether the response to a signal for a drug will be like the
reaction to the drug, or will oppose it, are not well understood.
His remark can be extended to the domain of classical conditioning
in general. As we have seen, sometimes the CR is a copy of the
UCR. In other instances it is a well-organized sequence of
appetitive behaviors, ensuring contact with the reinforcer, which
itself evokes a consummatory response. In still other instances
the CR opposes, and apparently compensates in advance for, the
impending onslaught of the UCS and its physiological effects.
Even these diverse examples do not do full justice to the diversity
of classically conditioned responses (cf. Konorski, 1967, Ch. 1).

REFERENCES

Annau, Z., & Kamin, L. J. The conditioned emotional response as a function of intensity of the US. JOURNAL OF COMPARATIVE AND PHYSIOLOGICAL PSYCHOLOGY, 1961, 54, 428-432.

Anrep, G. V. Pitch discrimination in the dog. JOURNAL OF PHYSIOLOGY, 1920, 53, 367-385.

Best, M. P. Conditioned and latent inhibition in taste-aversion learning: Clarifying the role of learned safety. JOURNAL OF EXPERIMENTAL PSYCHOLOGY: ANIMAL BEHAVIOR PROCESSES, 1975, 1, 97-113.

Bitterman, M. E. The CS-UCS interval in classical and avoidance conditioning. In W. F. Prokasy (Ed.), CLASSICAL CONDITIONING. New York: Appleton-Century-Crofts Inc., 1965.

Blackman, D. E. Response rate, reinforcement frequency, and conditioned suppression. JOURNAL OF THE EXPERIMENTAL ANALYSIS OF BEHAVIOR, 1968, 11, 503-516.

Blackman, D. E. Conditioned suppression and the effects of classical conditioning on operant behavior. In W. K. Honig & J. E. R. Staddon (Eds.), HANDBOOK OF OPERANT BEHAVIOR. Englewood Cliffs, New Jersey: Prentice-Hall, 1977.

Brown, P. L., & Jenkins, H. M. Auto-shaping of the pigeon's keypeck. JOURNAL OF THE EXPERIMENTAL ANALYSIS OF BEHAVIOR, 1968, 11, 1-8.

Estes, W. K., & Skinner, B. F. Some quantitative properties of anxiety. JOURNAL OF EXPERIMENTAL PSYCHOLOGY, 1941, 29, 390-400.

Gormezano, I. Classical conditioning. In J. B. Sidowski (Ed.), EXPERIMENTAL METHODS AND INSTRUMENTATION IN PSYCHOLOGY. New York: McGraw-Hill, 1966.

Gormezano, I. Investigations of defense and reward conditioning in the rabbit. In A. H. Black & W. F. Prokasy (Eds.), CLASSICAL CONDITIONING II. New York: Appleton-Century-Crofts, 1972.

Grant, D. A., & Adams, J. K. "Alpha" conditioning in the eyelid. JOURNAL OF EXPERIMENTAL PSYCHOLOGY, 1944, 34, 136-142.

Grether, W. F. Pseudo conditioning without paired stimulation encountered in attempted backward conditioning. JOURNAL OF COMPARATIVE PSYCHOLOGY, 1938, 25, 91-96.

Hammond, L. J. Retardation of fear acquisition when the CS has previously been inhibitory. JOURNAL OF COMPARATIVE AND PHYSIOLOGICAL PSYCHOLOGY, 1968, 66, 756-759.

Harlow, H. F., & Toltzien, F. Formation of pseudo-conditioned responses in the cat. JOURNAL OF GENERAL PSYCHOLOGY, 1940, 23, 367-375.

Harris, J. D. Studies of non-associative factors inherent in con-
    ditioning. COMPARATIVE PSYCHOLOGICAL MONOGRAPHS, 1943, 18
    (1, whole #33).

Hearst, E., & Jenkins, H. M. SIGN-TRACKING: THE STIMULUS-
    REINFORCER RELATION AND DIRECTED ACTION. Austin, Texas: The
    Psychonomic Society, 1974.

Hoffman, H. S. Stimulus generalization versus discrimination failure
    in conditioned suppression. In R. M. Gilbert & N. S. Suther-
    land (Eds.), ANIMAL DISCRIMINATION LEARNING. New York:
    Academic Press, 1969.

Hogan, J. A. Responses in Pavlovian conditioning studies. SCIENCE,
    1974, 186, 156-157.

Hull, C. L. PRINCIPLES OF BEHAVIOR. New York: Appleton-Century-
    Crofts, 1943.

Hull, C. L. A BEHAVIOR SYSTEM. New Haven: Yale University Press,
    1952.

Jenkins, H. M., & Moore, B. R. The form of the auto-shaped response
    with food or water reinforcers. JOURNAL OF THE EXPERIMENTAL
    ANALYSIS OF BEHAVIOR, 1973, 20, 163-181.

Kamin, L. J. Temporal and intensity characteristics of the condi-
    tioned stimulus. In W. F. Prokasy (Ed.), CLASSICAL CONDI-
    TIONING. New York: Appleton-Century-Crofts, 1965.

Kandel, E. R. CELLULAR BASIS OF BEHAVIOR. San Francisco: W. H.
    Freeman, 1976.

Konorski, J. INTEGRATIVE ACTIVITY OF THE BRAIN. Chicago: University
    of Chicago Press, 1967.

Konorski, J., & Szwejkowska, G. Chronic extinction and restoration
    of conditioned reflexes. I. Extinction against the excitatory
    background. ACTA BIOLOGIAE EXPERIMENTALIS, 1950, 15, 155-170.

Konorski, J., & Szwejkowska, G. Chronic extinction and resotration
    of conditioned reflexes. IV. The dependence of the course of
    extinction and restoration of conditioned reflexes on the
    "history" of the conditioned stimulus. ACTA BIOLOGIAE EXPERI-
    MENTALIS, 1952, 16, 95-113.

Lea, S. E. C., & Morgan, M. J. The measurement of rate-dependent
    changes in responding. In R. M. Gilbert & J. R. Millenson
    (Eds.), REINFORCEMENT: BEHAVIORAL ANALYSES. New York:
    Academic Press, 1972.

Lubow, R. E. Latent inhibition. PSYCHOLOGICAL BULLETIN, 1973, 79,
    398-407.

Lubow, R. E., Alek, M., & Arzy, J. Behavioral decrement following stimulus preexposure: Effects of number of preexposures, presence of a second stimulus, and interstimulus interval in children and adults. JOURNAL OF EXPERIMENTAL PSYCHOLOGY: ANIMAL BEHAVIOR PROCESSES, 1975, 1, 178-188.

Mackintosh, N. J. A theory of attention: Variations in the associability of stimuli with reinforcement. PSYCHOLOGICAL REVIEW, 1975, 82, 276-298.

Marchant, H. G. III, Mis, F. W., & Moore, J. W. Conditioned inhibition of the rabbit's nictitating membrane response. JOURNAL OF EXPERIMENTAL PSYCHOLOGY, 1972, 95, 408-411.

Marchant, H. G. III, & Moore, J. W. Below-zero conditioned inhibition of the rabbit's nictitating membrane response. JOURNAL OF EXPERIMENTAL PSYCHOLOGY, 1974, 102, 350-352.

Millenson, J. R., & deVilliers, P. A. Motivational properties of conditioned anxiety. In R. M. Gilbert & J. R. Millenson (Eds.), REINFORCEMENT: BEHAVIORAL ANALYSES. New York: Academic Press, 1972, 97-128.

Moore, B. R. On directed respondents. Unpublished Doctoral dissertation, Standord University, 1971.

Moore, B. R. The role of directed Pavlovian reactions in simple instrumental learning in the pigeon. In R. A. Hinde & J. Stevenson-Hinde (Eds.), CONSTRAINTS ON LEARNING. New York: Academic Press, 1973.

Newlin, R. J. The relation of Pavlovian and autoshaping outcomes. Unpublished Doctoral dissertation, the University of North Carolina at Chapel Hill, 1974.

Patterson, M. M. Classical conditioning of the rabbit's (*Oryctolagus cuniculus*) nictitating membrane response with fluctuating ISI and intracranial CS. JOURNAL OF COMPARATIVE AND PHYSIOLOGICAL PSYCHOLOGY, 1970, 72, 193-202.

Pavlov, I. P. CONDITIONED REFLEXES. Translated by G. V. Anrep. Oxford: Oxford University Press, 1927.

Pavlov, I. P. LECTURES ON CONDITIONED REFLEXES. Translated by W. H. Gantt. New York: International Publishers, 1928.

Reberg, D. Compound tests for excitation in early acquisition and after prolonged extinction of conditioned suppression. LEARNING AND MOTIVATION, 1972, 3, 246-258.

Reiss, S., & Wagner, A. R. CS habituation produces a "latent inhibition effect" but no active "conditioned inhibition." LEARNING AND MOTIVATION, 1972, 3, 237-245.

Rescorla, R. A.   Inhibition of delay in Pavlovian fear conditioning.
    JOURNAL OF COMPARATIVE AND PHYSIOLOGICAL PSYCHOLOGY, 1967, 64,
    114-120.

Rescorla, R. A.   Pavlovian conditioned fear in Sidman avoidance
    learning.   JOURNAL OF COMPARATIVE AND PHYSIOLOGICAL PSYCHOLOGY,
    1968, 65, 55-60.

Rescorla, R. A.   Pavlovian conditioned inhibition.   PSYCHOLOGICAL
    BULLETIN, 1969, 72, 77-94.

Rescorla, R. A.   Summation and retardation tests of latent inhibition.
    JOURNAL OF COMPARATIVE AND PHYSIOLOGICAL PSYCHOLOGY, 1971, 75,
    77-81.

Rescorla, R. A., & Furrow, D. R.   Stimulus similarity as a determi-
    nant of Pavlovian conditioning.   JOURNAL OF EXPERIMENTAL PSY-
    CHOLOGY: ANIMAL BEHAVIOR PROCESSES, 1977, 3, 203-215.

Rescorla, R. A., & Holland, P. C.   Associations in Pavlovian condi-
    tioned inhibition.   LEARNING AND MOTIVATION, 1977, 8, 429-447.

Rescorla , R. A., & LoLordo, V. M.   Inhibition of avoidance behavior.
    JOURNAL OF COMPARATIVE AND PHYSIOLOGICAL PSYCHOLOGY, 1965, 59,
    406-412.

Rodnick, E. H.   Does the interval of delay of conditioned responses
    possess inhibitory properties?   JOURNAL OF EXPERIMENTAL PSY-
    CHOLOGY, 1937, 20, 507-527.

Schwartz, B., & Gamzu, E.   Pavlovian control of operant behavior.
    In W. K. Honig & J. E. R. Staddon (Eds.), HANDBOOK OF OPERANT
    BEHAVIOR.   New York: Prentice-Hall, Inc., 1977.

Sears, R. R.   Effect of optic lobe ablation on the visuo-motor
    behavior of goldfish.   JOURNAL OF COMPARATIVE PSYCHOLOGY, 1934,
    17, 233-265.

Shapiro, K. L., Jacobs, W. J., & LoLordo, V. M.   Stimulus relevance
    in Pavlovian conditioning of pigeons.   Unpublished manuscript,
    1977.

Sheafor, P. J.   "Pseudoconditioned" jaw movements of the rabbit re-
    flect associations conditioned to contextual background cues.
    JOURNAL OF EXPERIMENTAL PSYCHOLOGY: ANIMAL BEHAVIOR PROCESSES,
    1975, 1, 245-260.

Sheafor, P. J., & Gormezano, I.   Conditioning the rabbit's jaw-
    movement responses: US magnitude effects on URs, CRs, and
    pseudo-CRs.   JOURNAL OF COMPARATIVE AND PHYSIOLOGICAL PSY-
    CHOLOGY, 1972, 81, 449-456.

Siegel, S.   Evidence from rats that morphine tolerance is a learned
    response.   JOURNAL OF COMPARATIVE AND PHYSIOLOGICAL PSYCHOLOGY,
    1975, 89, 498-506.

Siegel, S. Morphine analgesic tolerance: Its situation specificity supports a Pavlovian conditioning model. SCIENCE, 1976, 193, 323-325.

Siegel, S. Learning and psychopharmacology. In M. E. Jarvik (Ed.), PSYCHOPHARMACOLOGY IN THE PRACTICE OF MEDICINE. New York: Appleton-Century-Crofts, 1977.

Smith, M. C., Coleman, S. R., & Gormezano, I. Classical conditioning of the rabbit's nictitating membrane response at backward, simultaneous, and forward CS-UCS intervals. JOURNAL OF COMPARATIVE AND PHYSIOLOGICAL PSYCHOLOGY, 1969, 69, 226-231.

Szwejkowska, G., & Konorski, J. The influence of the primary inhibitory stimulus upon the salivary effect of excitatory conditioned stimuli. ACTA BIOLOGIAE EXPERIMENTALIS, 1959, 19, 161-174.

Testa, T. J. Effects of similarity of location and temporal intensity pattern of conditioned and unconditioned stimuli on the acquisition of conditioned suppression in rats. JOURNAL OF EXPERIMENTAL PSYCHOLOGY: ANIMAL BEHAVIOR PROCESSES, 1975, 1, 114-121.

Timberlake, W., & Grant, D. L. Auto-shaping in rats to the presentation of another rat predicting food. SCIENCE, 1975, 190, 690-692.

Wagner, A. R., Siegel, S., Thomas, E., & Ellison, G. D. Reinforcement history and extinction of the classical reward response. JOURNAL OF COMPARATIVE AND PHYSIOLOGICAL PSYCHOLOGY, 1964, 58, 354-358.

Wagner, A. R., Thomas, E., & Norton, T. Conditioning with electrical stimulation of the motor cortex: Evidence of a possible source of motivation. JOURNAL OF COMPARATIVE AND PHYSIOLOGICAL PSYCHOLOGY, 1967, 64, 191-199.

Wasserman, E. A. Pavlovian conditioning with heat reinforcement produces stimulus-directed pecking in chicks. SCIENCE, 1973, 181, 875-877.

Wessells, M. G. Autoshaping, errorless discrimination, and conditioned inhibition. SCIENCE, 1973, 182, 941-943.

Wickens, D. D., & Wickens, C. D. Some factors related to pseudo conditioning. JOURNAL OF EXPERIMENTAL PSYCHOLOGY, 1942, 31, 518-526.

Woodruff, G., & Williams, D. R. The associative relation underlying autoshaping in the pigeon. JOURNAL OF THE EXPERIMENTAL ANALYSIS OF BEHAVIOR, 1976, 26, 1-13.

3. CLASSICAL CONDITIONING: CONTINGENCY AND CONTIGUITY

V. M. LoLordo

Dalhousie University
Halifax, Nova Scotia, Canada

The last chapter described the sorts of experiments that led Pavlov to his conception of the conditions which are necessary and sufficient for the conditioning of excitation and inhibition. This chapter will begin by placing the sorts of relationships between CS and UCS that Pavlov explored within a broader framework, one which should help the reader organize a large body of data, and which should suggest a variety of new experiments on classical conditioning.

Contingency Space

This framework is called the contingency space, and it is a way of describing classical conditioning procedures in terms of the contingency or dependency between CS and UCS. Figure 1 illustrates a contingency space, which relates two conditional probabilities, the conditional probability of the occurrence of the UCS, given the occurrence of the CS or a trace of the CS, and the conditional probability of occurrence of the UCS, given the non-occurrence of the CS or its trace. The first conditional probability can be represented as p(UCS/CS), and the second, as p(UCS/$\overline{\text{CS}}$). The degree and direction of the contingency between a CS and the UCS in any classical conditioning experiment can be represented by a point in this contingency space (e.g. Catania, 1971; Gibbon, Berryman, & Thompson, 1974; Seligman, Maier, & Solomon, 1971).

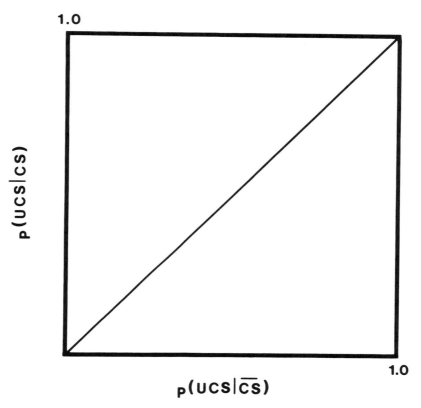

Figure 3.1   A contingency space, which plots the conditional proba-
             bility of the UCS in the presence of the trace of the
             CS [p(UCS/CS)] against the conditional probability of
             the UCS in the absence of the CS trace [p(UCS/C̄S)].

Let's consider how the contingencies are plotted. First, consider an experiment by Wagner, Siegel, Thomas, and Ellison (1964), in which the salivary response of the dog was conditioned to an auditory CS which preceded the presentation of a food pellet by 20 sec. If we divide the duration of the experimental session into intervals equal to the interval between the start of the CS and the termination of the UCS, in this instance 25 sec, we find that every interval which contained the CS also contained the UCS. Thus p (UCS/CS) = 1.0. The UCS never occurred at other times, thus p(UCS/$\overline{\text{CS}}$) = 0. This pair of conditional probabilities yields the limiting positive contingency, a point at the upper left corner of the contingency space. The same description can be applied to the experiment by Anrep (1920), described at the beginning of the last chapter, but that study included unreinforced test presentations of the CS roughly every tenth trial, so that p (UCS/CS) would be roughly 0.9, and the point describing the contingency would move down the left-hand edge of the square a little.

As a second sort of example, let's consider the Pavlovian differential conditioning, or "method of contrasts" procedure, in which one CS, CS+, is always followed by the UCS, but a different CS, CS-, is never followed by the UCS, and is separated from it by the relatively long intertrial interval. Computations would reveal that p(UCS/CS+) = 1.0, but p(UCS/$\overline{\text{CS}}$+) = 0. Because CS- is never followed by the UCS, p(UCS/CS-) = 0. However, because the UCS does occur at other times, specifically during or soon after CS+, p(UCS/$\overline{\text{CS}}$-) is greater than zero. How much greater depends upon the number of UCSs per unit of time; a typical value might be 0.1. Thus, the contingency between CS- and the UCS in a differential conditioning experiment can be represented by a point somewhere on the bottom edge of the square.

In the aforementioned studies, a high positive contingency between a CS and a UCS led to an increase in the CS's capacity to evoke a salivary CR, which we have called excitatory conditioning. When food was negatively contingent on CS-, however, the asymptotic level of salivation was near zero, and summation or retardation of acquisition tests revealed an active tendency not to salivate, which we have called inhibitory conditioning. Thus, these few outcomes suggest that the contingency between CS and UCS might be an important determinant of the outcome of a classical conditioning experiment. Ten years ago Rescorla (1967a) rekindled interest in classical conditioning by proposing that the contingency between a CS and the UCS is a fundamental determinant of classical conditioning (cf. Prokasy, 1965). Rescorla's viewpoint will be evaluated in this chapter. But first, some of the wealth of questions that arose from Rescorla's reformulation of classical conditioning will

be posed, and some tentative answers will be suggested.   These
questions have had a profound impact on research and theory-building in classical conditioning in the last ten years.

First, although positive contingencies between CS and UCS resulted in excitatory conditioning, and a negative contingency, in
inhibitory conditioning, in the examples just presented, will this
always be the case? Across diverse species, unconditioned stimuli
and response systems?

Second, do animals behave as though they calculate the two conditional probabilities and form associations between CS and UCS
On the basis of some relationship between the two probabilities,
e.g., their difference? Put another way, will researchers be able
to make some sense of the locus of points in the contingency space
that yield the same outcome?

Third, what will be the effect of procedures in which the probability of the UCS is unaffected by the presence or absence of a
CS or its trace, i.e., when $p(UCS/CS) = p(UCS/\overline{CS})$? Such procedures
will be called zero-contingency procedures, and they can be represented by points on the positive diagonal of the contingency space.

Finally, how will the answers to these and other questions reflect upon the adequacy of the Pavlovian account of classical conditioning, which was described in the last chapter?

Data which bear on these questions will be presented in the
next few sections of this chapter.

Effects of Positive and Negative Contingencies

In this section the question whether positive contingencies
inevitably result in excitatory classical conditioning, and negative
contingencies, in inhibitory conditioning, will be considered.

Effects of CSs from Defensive Conditioning Procedures. Rescorla
and LoLordo (1965) studied the effects of positive and negative contingencies between auditory CSs and electric shock, using an indirect method related to the Conditioned Emotional Response paradigm. Their procedure had three phases. First, dogs were trained
to jump a hurdle in a two-compartment chamber called a shuttlebox
in order to avoid unsignalled electric shock delivered through the
grid floor. The unsignalled avoidance schedule (Sidman, 1953) resulted in a fairly steady rate of jumping at 5 to 8 responses per
minute.

Second, a classical conditioning treatment was administered while the hurdle-jump could not be performed. During several sessions the dogs were confined in one compartment of the shuttlebox, and received classical conditioning trials separated by intervals of 1-2 min. The studies to be described included various arrangements of 5-sec auditory CSs and 5-sec electric shock UCSs during classical conditioning.

Finally, there was a test session, during which the dogs were permitted to jump the hurdle repeatedly. During this session the CSs were each presented 60 times, independent of the dog's behavior, and no shocks were administered. It was assumed that the rate of avoidance is a monotonically increasing function of the strength of a classically conditioned fear response, so that the superimposition of a classically conditioned excitor of fear would produce an increase in the rate of jumping, but the superimposition of an inhibitor of fear would produce a decrease in the rate of jumping. Formally, this test procedure is like the summation test procedures discussed in the last chapter.

In the first experiment there were two kinds of conditioning trials. On half the trials, a 5-sec, auditory CS+ occurred, and after a brief trace interval a 5-sec, electric shock UCS was presented. On the other trials the CS+ and trace interval again occurred, but a different auditory stimulus, CS-, occured instead of the UCS. In Pavlovian terms this is a conditioned inhibition procedure, because a CS+ is followed by a UCS unless that CS+ is accompanied by another CS (CS-). Calculating the various conditional probabilities, we find that p(UCS/trace of CS+) = 0.5, and p(UCS/no trace of CS+) = 0, i.e., there is a positive contingency between CS+ and the UCS. Further, p(UCS/CS-) = 0, but p(UCS/CS-) = roughly 0.05, i.e., there is a negative contingency between CS- and the shock.

Observations of the dogs during classical conditioning were consistent with the inference that the CS+ aroused fear, but the CS- was an inhibitor of fear:

> Upon the onset of CS_1 (the CS+) <u>S</u> typically showed signs of fear and agitation--barking, crouching, running around with its ears back and its tail between its legs, etc.; however, by the last conditioning session, CS_2 (CS-) clearly produced termination of this behavior; during CS_2 (CS-) <u>S</u> seemed to relax, cocking its head at the sound of the stimuli. Rescorla and LoLordo, 1965; p. 408.

Further, as Figure 2 illustrates, the presentation of CS+ resulted in a marked increase in the rate of jumping, which was sustained for the 5-sec period immediately after CS+ terminated.

This result was consistent with the assertion that fear had been classically conditioned to the trace of CS+. When CS- was superimposed on the baseline, on the other hand, the rate of avoidance dropped nearly to zero, and remained low for the 5 sec immediately after termination of CS-. Thus the CS- from this conditioned inhibition procedure became an inhibitor of fear.

In a second experiment two groups of dogs received repeated pairings of a 5-sec CS+ and a 5-sec shock UCS, which began as the CS+ ended. On the other half of the trials Group CI received a 5-sec CS- immediately followed by the CS+, but no shock. Again there was a positive contingency between CS+ and shock and a negative contingency between CS- and shock in a conditioned inhibition procedure. Group DI received only the CS- on non-reinforced trials. This differential conditioning procedure was like the conditioned inhibition procedure, except that the CS+ was omitted on non-reinforced trials.

Behavior of the two groups did not differ. Figure 3 illustrates that the CS+ again produced a marked increase in the rate of avoidance. As in the previous experiment, the CS- produced virtual cessation of responding, and the rate remained low for five sec after the CS- terminated. These experiments varied the temporal relationship between CS+ and CS-, and the relationship of CS+ to shock, without affecting the strong inhibitory effect of the CS-. However, the probability of shock in the presence of the trace of CS- was always zero, and shocks did sometimes occur in the absence of CS-. Thus there was a negative contingency between CS- and the UCS. The next experiment further simplified the classical conditioning procedure, by presenting only the negative contingency between an auditory CS- and electric shock. During each session tones and shocks occurred in a mixed order, and the interval between successive events was 1-2 min, as before. Thus shock could be presented only if there had been no tone for at least 1 min.

Figure 4 illustrates the results of the test session. Presentation of the CS- reduced the rate at which the dogs jumped to nearly zero, and the rate remained low, though recovering somewhat, during the 5 sec immediately following presentation of CS-. Thus separate, unpaired presentations of CS and UCS, a procedure which had been used as a control for non-associative factors in classical conditioning, yield a conditioned effect, albeit an inhibitory one. The implications of this finding for the design of control groups will be discussed later in this chapter.

The outcomes of the experiments which were just described are consistent with the assertion that a negative contingency between

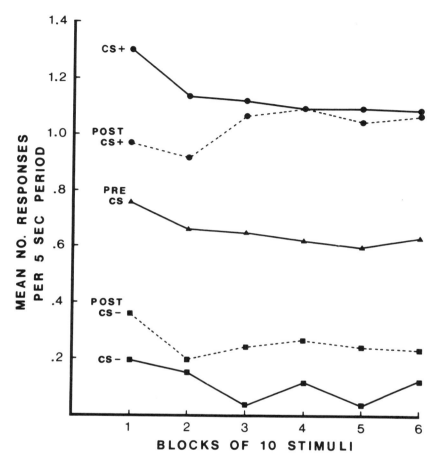

Figure 3.2.   Mean number of avoidance responses in 5-sec periods
before, during, and following test presentations of
CS+ and CS- in Experiment 1 of Rescorla and LoLordo
(1965).

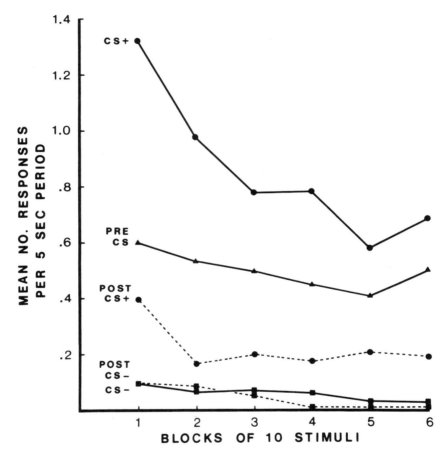

Figure 3.3   Mean number of avoidance responses in 5-sec periods
before, during, and following test presentations of
CS+ and CS- in Experiment 2 of Rescorla and LoLordo
(1965).   Data from groups CI and DI have been combined.

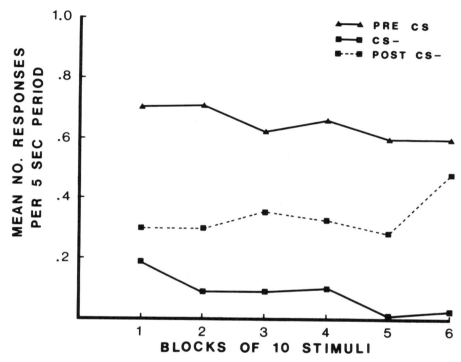

Figure 3.4 Mean number of avoidance responses in 5-sec periods
before, during, and following test presentations of
CS- in Experiment 3 of Rescorla and LoLordo (1965).

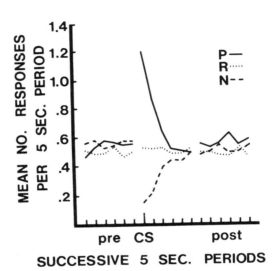

Figure 3.5   Mean number of avoidance responses in successive 5-sec periods before, during, and following test presentations of the CS for groups which had received positive (P), negative (N), or zero (R) contingencies between that CS and electric shock (Rescorla, 1966).

CS and UCS is sufficient for the conditioning of inhibition of a CR to that CS. In these studies not only was the CS- non-reinforced but in addition the UCS occurred in the absence of the CS-. Suppose the CS occurred in the absence of shock, and that no shocks occurred at other times. Would the CS become an inhibitor? In the last chapter we concluded that the latent inhibition procedure, which entailed simple non-reinforcement of a stimulus, did not produce an inhibitor. Rescorla and LoLordo (1965) obtained data in accord with this conclusion; two auditory stimuli which had been presented repeatedly in sessions in which no shocks occurred subsequently had no effect upon the rate at which dogs jumped.

Other studies which have inferred the excitatory and inhibitory characters of Pavlovian CSs from their effects upon avoidance behavior have provided additional information about the effects of contingencies upon conditioning. Rescorla (1966) presented dogs with tones and shocks at varying intervals within classical conditioning sessions, so that for Group R (Random) tones and shocks were independent, i.e., $p(UCS/CS) = p(UCS/\overline{CS})$. Two other groups of dogs received a modified pattern of shocks. For Group P (Positive), only shocks programmed to occur within 30 sec after the onset of the CS actually occurred. For Group N (Negative), only the shocks that were programmed to occur more than 30 sec after the onset of the. CS occurred. Thus Group P received a positive contingency between CS and UCS, Group N, a negative contingency, and Group R, a zero contingency.

Figure 5 illustrates the effects of the CS when it was superimposed upon the avoidance baselines of the three groups of dogs. The CS markedly increased the jumping rate of Group P, but produced a large decrease in rate for Group N. In both cases the rate of avoidance gradually returned to baseline levels after the CS terminated. The behavior of dogs in Group R was unaffected by the presentation of the CS. It is noteworthy that Groups P and R received the same number of pairings of CS and UCS, i.e., occasions on which the UCS occurred less that 30 sec after the onset of the 5-sec CS, yet excitatory conditioning occurred only in Group P. Rescorla (1966) interpreted this result to mean that contiguity of CS and UCS was not sufficient for excitatory classical conditioning, and that a positive contingency between the two events was necessary for such conditioning.

To complete the review of studies which have varied the temporal relationship between CS and electric shock, and examined the subsequent effect of the CS upon avoidance behavior, two additional sets of studies will be mentioned. Rescorla (1967b, 1968b) has obtained evidence for inhibition of delay in aversive conditioning of dogs. The rate of avoidance responding was markedly reduced during the first 5 sec of a long-delay or long-trace CS that preceded shock, but increased to greater than baseline rate towards the end of the CS. If we grant organisms the capacity to discriminate the early part of the CS from the later part, then there is a negative contingency between the early part of the CS and shock, but a positive contingency between the later part and shock, and these data are consistent with others illustrating the importance of contingencies. Nonetheless, this account can only explain graded increases in the strength of excitatory conditioning as time passes if additional assumptions about "timing mechanisms" are added (e.g. Anger, 1963; Bitterman, 1965; Killeen, 1975).

The last paradigm which will be considered in this survey is backward conditioning, in which the onset of the UCS precedes the onset of the CS. Researchers were first interested in the effects of backward conditioning because of the possibility that excitatory conditioning might occur, an outcome which would have been difficult to reconcile with the Pavlovian view that the CS+ comes to elicit an excitatory CR because the trace of the CS is present when the UCS begins (Pavlov, 1927; Razran, 1956). This view was so strong that, in the absence of impressive demonstrations of backward excitatory conditioning, the backward conditioning procedure came to be used as a control for non-associative effects (cf. the previous chapter). More recently, however, there has been considerable interest in backward conditioning as a procedure which generates conditioned inhibition (e.g., Konorski, 1948; Moscovitch & LoLordo, 1968).

Moscovitch and LoLordo (1968) presented dogs with a CS which began 1 sec after shock ended on each conditioning trial, with trials separated by 2-3 min. Following conditioning, the CS was superimposed upon the avoidance baseline in a test session. Figure 6 shows that the CS produced a marked decrease in the rate of avoidance. The rate recovered gradually after CS termination, but recovery was incomplete even after 30 sec.

As most experiments in classical conditioning do, the experiment by Moscovitch and LoLordo (1968) included a relatively long minimum interval between trials. Given the foregoing discussion of contingencies and their effects, this feature immediately suggests the possibility that the inhibitory effect of a CS from these experiments is totally, or in part, a result of the between-trials negative contingency between CS and UCS which they include. Moscovitch and LoLordo (1968) assessed the contributions of within-trials and between-trials factors in a second experiment, which included two additional groups treated just like the first one except during classical conditioning. Group B15 received backward

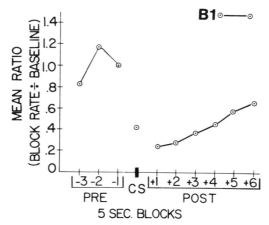

Figure 3.6   The mean ratios of the rates of avoidance responding
in successive 5-sec periods before, during, and following test presentations of a CS to the average baseline rate.   Group B1 had received a backward conditioning treatment (Moscovitch & LoLordo, 1968; Experiment
1).

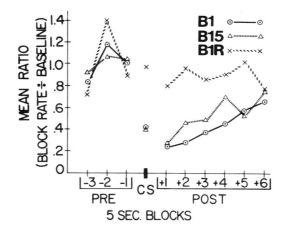

Figure 3.7   The mean ratios of the rates of avoidance responding
in successive 5-sec periods before, during, and following test presentations of a CS to the average baseline rate of avoidance for Group B1 (Experiment 1) and
Groups B15 and B1R (Experiment 2, Moscovitch & LoLordo,
1968).

conditioning trials in which the CS began 15 sec after shock ended. In the test session (see Figure 7) the CS had the same pronounced suppressive effect upon jumping that it had in the first experiment, suggesting that lengthening the interval between UCS termination and CS onset on a backward conditioning trial from 1-15 sec had no effect upon the strength of the resulting conditioned inhibition. Recently, however, Maier, Rapaport, and Wheatley (1976) found that a similar manipulation with rat subjects resulted in strong inhibitory effects when the CS followed shock by only 3 sec, but no inhibition when the UCS-CS interval was 30 sec, even though the minimum intertrial interval was 2.5 min (cf. Plotkin & Oakley, 1975). This discrepancy has not been resolved.

Group B1R received conditioning trials in which the CS began 1 sec after shock ended on each trial. However, the intertrial intervals were allowed to vary from 0-15 min, around a mean of 2.5 min. Thus for Group B1R the CS did not signal a long minimum time to the next UCS. In the test session (Figure 7), the CS had no effect upon the jumping rate of dogs in this group, suggesting that the CS inhibited fear in Groups B1 and B15 because it signalled that shock would not occur for a while, i.e., because of the strong negative contingency between CS and shock. These experiments provided no evidence for backward association, of either an excitatory or an inhibitory sort (but see Heth, 1976; Mahoney & Ayres, 1976).

The foregoing discussion has focused on the results obtained from one paradigm, which has assessed classical conditioning to stimuli placed in various relations to shock by measuring the effects of those stimuli upon the rate of shock avoidance. This particular paradigm has been the focus of attention, not because it is one of the most commonly studied, but because it provided much of the impetus for the view that the contingency between CS and UCS is a fundamental determinant of classical conditioning. The data obtained from this indirect method of studying classical aversive conditioning paint a reasonably clear picture. They suggest that when there is a positive contingency between CS and UCS the CS will become excitatory, when there is a negative contingency the CS will become inhibitory, and when there is a zero contingency, the CS will remain associatively neutral. But how general are these effects?

Contingencies as "Causal." Evidence bearing on the necessity of a positive contingency between CS and UCS for excitatory classical conditioning has been obtained in studies of the effects of zero-contingencies. If exposure to a zero-contingency between CS and UCS does not result in excitatory conditioning to the CS, despite a moderate probability of the UCS in the presence of the trace of the CS, but the same p(UCS/CS) results in excitatory conditioning when the overall contingency between the two events is positive, then the contiguity of the trace of the CS with the UCS is not a sufficient condition for excitatory conditioning. Further, a positive contingency would seem to be necessary. The comparison of Groups P and R in the study by Rescorla (1966) yielded such an outcome; excitatory conditioning occurred only when the overall contingency between CS and shock was positive. Similar results have been obtained in studies of the CER in rats (Hammond, 1967; Keller, Ayres, & Mahoney, 1977; Rescorla, 1967b) and of autoshaped pecking in pigeons (Gamzu & Williams, 1971). On the other hand, a number of studies using variations of the conditioned emotional response procedure in rats have obtained at least pre-asymptotic excitatory conditioned effects of zero-contingencies (e.g. Benedict & Ayres, 1972; Keller et al., 1977; Kremer, 1974). This issue will be treated in more depth later in this chapter, but for the moment must be left unresolved.

Evidence that a negative contingency between CS and UCS leads to inhibitory conditioning to the CS comes mostly from the differential conditioning and conditioned inhibition paradigms. In a pair of papers reporting both summation and retardation tests, Hammond (1967, 1968) showed that the CS- in a differential conditioned emotional response procedure became a conditioned inhibitor of response suppression. During each classical conditioning session two groups of rats received three presentations of a tone CS+, which always terminated with electric shock delivered through the grid floor. There were also three presentations of a light during these sessions, but the temporal relationship between presentations of the light and the tone-shock pairing differed for the groups. For Group I the light was a CS-, and was never presented during a tone-shock trial or the 3 min immediately preceding such a trial. Thus light and shock were related by a negative contingency. For group R, on the other hand, the light CS was scheduled independently of tone-shock pairings, and could occur at any time (even during a tone-shock sequence).

Following conditioning, summation tests were carried out, in an attempt to detect differential effects of the light CS in the two groups. During each session a simultaneous compound of light and tone was presented three times without shock. Figure 8 illustrates the results of these tests. There was less suppression of lever-pressing during the compound CS in group I than in group R. Since suppression is identified with conditioned excitation in the CER procedure, this outcome suggests that the CS- in group I had become a conditioned inhibitor of conditioned suppression of lever-pressing.

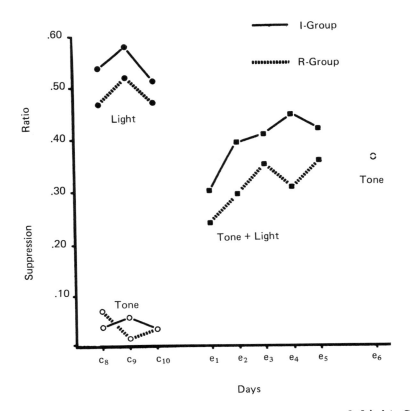

Figure 3.8   Daily suppression ratios for the tone and light CSs
             during Groups I and R in an experiment by Hammond
             (1967).   The suppression ratio equals the number of
             responses during the stimulus divided by the sum of
             the number of responses during the stimulus and the
             number during an equally long period just prior to
             the stimulus.

This conclusion was confirmed in a second study which administered a retardation of acquisition test following conditioning (Hammond, 1968). It would be difficult to argue that the inhibitory effect of the CS- in these experiments results from some non-associative process, because in both experiments the experimental and control groups differed only in the contingency between the visual CS and electric shock.

Several other experiments which have included both summation and retardation-of-acquisition tests have obtained analogous results. In an autoshaping experiment Wessells (1973) found that pigeons pecked at a localized visual CS+ upon which the presentation of grain was positively contingent, but would inhibit pecking in the presence of a different localized visual stimulus, a CS- from a discriminative conditioning procedure. Further, Wasserman, Franklin, and Hearst (1974) showed that pigeons would even move away from such a CS-.

Evidence from the Pavlovian conditioned inhibition procedure, in which CS+ is reinforced unless it is compounded with CS-, generally supports the data from studies of discriminative conditioning, demonstrating that a negative contingency between CS and UCS results in the conditioning of inhibition. The CS- from conditioned inhibition procedures has been shown to be inhibitory in salivary conditioning (Pavlov, 1927), the CER procedure with rats (Rescorla, 1971), nictitating membrane conditioning of rabbits (Marchant, Mis, & Moore, 1972; Marchant & Moore, 1974), and the conditioning of aversions to taste cues in rats (Best, 1975; Taukulis & Revusky, 1975).

Backward conditioning procedures have resulted in the conditioning of inhibition to the backward CS in studies of the CER of rats (Siegel & Domjan, 1971), as well as in studies of the rabbit's conditioned eyeblink (Siegel & Domjan, 1971) and nictitating membrane (Plotkin & Oakley, 1975). In each of these experiments animals which had received backward conditioning, including a negative contingency between the trace of the CS and UCS onset, were slower than control groups, e.g., groups which had received independent CSs and UCSs, to acquire an excitatory CR when the CS was subsequently paired with the UCS.

In summary, the balance of the evidence suggests that a negative contingency between CS and UCS results in conditioning of inhibition to that CS. However, very little of this evidence has come from the procedure in which only CS- and the UCS occur during the session, and there is a negative contingency between them. In addition to the work of Rescorla and LoLordo (1965), cited earlier, evidence that a negative contingency of the UCS upon CS-

in the absence of other stimuli yields conditioned inhibition has
been obtained by Rescorla (1969) in the CER procedure with rats,
and by Wasserman, Franklin, and Hearst (1974; Hearst & Franklin,
1977), when the presentation of grain was negatively contingent
upon illumination of a pigeon's pecking key. In the latter case,
the birds acquired a tendency to withdraw from the CS-. Further,
when a positive contingency was subsequently arranged between the
former CS- and food in a retardation-of-acquisition test, the birds
acquired an approach response to the CS more slowly than pigeons
which had formerly been exposed to a zero contingency between the
CS and food (Hearst & Franklin, 1977).

On the other hand, Plotkin and Oakley (1975) found that a
group of rabbits that received alternating CS-only and UCS-only
trials, with an intertrial interval of 55 sec, subsequently acquired the nictitating membrane response in a retardation test as
rapidly as a control group that had merely been placed in the
restraining apparatus, and more rapidly than backward conditioning
groups.

These data suggest that unpaired presentations of a CS and UCS
that evokes extension of the rabbit's nictitating membrane do not
endow the CS with the capacity to inhibit extension of the membrane,
and thus that a negative contingency between CS and UCS is not
sufficient for the conditioning of inhibition in this response
system (cf. Furedy, Poulos, & Schiffmann, 1975; Furedy & Schiffmann, 1973 for a related discussion re the human galvanic skin
response (GSR) and other human autonomic responses). Plotkin &
Oakley's result appears to contradict that of Rescorla and LoLordo
(1965), described earlier, although the two studies differ in many
details. Perhaps Plotkin and Oakley failed to find inhibition of
the nictitating membrane CR when CS- simply preceded a long shock-
free period because non-reinforcement of the CS- must occur in the
presence of excitation in order for inhibition to develop, and excitation of the nictitating membrane response was not conditioned to
background cues (see Maier, et al., 1976). Excitation would have
been present when the UCS had just occurred in backward conditioning (Plotkin & Oakley, 1975), and in that case conditioned inhibition of the rabbit's nictitating membrane response did occur. One
testable implication of this speculation is that inhibition of
other CRs, which are conditioned to background cues, would occur
in the very same experiments which failed to demonstrate inhibition
of the nictitating membrane response. The possible dissociation
of several CRs in a given experiment points to the importance of
distinguishing learning from performance, i.e., the formation of
associations from their manifestation in changes in a given response
system. To the extent that one is interested in what has been

learned, rather than in the performance of some model response system, then the measurement of only a single response may be misleading (cf. Holland, 1977).

Finally, we can ask whether a negative contingency between CS and UCS is necessary for the conditioning of inhibition. We have already seen that simple non-reinforcement in the latent inhibition procedure does not yield conditioned inhibition (Rescorla, 1971). Nor does non-reinforcement in the standard extinction procedure produce an inhibitory stimulus (Reberg, 1972). These data suggest that simple non-reinforcement is not sufficient for the conditioning of inhibition, and at this point negative contingency seems to be the critical determinant of inhibition more or less by default. However, in the next chapter we will return to this question, and discuss an alternative formulation of the empirical basis of inhibition. In the meantime, the other questions posed at the beginning of this chapter will be answered.

Magnitude of the contingency. Our second question was "do animals behave as though they calculate the two conditional probabilities and form associations between CS and UCS on the basis of some relationship between those probabilities, e.g., their difference?" Data pertinent to this question have been obtained in studies which hold one of the two conditional probabilities constant at zero, and vary the other across groups. When $p(UCS/\overline{CS})$ is zero, but $p(UCS/CS)$ varies, either the rate of acquisition of the excitatory CR (e.g., Gormezano & Coleman, 1975) or the asymptotic level of excitatory conditioning (Gonzalez, Longo, & Bitterman, 1961; Ost & Lauer, 1965; Wagner, et al., 1964) increases as $p(UCS/CS)$ increases. Thus, the strength of excitatory conditioning does increase with the magnitude of the positive contingency, i.e., as the point representing the contingency moves up the left-hand edge of the contingency space. Only two experiments have varied the magnitude of the negative contingency by holding $p(UCS/CS) = 0$ and varying $p(UCS/\overline{CS})$. Both Rescorla (1969), using the CER procedure with rats, and Hearst and Franklin (1977), studying pigeons' conditioned withdrawal from a localized visual cue upon which the presentation of food was negatively contingent, have found that asymptotic level of inhibitory conditioning increases with the magnitude of $p(UCS/\overline{CS})$, or average density of reinforcement in the absence of the CS. Thus the strength of inhibitory conditioning increases with the magnitude of the negative contingency, as the point representing the contingency moves to the right along the lower edge of the contingency space.

Additional data bearing on the question that began this section come from experiments which varied both $p(UCS/CS)$ and $p(UCS/\overline{CS})$.

In a CER experiment Rescorla (1968a) presented ten groups of rats
with the following pairs of conditional probabilities of electric
shock in the presence and absence of a 2-min auditory CS, respectively: .4/2 min-.4/2 min, .2-.2, .1-.1, 0-0, .4-.2, .4-.1, .4-0,
.2-.1, .2-0, and .1-0. Figure 9 illustrates suppression ratios
over six extinction test sessions which followed conditioning.
The CS produced little suppression in groups which received a zero-
contingency between CS and UCS (Groups .4-.4, .2-.2, .1-.1. and
0-0). Second, with p(UCS/CS) held constant, the amount of conditioned suppression decreased in an orderly fashion as p(UCS/$\overline{CS}$)
approached p(UCS/CS). Keller, et al., (1977) obtained a similar result when p(UCS/CS) was held constant at 1.0.

Comparison of the various panels in the figure indicates that
the magnitude of the contingency was a much better predictor of
suppression than the absolute number of shocks which occurred during the CS. For example, the CS evoked much more suppression in
Group .1-0 than in Group .4-.4, although four times as many shocks
occurred during CS in the latter group. Moreover, the difference
between p(UCS/CS) and p(UCS/$\overline{CS}$) was not a very good predictor of
the amount of conditioned suppression. For example, on the first
test day, the suppression ratio for Group .4-.2 was roughly .33,
i.e., the presentation of the CS reduced the lever-pressing rate
by half. Group .2-0 received the same value of p(UCS/CS) minus
p(UCS/$\overline{CS}$), but the CS reduced the lever-pressing rate of this group
to roughly a tenth of its baseline value. When the phi-coefficient,
a measure of contingency which takes into account the proportion of
the session during which the CS is present as well as the two conditional probabilities (Gibbon, Berryman, & Thompson, 1974), was
computed for each group in this study, the obtained suppresion
ratio was linear with the value of phi.

These data, along with those presented in previous sections,
suggest the importance of the contingency between CS and UCS for
the outcome of a Pavlovian conditioning experiment, although they
also reveal that more parametric studies must be done if we hope to
understand the effects of contingency manipulation in any quantitative way. In any case, consideration of the data which have been
presented in the last few sections should make the view of conditioning which will be presented in the next section at least plausible.

A Contingency View of Conditioning

On the basis of some of the data which have been presented in
this chapter, Rescorla (1967a) proposed a model of classical conditioning which was based on the notion of contingency. According to
this viewpoint, the conditioned outcome of a classical conditioning

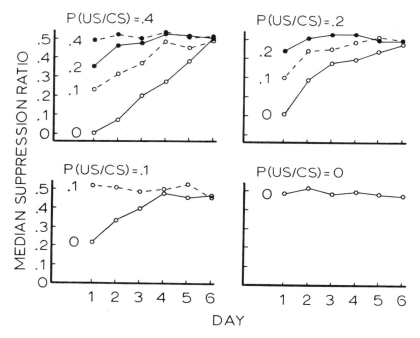

Figure 3.9 Median suppression ratios for each group of rats over
six test sessions in an experiment by Rescorla (1968a).
Within each panel p(UCS/CS) is constant, and p(UCS/$\overline{CS}$)
varies across groups.

experiment depends upon the two conditional probabilities we have been considering, p(UCS/CS) and p(UCS/$\overline{CS}$). Rescorla suggested that zero-contingencies should result in no associative effect, that excitatory conditioning should occur when p(UCS/CS) > p(UCS/$\overline{CS}$), and that inhibitory conditioning should occur when p(UCS/CS) < p(UCS/$\overline{CS}$). The mathematical function which should be applied to the two conditional probabilities to produce a single number representing the contingency was not specified by Rescorla. As was noted earlier, Gibbon, Berryman, and Thompson (1974) discussed several possible functions, and ultimately suggested the phi-coefficient.

On Rescorla's view the number of pairings of the CS and the UCS, i.e., the number of occurrences of contiguity between CS and UCS, is an important determinant of conditioning only insofar as it affects the contingency between CS and UCS. Contiguity was held not to be sufficient for classical conditioning, largely because of Rescorla's (1966) own finding that zero-contingency procedures produced no conditioning, despite the occurrence of chance pairings of CS and UCS. Rescorla did not deny the necessity of contiguity of CS trace and UCS for excitatory classical conditioning, but his emphasis on the two conditional probabilities p(UCS/CS) and p(UCS/$\overline{CS}$) did not suggest a treatment of effects that are graded with time. For example, so long as the UCS occurs during CS, it should not matter whether it occurs immediately after the termination of the CS, or just before the next CS, according to a strict interpretation of the conditional probabilities, though the two cases would differ greatly in the contiguity of the CS trace with UCS onset.

In the previous chapter, the problem of controlling for non-associative effects in classical conditioning was discussed from the Pavlovian point of view. Given the view that contiguity between the trace of the CS and the UCS is necessary for conditioning, then procedures which include the same number of stimulus events as the experimental group receives, but include no contiguous occurrences of CS and UCS, should yield no associative effects. Consequently, any responses to the CS in these groups should represent the net effect of non-associative factors, e.g., sensitization and pseudoconditioning. This view suggests that the level of responding to the CS in a backward conditioning procedure or a procedure in which CS and UCS are always separated by an intertrial interval should be subtracted from the level of responding to the CS in the group which received CS-UCS pairings, and the difference should be identified with the amount of excitatory conditioning (cf. Rescorla, 1967a).

Now, reconsider the validity of this approach, in light of data presented in this chapter. When backward conditioning trials or unpaired CSs and UCSs are presented to an organism, summation and retardation-of-acquisition tests reveal that the CS does not become associatively neutral, but rather becomes a conditioned inhibitor of the CR. Insofar as negative contingencies between CS and UCS do result in inhibition within a given response system, then comparison of the levels of responding in groups that received paired vs. unpaired CSs and UCSs will reveal the sum of excitatory and inhibitory conditioned effects, rather than excitatory effects alone.

Rescorla's contingency viewpoint suggests that the most appropriate control for non-associative effects is a procedure which includes the same number of CSs and UCSs as the experimental group receives, but removes the contingency between CS and UCS. The zero-contingency procedure should provide an associatively neutral reference point, against which the effects of both positive and negative contingencies can be assessed. For example, suppose a rat acquires conditioned suppression to a tone CS paired with shock more slowly if the shock has previously been negatively contingent upon the tone than if neither event has been presented before. In isolation, this outcome might be attributed to the conditioning of inhibition (of suppression) to the tone CS in the experimental group. Alternatively, perhaps the retardation of excitatory conditioning in the experimental group is attributable to some non-associative effect, e.g., latent inhibition, and is not dependent upon a prior negative contingency between CS and UCS. These possibilities can be teased apart by the addition of a zero-contingency group. To the extent that a prior negative contingency between CS and UCS results in a decrement in the rate of subsequent excitatory conditioning, relative to a prior zero-contingency, the negative contingency can be said to have produced conditioned inhibition.

The virtues of the zero-contingency control procedure are the virtues of the contingency veiwpoint on which it is based, chiefly the recognition that a wide variety of procedures which entail negative contingencies between CS and UCS yield inhibitory conditioning, and that inhibitory effects of UCSs which occur in the absence of CS somehow combine with excitatory effects of UCSs which occur in presence of a CS to produce some net conditioned effect. The special rationale for the zero-contingency procedure as a control for non-associative, incremental effects disappears for response systems which do not yield conditioned inhibition. In such cases the negative contingency procedure would be the best control for sensitization, since it differs from the positive contingency procedure only

in eliminating CS-UCS pairings, presumably the critical factor in
the absence of inhibition.

The contingency viewpoint also has important implications for
the study of extinction. First, it suggests that the most appropriate extinction procedure following excitatory conditioning is one
which removes the positive contingency between CS and UCS, but
holds the number of CS and UCS events per session constant (Rescorla
1967a). This procedure should provide more straightforward information about associative effects than the standard "CS-alone
extinction" procedure discussed in the previous chapter, which allows
non-associative factors to contribute to the decrement in performance across trials. Total removal of the UCS from the session
might remove a source of drive which facilitates performance, particularly in the case of aversive conditioning (e.g., Spence, 1966).
Removal of the UCS from the session changes the stimulus context
markedly, and might result in a performance decrement on that
basis. Further, Rescorla and Heth (1975) have argued that the
memorial representation of the UCS might be modified during extinction, i.e., the animal might treat the CS-alone extinction procedure as a drastic reduction in the intensity of the UCS. All
these putative non-associative factors should result in fewer CRs,
at least during the early trials, in the CS-alone procedure than in
a zero-contingency procedure which presented the same number of CSs
and UCSs per session as in acquisition.

If we reasonably extend the contingency viewpoint to permit
"computation" of contingencies across sessions, then in both of
the aforementioned procedures the overall positive contingency
between CS and UCS should diminish in magnitude as the extinction
treatment progresses. The relative associative decrements produced
by the two procedures would depend upon the particular quantitative
model applied to the two conditional probabilities (e.g., the phi-
coefficient), but in neither case should $p(UCS/CS)$ be less than
$p(UCS/\overline{CS})$. Thus, neither procedure should result in inhibition.
As we have seen, in the standard CS-alone extinction procedure the
latter prediction has been borne out by Reberg's (1972) data; an
extinguished CS seems not to become a net inhibitor of the CR.
Nonetheless, the former prediction, that an associative process
contributes to the decrement in the CR across CS-alone trials, remains plausible. For example, non-reinforcement of the CS might
produce a partial loss of excitation. Spontaneous recovery is
rarely complete, and thus implies only that excitatory conditioning
is not totally obliterated by subsequent unreinforced presentations
of the CS. On the other hand, the performance decrement in extinction might in part reflect the build-up of an inhibitory process
which opposes the previously conditioned excitation, but which is

parasitic on, and can never be stronger than, excitation. The contingency viewpoint is silent on questions of mechanism, and is thus compatible with either of these possibilities.

Although there have been very few comparisons of CS-alone extinction procedures with zero-contingency procedures that matched the number of UCSs per session with acquisition (Ayres & DeCosta, 1971; Ayres, Mahoney, Proulx, & Benedict, 1976), several studies have compared the CS-alone procedure with procedures in which: (a) the UCS was delayed on extinction trials, so that the CS-UCS interval would not have produced acquisition of that response (e.g., Frey & Butler, 1977; Leonard, 1975; McAllister, 1953; Spence, 1966), or (b) the CS and UCS occurred on separate trials (Frey & Butler, 1977; Leonard, 1975; Spence, 1966). In all of these studies of eye-blink conditioning in humans and rabbits there were more CRs during the extinction treatments which included UCS presentation than in the CS-alone treatments, suggesting that removal of the UCS from the situation in the CS-alone procedure removed some non-associative support for the CR, e.g., motivational, stimulus, or memorial support.

In one of these studies Frey and Butler reconditioned the sub-jects, i.e., again presented CS-UCS pairings, following exposure to the extinction treatments. During reconditioning they obtained fewer CRs from rabbits that had received the CS and UCS unpaired during extinction than from rabbits that had received the CS-alone procedure. These results suggest that larger associative decre-ments occurred in the negative contingency and delayed UCS treat-ments than in the CS-alone procedure. The contingency viewpoint predicts greater associative decrements in the negative contin-gency treatment than in the CS-alone procedure, because averaging a positive contingency with a negative contingency should reduce the overall contingency between CS and UCS to zero more rapidly than should averaging a positive contingency with a zero-contingency in which $p(UCS/CS) = p(UCS/\overline{CS}) = 0$.

Rescorla's contingency view of classical conditioning was a very useful one, primarily because it suggested a variety of funda-mentally important experiments that would not have been suggested by the Pavlovian view. Further, it allowed researchers to inte-grate a much larger body of data, particularly data on inhibitory conditioning, than could be explained by the Pavlovian account. For these reasons it was the dominant theoretical impetus to re-search in classical conditioning in the late 1960s. However, sev-eral limitations of the theory, along with the growth of interest in questions about the effects of presenting multiple stimuli on a conditioning trial, have led to the development of a quite different model (Rescorla & Wagner, 1972; Wagner & Rescorla, 1972), which will be the focus of the next chapter. The next section of the present chapter will discuss some limitations of the contingency view, and in so doing point out requirements that any new model of classical conditioning should fulfill.

Limitations of the contingency model. One characteristic of the contingency model has not been widely discussed, and should be noted. The contingency view cannot escape contiguity, but must assume the necessity of the contiguity of the trace of the CS with the UCS for the occurrence of excitatory conditioning. This assumption becomes explicit if we speak of the conditional probabilities of the UCS in the presence and absence of the trace of the CS. If the latter formulation is not used, then a trace conditioning procedure with a very brief gap between CS and UCS becomes an example of a negative contingency, which should produce inhibition; often it does not (e.g., Ellison, 1964; Kamin, 1965). The inclusion of contiguity as a necessary condition for excitatory conditioning does not blunt the main thrust of the contingency view, that a positive contingency is necessary for excitatory conditioning, and contiguity of CS trace and UCS is not sufficient. However, necessity of contiguity implies that global statements like "positive contingencies invariably lead to excitatory conditioning, and negative contingencies invariably lead to inhibition" may run afoul of the temporal characteristics of particular response systems. For example, suppose a half-sec tone reliably precedes a brief shock by 10 sec. Some responses, e.g., heart rate changes and the subsequent punishing effect of the tone upon instrumental appetitive behaviors, would very likely reveal excitatory conditioning, but others, e.g., the rabbit's eyeblink, would reveal no excitatory conditioning (VanDercar & Schneiderman, 1967). Thus, if a contingency between CS and UCS is described without specifying the response system to be studied, the effect of this contingency may depend on the response system.

One limitation of the contingency view has been its failure to make predictions about the course of acquisition of excitation and inhibition. It predicts the net conditioned effects of treatments involving presentations of CSs and UCSs, as though organisms calculate $p(UCS/CS)$ and $p(UCS/\overline{CS})$ and somehow compare the two values to arrive at the contingency, but it does not specify the conditioned effects of reinforced and non-reinforced presentations of CSs on particular trials.

A second limitation of the contingency viewpoint arises from the first. The absence of statements about the underlying processes, or at least of rules which describe trial-by-trial changes

in the strength of associations as function of reinforcement or
nonreinforcement, has led to a lack of rules which predict the
net associative effects of several different treatments given in
succession. In the absence of specific rules in the discussion of
extinction earlier in this chapter, it was simply assumed that
across-sessions p (UCS/CS) and p(UCS/$\overline{\text{CS}}$) were computed by means of
arithmetic averaging of the rates of UCS per unit time in the
presence/absence of the CS in the various conditioning sessions.
By default, such an assumption predicts no order effects when conditioning treatments A and B are administered in the orders AB and
BA. If the A and B treatments are, e.g., CS-UCS pairings and presentations of the UCS alone, then the prediction must be that the
strength of conditioning would be equal following the two pairs
of treatments, and less than in a group that had received only the
CS-UCS pairings. Experimental disconfirmations, as well as con-
firmations, of this prediction abound (e.g., Rescorla, 1974; Riley,
Jacobs, & LoLordo, 1976). More important, there are other theoretical positions which make positive predictions, based on postulated
mechanisms, about these and other experiments, and are therefore
likely to be theoretically "richer."

     These limitations upon the utility of the contingency model
provided part of the impetus for the development of more recent
models (e.g., Rescorla & Wagner, 1972). Another impetus for such
developments came from the results of a number of zero-contingency
experiments, which will be discussed in the next section.

Effects of a Zero Contingency Between CS and UCS

     The zero-contingency procedure arose from the contingency view
as the most appropriate procedure for assessing non-associative contributions to the outcomes of classical conditioning experiments.
At first, the zero-contingency procedure was studied primarily because of its potential as a control procedure. Recently, however,
it has been seen as important in its own right.

     In several experiments described earlier, a CS from the zero-
contingency procedure has: (1) become less excitatory than a CS
upon which a UCS had been positively contingent (Gamzu & Williams,
1971; Rescorla, 1968a); (2) had no effect upon the response to a
CS+ in summation tests (Hammond, 1967; Rescorla, 1966; 1967b, 1968a),
and (3) been transformed into an excitatory CS in a retardation-of-
acquisition test more rapidly than stimuli on which the UCS had
been negatively contingent (Hammond, 1968). Taken together, these
results suggest that the zero-contingency procedure produces no
associative effects. On the other hand, a number of studies using

variations of the conditioned emotional response procedure with rats have obtained excitatory conditioned effects with the zero-contingency procedure (Benedict & Ayres, 1972; Kremer, 1971, 1974; Kremer & Kamin, 1971; Quinsey, 1971). Several of these studies found that the strength of excitatory conditioning produced by a zero-contingency procedure was directly related to the number of chance pairings that occurred (Kremer, 1971; Kremer & Kamin, 1971; Quinsey, 1971), as an orthodox Pavlovian view would have predicted.

Benedict and Ayres (1972), using the CER procedure with rats, varied the number of randomly distributed CSs and UCSs per hr across a number of zero-contingency groups, thereby varying the number of chance pairings of CS and UCS. All groups received relatively high densities of CSs and/or UCSs in this experiment. In the test session, the magnitude of conditioned suppression evoked by the CS in the various groups was not directly related to the number of chance pairings of CS and UCS received by those groups. On the contrary, the two groups which showed considerable conditioned suppression in the test had received low or moderate numbers of chance pairings of CS and UCS. However, a closer look at the data did reveal an orderly relationship; the two groups that ultimately showed conditioned suppression to the CS experienced a positive contingency between CS and UCS during the first 100 sec of conditioning, but other groups experienced a negative contingency.

Later experiments (Ayres, Benedict, & Witcher, 1975; Benedict & Ayres, 1972) pointed out an additional complication. Zero-contingency procedures using massed trials resulted in excitatory conditioning when "chance pairings" occurred very early in the first conditioning session or when they occurred later but were not preceded by many unsignalled UCSs. However, when several unsignalled UCSs preceded the first pairing of CS and UCS in a zero-contingency procedure, excitatory conditioning did not occur.

Thus, when trials are massed in a zero-contingency procedure, the CS from that procedure will acquire the capacity to suppress appetitive behaviors, unless the first chance pairings of CS and UCS have been preceded by several unsignalled UCSs. This effect is very likely associative. Recent data by Keller, Ayres, and Mahoney (1977) suggest that the effect is preasymptotic (but see Kremer, 1974). They found that, with additional hours of exposure to a zero-contingency procedure which initially resulted in strong excitatory conditioning, the amount of conditioned suppression progressively diminished, until there was none.

What are the implications of this research? First, it suggests that an overall positive contingency between CS and UCS is not a necessary condition for the formation of an excitatory CR. Second, the conditions under which a zero-contingency will yield an excitatory CS at least pre-asymptotically can be specified, i.e., only sequences of events with specifiable properties will result in excitatory conditioning. The latter fact again points to the need for a trial-by-trial model of conditioning, which can explain why certain sequences of CSs, CS-UCS pairings, and UCSs result in excitatory conditioning, and others do not.

Mackintosh (1973, 1975) has observed another outcome of zero-contingency procedures which was not anticipated by the contingency viewpoint. Briefly, in both the CER procedure and in appetitive conditioning, when a CS and a UCS had been presented in a zero-contingency procedure, subsequent excitatory classical conditioning in a retardation-of-acquisition test occurred more slowly than in naive subjects or subjects that had received CS-alone or UCS-alone treatments. Figure 10 illustrates this result for groups of rats that were given the CER test procedure. Over the first four days of conditioning, the group that had previously received a zero-contingency between CS and shock exhibited less conditioned suppression than any of the other groups, which did not differ from each other. Figure 11 illustrates that an analogous pattern of results occurred in the appetitive test procedure. Over the six sessions, the group that had previously received a zero-contingency between the CS and water did less conditioned licking than the other groups, when the dependent variable was the difference between the rates of licking during the first 15 sec of the CS and the 5 sec preceding each CS. The same result has been observed in autoshaping of the pigeon's keypeck (Mackintosh, 1973), eyeblink conditioning of rabbits (Siegel & Domjan, 1971), as well as the CER (Baker, 1976, Experiment 1) and conditioned licking (Baker, 1974, Experiment 3). Mackintosh, who called this phenomenon "learned irrelevance." reasoned that the arrangement of a zero-contingency between a CS and a UCS resulted in retarded development of an excitatory CR when the CS was subsequently paired with the UCS because the rats first learned that the CS and UCS were independent of each other, or irrelevant for each other, and this learning interfered with the learning of a different relationship, e.g., a positive contingency between the same two events.

Another interpretation of Mackintosh's results can be offered. The zero-contingency procedure might be producing conditioned inhibitory effects, which would yield retardation of excitatory conditioning. That is, the deficit might be attributable to the conditioning of a response during the zero-contingency phase which must be overcome in the subsequent positive contingency phase, rather than to the learning of a relationship which is incompatible

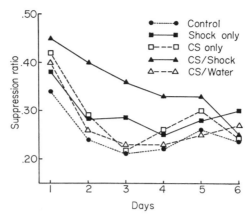

Figure 3.10   Mean suppression ratios across six days of a condi-
              tioned emotional response procedure in which a CS was
              paired with electric shock for various groups in an
              experiment by Mackintosh (1973). Various groups had
              previously received no treatment, only CSs, only UCSs,
              a zero-contingency between CS and shock (Group CS/
              Shock), or a zero-contingency between CS and water
              (Group CS/Water).

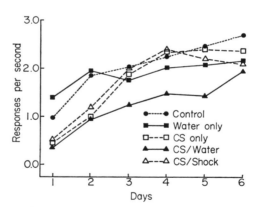

Figure 3.11   Mean lick CRs (i.e., the number of licks/sec during
              the CS minus the number of licks/sec during the 5
              sec immediately prior to the CS) for groups of rats
              in an experiment by Mackintosh (1973). Various
              groups had previously received no treatment, only CSs,
              only (water) UCSs, a zero-contingency between CS and
              water (Group CS/Water), or a zero-contingency between
              CS and shock (Group CS/Shock).

with the relationship to be learned in the subsequent phase. In cases where it's reasonable to suspect that the zero-contingency procedure might have produced conditioned inhibition, an experiment in which the test phase involves inhibitory rather than excitatory conditioning should permit rejection of one of the alternative explanations. Learned irrelevance arising from a zero-contingency procedure should lead to retarded acquisition of inhibitory conditioning, since the latter also requires learning of a new relationship between CS and UCS. On the other hand, if the CS from a zero-contingency procedure becomes inhibitory, then subsequent inhibitory conditioning should be facilitated.

The "learned irrelevance" effect should have a profound impact upon theory building in classical conditioning, because it suggests that there is a form of learning occurring when CS and UCS are independent that is orthogonal to the associative learning continuum which has been the basis of both the Pavlovian and the contingency viewpoints. It remains to be seen how the two forms will be integrated.

Conclusions

Despite the limitations which have been outlined in the last few pages, the introduction of the concept of contingency between CS and UCS has had a beneficial impact on the study of learning. It rearoused interest in classical conditioning, and led researchers to consider a much broader range of problems. Further, by emphasizing the symmetry of positive and negative contingencies around a zero point, it led to a new focus on the conditions which lead to the development of inhibition, thereby restoring a balance that did characterize Pavlov's and Konorski's thinking, but not earlier North American work.

Nonetheless, by the late 1960's it had become clear that a model of classical conditioning that made trial-by-trial predictions of the effects of reinforcement and non-reinforcement was needed. Such a model would be required to account for the excitatory and inhibitory effects of diverse manipulations of the contingency between CS and UCS. Further, it would have to explicitly acknowledge the importance of contiguity in determining the effects of such manipulations. The next chapter will focus on a model which meets these requirements.

REFERENCES

Anger, D. The role of temporal discriminations in the reinforcement of Sidman avoidance behavior. JOURNAL OF THE EXPERIMENTAL ANALYSIS OF BEHAVIOR, 1963, 6, 477-506.

Anrep, G. V. Pitch discrimination in the dog. JOURNAL OF PSYCHOLOGY, 1920, 53, 367-385.

Ayres, J. J. B., Benedict, J. O., & Witcher, E. S. Systematic manipulation of individual events in a truly random control in rats. JOURNAL OF COMPARATIVE AND PHYSIOLOGICAL PSYCHOLOGY, 1975, 88, 97-103.

Ayres, J. J. B., & DeCosta, M. J. The truly random control as an extinction procedure. PSYCHONOMIC SCIENCE, 1971, 24, 31-33.

Ayres, J. J. B., Mahoney, W. J., Proulx, D. T., & Benedict, J. O. Backward conditioning as an extinction procedure. LEARNING AND MOTIVATION, 1976, 7, 368-381.

Baker, A. G. Rats learn that events, be they stimuli or responses, bear no relation to one another. Unpublished Doctoral dissertation, Dalhousie University, 1974.

Baker, A. G. Learned irrelevance and learned helplessness: Rats learn that stimuli, reinforcers, and responses are uncorrelated. JOURNAL OF EXPERIMENTAL PSYCHOLOGY: ANIMAL BEHAVIOR PROCESSES, 1976, 2, 130-131.

Benedict, J. O., & Ayres, J. J. B. Factors affecting conditioning in the truly random control procedure in the rat. JOURNAL OF COMPARATIVE AND PHYSIOLOGICAL PSYCHOLOGY, 1972, 78, 323-330.

Best, M. P. Conditioned and latent inhibition in taste-aversion learning: Clarifying the role of learned safety. JOURNAL OF EXPERIMENTAL PSYCHOLOGY: ANIMAL BEHAVIOR PROCESSES, 1975, 1, 97-113.

Bitterman, M. E. The CS-UCS interval in classical and avoidance conditioning. In W. F. Prokasy (Ed.), CLASSICAL CONDITIONING. New York: Appleton-Century-Crofts, Inc., 1965.

Catania, A. C. Elicitation, reinforcement, and stimulus control. In R. Glaser (Ed.), THE NATURE OF REINFORCEMENT. New York: Academic Press, 1971.

Coleman, S. R., & Gormezano, I. Classical conditioning of the rabbit's (Oryctolagus cuniculus) nictitating membrane response under symmetrical CS-UCS interval shifts. JOURNAL OF COMPARATIVE AND PHYSIOLOGICAL PSYCHOLOGY, 1971, 77, 447-455.

Ellison, G. D. Differential salivary conditioning to traces. JOURNAL OF COMPARATIVE AND PHYSIOLOGICAL PSYCHOLOGY, 1964, 57, 373-380.

Frey, P. W., & Butler, C. S. Extinction after aversive conditioning: An associative or non-associative process? LEARNING AND MOTIVATION, 1977, 8, 1-17.

Fureday, J. J., Poulos, C. X., & Schiffmann, K. Contingency theory and classical autonomic excitatory and inhibitory conditioning: Some problems of assessment and interpretation. PSYCHOPHYSIOLOGY, 1975, 12, 98-105.

Fureday, J. J., & Schiffmann, K. Concurrent measurement of autonomic and cognitive processes in a test of the traditional discriminative control procedure for Pavlovian electroderman conditioning. JOURNAL OF EXPERIMENTAL PSYCHOLOGY, 1973, 100, 210-217.

Gamzu, E., & Williams, D. R. Classical conditioning of a complex skeletal response. SCIENCE, 1971, 171, 923-925.

Gibbon, J., Berryman, R., & Thompson, R. L. Contingency spaces and measures in classical and in instrumental conditioning. JOURNAL OF THE EXPERIMENTAL ANALYSIS OF BEHAVIOR, 1974, 21, 585-605.

Gonzalez, R. C., Longo, N., & Bitterman, M. E. Classical conditioning in the fish: Exploratory studies of partial reinforcement. JOURNAL OF COMPARATIVE AND PHYSIOLOGICAL PSYCHOLOGY, 1961, 54, 453-456.

Gormezano, I., & Coleman, S. R. Effects of partial reinforcement on conditioning, conditional probabilities, asymptotic performance, and extinction of the rabbit's nictitating membrane response. THE PAVLOVIAN JOURNAL OF BIOLOGICAL SCIENCE, 1975, 10, 13-22.

Hammond, L. J. A traditional demonstration of the active properties of Pavlovian inhibition using differential CER. PSYCHONOMIC SCIENCE, 1967, 9, 65-66.

Hammond, L. J. Retardation of fear acquisition when the CS has previously been inhibitory. JOURNAL OF COMPARATIVE AND PHYSIOLOGICAL PSYCHOLOGY, 1968, 66, 756-759.

Hearst, E., & Franklin, S. R. Positive and negative relations between a signal and food: Approach-withdrawal behavior to the signal. JOURNAL OF EXPERIMENTAL PSYCHOLOGY: ANIMAL BEHAVIOR PROCESSES, 1977, 3, 37-52.

Heth, C. D. Simultaneous and backward fear conditioning as a function of number of CS-UCS pairings. JOURNAL OF EXPERIMENTAL PSYCHOLOGY: ANIMAL BEHAVIOR PROCESSES, 1976, 2, 117-129.

Holland, P. C. Conditioned stimulus as a determinant of the form of the Pavlovian conditioned response. JOURNAL OF EXPERIMENTAL PSYCHOLOGY: ANIMAL BEHAVIOR PROCESSES, 1977, 3, 77-104.

Kamin, L. J. Temporal and intensity characteristics of the conditioned stimulus. In W. F. Prokasy (Ed.), CLASSICAL CONDITIONING. New York: Appleton-Century-Crofts, 1965.

Keller, R. J., Ayres, J. J. B., & Mahoney, W. J. Brief versus ex-
    tended exposure to truly random control procedures. JOURNAL OF
    EXPERIMENTAL PSYCHOLOGY: ANIMAL BEHAVIOR PROCESSES, 1977, 3,
    53-65.

Killeen, P. On the temporal control of behavior. PSYCHOLOGICAL
    REVIEW, 1975, 82, 89-115.

Konorski, J. CONDITIONED REFLEXES AND NEURON ORGANIZATION. Cam-
    bridge: Cambridge University Press, 1948.

Kremer, E. F. Truly random and traditional control procedures in
    CER conditioning in the rat. JOURNAL OF COMPARATIVE AND PHY-
    SIOLOGICAL PSYCHOLOGY, 1971, 76, 441-448.

Kremer, E. F. The truly random control procedure: Conditioning to
    the static cues. JOURNAL OF COMPARATIVE AND PHYSIOLOGICAL
    PSYCHOLOGY, 1974, 86, 700-707.

Kremer, E. F., & Kamin, L. J. The truly random control procedure:
    Associative or non-associative effects in rats. JOURNAL OF
    COMPARATIVE AND PHYSIOLOGICAL PSYCHOLOGY, 1971, 74, 203-210.

Leonard, D. W. Partial reinforcement effects in classical aversive
    conditioning in rabbits and human beings. JOURNAL OF COMPARA-
    TIVE AND PHYSIOLOGICAL PSYCHOLOGY, 1975, 88, 596-608.

Mackintosh, N. J. Stimulus selection: Learning to ignore stimuli
    that predict no change in reinforcement. In R. A. Hinde &
    J. Stevenson-Hinde (Eds.), CONSTRAINTS ON LEARNING. New York:
    Academic Press, 1973.

Mackintosh, N. J. A theory of attention: Variations in the associ-
    ability of stimuli with reinforcement. PSYCHOLOGICAL REVIEW,
    1975, 82, 276-298.

Mahoney, W. J., & Ayres, J. J. B. One trial simultaneous and back-
    ward fear conditioning in rats. ANIMAL LEARNING & BEHAVIOR,
    1976, 4, 357-362.

Maier, S. F., Rapaport, P., & Wheatley, K. L. Conditioned inhibition
    and the UCS-CS interval. ANIMAL LEARNING & BEHAVIOR, 1976, 4,
    217-220.

Marchant, H. G., III, Mis, F. W., & Moore, J. W. Conditioned inhi-
    bition of the rabbit's nictitating membrane response. JOURNAL
    OF EXPERIMENTAL PSYCHOLOGY, 1972, 95, 408-411.

Marchant, H. G., III, & Moore, J. W. Below-zero conditioned inhibi-
    tion of the rabbit's nictitating membrane response. JOURNAL OF
    EXPERIMENTAL PSYCHOLOGY, 1974, 102, 350-352.

McAllister, W. R. The effect on eyelid conditioning of shifting the
    CS-UCS interval. JOURNAL OF EXPERIMENTAL PSYCHOLOGY, 1953, 45,
    423-428.

Moscovitch, A., & LoLordo, V. M. Role of safety in the Pavlovian backward fear conditioning procedure. JOURNAL OF COMPARATIVE AND PHYSIOLOGICAL PSYCHOLOGY, 1968, 66, 673-678.

Ost, J. W. P., & Lauer, D. W. Some investigations of classical salivary conditioning in the dog. In W. F. Prokasy (Ed.), CLASSICAL CONDITIONING: A SYMPOSIUM. New York: Appleton-Century-Crofts, 1965.

Pavlov, I. P. CONDITIONED REFLEXES. Translated by G. V. Anrep. Oxford: Oxford University Press, 1927.

Plotkin, H. C., & Oakley, D. A. Backward conditioning in the rabbit (*Oryctolagus cuniculus*). JOURNAL OF COMPARATIVE AND PHYSIOLOGICAL PSYCHOLOGY, 1975, 88, 586-590.

Prokasy, W. F. Classical eyelid conditioning: Experimenter operations, task demands, and response shaping. In W. F. Prokasy (Ed.), CLASSICAL CONDITIONING: A SYMPOSIUM. New York: Appleton-Century-Crofts, 1965.

Quinsey, V. L. Conditioned suppression with no CS-UCS contingency in the rat. CANADIAN JOURNAL OF PSYCHOLOGY, 1971, 25, 69-82.

Razran, G. Backward conditioning. PSYCHOLOGICAL BULLETIN, 1956, 63, 55-69.

Reberg, D. Compound tests for excitation in early acquisition and after prolonged extinction of conditioned suppression. LEARNING AND MOTIVATION, 1972, 3, 246-258.

Rescorla, R. A. Predictability and number of pairings in Pavlovian fear conditioning. PSYCHONOMIC SCIENCE, 1966, 4, 383-384.

Rescorla, R. A. Pavlovian conditioning and its proper control procedures. PSYCHOLOGICAL REVIEW, 1967, 74, 71-80. (a)

Rescorla, R. A. Inhibition of delay in Pavlovian fear conditioning. JOURNAL OF COMPARATIVE AND PHYSIOLOGICAL PSYCHOLOGY, 1967, 64, 114-120. (b)

Rescorla, R. A. Probability of shock in the presence and absence of shock in fear conditioning. JOURNAL OF COMPARATIVE AND PHYSIOLOGICAL PSYCHOLOGY, 1968, 66, 1-5. (a)

Rescorla, R. A. Pavlovian conditioned fear in Sidman avoidance learning. JOURNAL OF COMPARATIVE AND PHYSIOLOGICAL PSYCHOLOGY, 1968, 65, 55-60. (b)

Rescorla, R. A. Conditioned inhibition of fear resulting from negative CS-UCS contingencies. JOURNAL OF COMPARATIVE AND PHYSIOLOGICAL PSYCHOLOGY, 1969, 67, 504-509.

Rescorla, R. A. Summation and retardation tests of latent inhibition. JOURNAL OF COMPARATIVE AND PHYSIOLOGICAL PSYCHOLOGY, 1971, 75, 77-81.

Rescorla, R. A.   Effect of inflation of the unconditioned stimulus value following conditioning. JOURNAL OF COMPARATIVE AND PHYSIOLOGICAL PSYCHOLOGY, 1974, 86, 101-106.

Rescorla, R. A., & Heth, C. D.   Reinstatement of fear to an extinguished conditioned stimulus. JOURNAL OF EXPERIMENTAL PSYCHOLOGY: ANIMAL BEHAVIOR PROCESSES, 1975, 1, 88-96.

Rescorla, R. A., & LoLordo, V. M.   Inhibition of avoidance behavior. JOURNAL OF COMPARATIVE AND PHYSIOLOGICAL PSYCHOLOGY, 1965, 59, 406-412.

Rescorla, R.A., & Wagner, A. R.   A theory of Pavlovian conditioning: Variations in the effectiveness of reinforcement and non-reinforcement. In A. H. Black & W. F. Prokasy (Eds.), CLASSICAL CONDITIONING II. THEORY AND RESEARCH. New York: Appleton-Century-Crofts, 1972.

Riley, A. L., Jacobs, W. J., & LoLordo, V. M.   Drug exposure and the acquisition and retention of a conditioned taste aversion. JOURNAL OF COMPARATIVE AND PHYSIOLOGICAL PSYCHOLOGY, 1976, 90, 799-807.

Seligman, M. E. P., Maier, S. F., & Solomon, R. L.   Unpredictable and uncontrollable aversive events. In F. R. Brush (Ed.), AVERSIVE CONDITIONING AND LEARNING. New York: Academic Press, 1971.

Sidman, M.   Avoidance conditioning with brief shock and no exteroceptive warning signal. SCIENCE, 1953, 118, 157-158.

Siegel, S., & Domjan, M.   Backward conditioning as an inhibitory procedure. LEARNING AND MOTIVATION, 1971, 2, 1-11.

Spence, K. W.   Extinction of the human eyelid CR as a function of presence or absence of the UCS during extinction. JOURNAL OF EXPERIMENTAL PSYCHOLOGY, 1966, 71, 642-648.

Taukulis, H. K., & Revusky, S. H.   Odor as a conditioned inhibitor: Applicability of the Rescorla-Wagner model to feeding behavior. LEARNING AND MOTIVATION, 1975, 6, 11-27.

Van Dercar, D. H., & Schneiderman, N.   Interstimulus interval functions in different response systems during classical discrimination conditioning of rabbits. PSYCHONOMIC SCIENCE, 1967, 9, 9-10.

Wagner, A. R., & Rescorla, R. A.   Inhibition in Pavlovian conditioning: Application of a theory. In R. A. Boakes & M. S. Halliday (Eds.), INHIBITION AND LEARNING. London: Academic Press, 1972.

Wagner, A. R., Siegel, S., Thomas, E., & Ellison, G. D.   Reinforcement history and extinction of the classical reward response. JOURNAL OF COMPARATIVE AND PHYSIOLOGICAL PSYCHOLOGY, 1964, 58, 354-358.

Wasserman, E. A., Franklin, S. R., & Hearst, E.   Pavlovian appetitive contingencies and approach vs. withdrawal to conditioned stimuli in pigeons.   JOURNAL OF COMPARATIVE AND PHYSIOLOGICAL PSYCHOLOGY, 1974, 86, 616-627.

Wessells, M. G.   Errorless discrimination, autoshaping, and conditioned inhibition.   SCIENCE, 1973, 182, 941-943.

4. CLASSICAL CONDITIONING: COMPOUND CSs AND THE RESCORLA-WAGNER MODEL

Vincent M. LoLordo

Dalhousie University
Halifax, Nova Scotia, Canada

Chapter Three concluded with a statement of the need for a model of classical conditioning that predicts the effects of re-inforcement and non-reinforcement on a trial-by-trial basis, accounts for the outcomes of various manipulations of the contingency between CS and UCS, and predicts the net associative effects of several different conditioning treatments given in succession. Such a model was devised by Rescorla and Wagner (1972; Wagner & Rescorla, 1972) to meet these needs, but also to explain provocative new variations in the effects of reinforcement and non-reinforcement in cases where more than one CS was presented on a trial.

Why should a single model of conditioning be emphasized in a relatively brief treatment of classical conditioning? Not because this model explains every known fact of classical conditioning. Later in this chapter we shall see that it does not address certain important issues, and makes incorrect predictions in some other instances. Rather, this chapter will focus on the Rescorla-Wagner model because it has been one of the major foci of research and theory-building in classical conditioning during the last five years. Moreover, even someone who is taking his first careful look at research in classical conditioning should, in understanding the model, be able to classify a great many experiments and remember their results. Since the model emphasizes certain features of classical conditioning experiments the reader should be alerted to these features in new experiments, and should be able to fit the experiments into a coherent framework and predict their outcomes.

REINFORCEMENT AND NON-REINFORCEMENT OF COMPOUND CSs

The impetus for the Rescorla-Wagner model came largely from demonstrations of variations in the effectiveness of reinforcement or non-reinforcement in conditioning a response to a given stimulus element of a compound CS, depending upon the separate reinforcement or non-reinforcement of the other elements. The first two examples will describe variations in the effectiveness of reinforcement.

Wagner and Saavedra (in Wagner, 1969) conditioned the eyeblink response of three groups of rabbits to brief auditory and visual CSs that began 1 sec before, and overlapped, para-orbital electric shock. During the first phase of the study, all three groups received 200 trials on which a simultaneous compound of a tone (call it A) and a flashing light (call it X) were followed by the UCS. The control group, Group AX+, received only these trials. However, the other two groups also received 200 presentations of X interspersed among compound trials. For Group AX+/X+, the X trials were always reinforced, but for Group AX+/X-, they were never reinforced. Thus, X should have been strongly excitatory in Group AX+/X+ and much less excitatory in Group AX+/X-. Wagner and Saavedra asked whether these treatments of X alone would affect the amount of excitatory conditioning to A, which was treated identically for the three groups. They answered this question by presenting all rabbits with 16 reinforced presentations of A alone in the test phase.

Figure 1 illustrates the median percent eyeblink CRs to the CSs in both the conditioning and test phases of the experiment. The left-hand bar in each panel illustrates responding to the reinforced AX compound, which was equivalent for the three groups. The middle bars illustrate responding to X. When it was always reinforced, X became strongly excitatory, but when it was never reinforced, it evoked only 10% CRs. The right-hand bars illustrate test responding to the tone, A. The excitatory strength of A was a function of the treatment of X. Reinforcement of X alone decreased the amount of excitatory conditioning to A on reinforced AX trials, whereas non-reinforcement of X alone increased the amount of excitatory conditioning to A on reinforced AX trials, relative to the control value (middle panel).

In a series of experiments using the CER procedure and rat subjects, Kamin (1969) demonstrated that prior reinforced presentations of CS X markedly attenuated excitatory conditioning to CS A when A was added to X, and the compound reinforced as before. Kamin used as CSs long-duration auditory and visual cues which

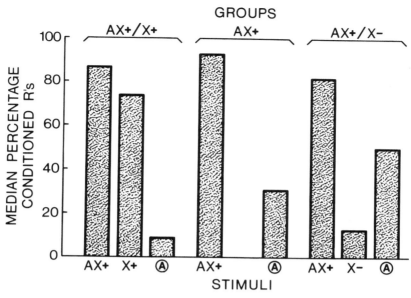

Figure 4.1 Median percentage conditioned eyeblink responses to an AX compound and to X alone in phase one, and to A alone in the test phase of the study by Wagner and Saavedra. The left-hand panel presents the data for Group AX+/X+; the middle panel, for Group AX+; and the right-hand panel for Group AX+/X- (adapted from Rescorla & Wagner, 1972).

TABLE 1

Design of a Typical Blocking Experiment
(after Kamin, 1969)

	STAGE 1	STAGE 2	TEST	SUPPRESSION RATIO
Group AX+, X+	AX(8)	X(16)	A	.25
Group X+, AX+	X(16)	AX(8)	A	.45
Group AX+		AX(8)	A	.05
Group X+		X(24)	A	.44

yielded no cross-modal stimulus generalization. In one experiment, there were four groups of rats as shown in Table 1. Group X+,AX+ received 16 reinforced presentations of a white noise CS (X) in the first phase. This treatment produced complete suppression of food-reinforced lever-pressing. In the second phase of the study, CS A, a light, was added to X to form a simultaneous compound. This compound was reinforced on eight occasions. In order to determine the amount of excitatory conditioning to the added element A, it was then presented alone on a single test trial. As the table illustrates, the suppression ratio was 0.45 on this trial, i.e., the CS produced very little suppression. In fact, it produced no more suppression than in Group X+, which had received 24 pairings of CS X with electric shock but had never received a pairing of A with shock. The suppression ratio to CS A was 0.05 in control Group AX+, indicating that A produced nearly complete suppression of appetitive behavior after it had been reinforced eight times in compound with X, in the absence of prior reinforcement of noise alone. Finally, the suppression ratio to CS A was 0.25 in control Group AX+,X+ which received eight reinforced presentations of the compound CS prior to 16 reinforced presentations of CS X. The data from this group indicate that the order in which the blocks of $AX+$ and $X+$ trials were presented was a critical determinant of conditioning to CS A. When all the $AX+$ trials came first, subsequently there was moderate conditioned suppression to A, despite a five-day retention interval between the last reinforced presentation of A and the test trial. On the other hand, when $X+$ trials preceded $AX+$ trials, a single test trial revealed no evidence of excitatory conditioning to A. Kamin called the latter outcome "blocking." Apparently prior asymptotic conditioning to one CS blocked conditioning to a second CS which was added as a redundant cue, preceding the same UCS. On the basis of these results, Kamin argued that a UCS must be surprising to promote classical conditioning and that it was no longer surprising by the seventeenth trial when CS A was added for Group X+,AX+.

The experimental analysis of blocking has progressed very rapidly in the last few years, and the most recent research will be discussed in detail later in this chapter. For the present, Kamin's blocking effect and the data from Wagner and Saavedra's Groups AX+/X+ versus AX+ can be summarized in one statement. The effect of reinforcement of a simultaneous compound of CSs A and X in conditioning excitatory strength to A is diminished when the excitatory strength of X has been increased by means of separate reinforcement of X. This outcome was obtained in experiments with different species, different responses, and vastly different temporal parameters, and it implies that neither contiguity of the CS trace with the UCS nor a positive contingency between CS and UCS is sufficient for excitatory conditioning to a given cue.

A second general statement can be drawn from the comparison of Wagner and Saavedra's Groups AX+/X- versus AX+. The effect of reinforcement of a simultaneous compound of CSs A and X in conditioning excitatory strength to A is enhanced when the excitatory strength of X has been reduced by separate non-reinforcement of X. The theoretical implications of these results will be considered in more detail later, but first some variations in the effectiveness of non-reinforcement will be considered.

In an eyeblink conditioning study by Wagner, Saavedra, and Lehman (in Wagner, 1969) two groups of rabbits received 224 reinforced presentations of CS A, 28 of CS B, and 224 of CS X, in an irregular order. This procedure established CSs A and X as strong excitors, and CS B as a weak one. Next, half the rabbits received 32 non-reinforced presentations of the AX compound, and the rest received 32 non-reinforced presentations of BX. Wagner, et al., wanted to assess the relative decrements in the excitatory strength of CS X produced by non-reinforcement of X in compound with a strong (A) versus a weak (B) excitatory CS. These effects were assessed by reconditioning X; all rabbits received 32 reinforced presentations of X in the test phase.

Figure 2 illustrates the percent CRs to the various stimuli in the three phases of this experiment. In the first phase, conditioned responding to CSs A and X reached a high level, and responding to B, a moderate level. In the second phase, responding initially occurred more often to the compound of two strong excitors (AX) than to the compound of a strong and a weak excitor (BX), but the two compounds ultimately evoked the same level of responding. These data suggest that non-reinforcement produced greater decrements in the excitatory strength of the more excitatory compound. Furthermore, the rabbits that had received extinction of X in compound with the stronger excitor A made fewer responses during reconditioning to X than did rabbits that had received extinction of BX. Non-reinforcement of a CS reduces its excitatory strength more when that non-reinforcement occurs in the presence of a strong excitor than when it occurs in the presence of a weak excitor.

Wagner and Saavedra (in Rescorla & Wagner, 1972) extended the preceding study to the conditioned inhibition paradigm, asking whether the excitatory strength of a CS+ determined the amount of inhibition conditioned to a novel CS which was subsequently non-reinforced in compound with the CS+. In the first phase of an eyeblink conditioning study, rabbits received 240 reinforced presentations of CS A, eight reinforced presentations of CS B, and 548 reinforcements of CS C in a mixed order. Then the rabbits were

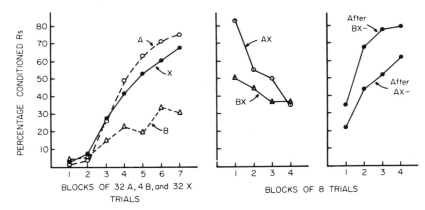

Figure 4.2   Mean percentage conditioned eyeblink responses to the
              CS elements and compounds in the three phases of the
              study by Wagner, Saavedra, and Lehman (in Wagner, 1969).
              The left-hand panel represents initial acquisition to
              CSs *A*, *B*, and *X* for all rabbits. The middle panel rep-
              resents extinction to the *AX* and *BX* compounds for sep-
              arate groups of rabbits. The right-hand panel repre-
              sents reconditioning to CS *X* for the same two groups.

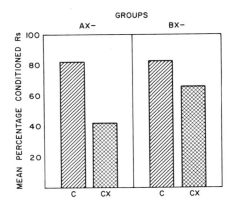

Figure 4.3   Mean percentage conditioned eyeblink responses to a
              strong excitatory CS (*C*) and to a simultaneous com-
              pound of that CS and another CS, *X*. CS *X* had previ-
              ously been non-reinforced in compound with a strong
              excitor in Group AX (left-hand panel), or in compound
              with a weak excitor (right-hand panel). Data are from
              Wagner and Saavedra, in Rescorla and Wagner (1972).

assigned to two groups, both of which received a Pavlovian conditioned inhibition treatment. Group AX- received 64 non-reinforced presentations of a novel CS X in compound with the strong conditioned excitor A. Interspersed among these trials were 64 reinforced presentations of A alone; these were included to maintain the excitatory strength of A. Group BX- received the weak excitor B instead of CS A. To ask which treatment conditioned the most inhibition to X, Wagner and Saavedra conducted a summation test in which reinforced compounds of X and the strong excitor C were interspersed among reinforced presentations of C alone. Figure 3 shows that both groups of rabbits responded less frequently to the CX compound than to C alone, i.e., CS X had become a conditioned inhibitor of the eyeblink in both groups. Moreover, the inhibitory effect was larger in Group AX-. The amount of inhibition conditioned to a novel CS as a result of non-reinforcement of that CS in compound with an excitatory CS is directly related to the strength of the excitor.

The results of these studies of variations in the effectiveness of non-reinforcement can be summarized in a way which reveals the parallels with the outcomes of the studies of variations in the effectiveness of reinforcement. The effect of non-reinforcement of a simultaneous compound of CSs A and X in conditioning inhibition to A, or in conditioning decrements in excitatory strength to A, is increased when the excitatory strength of X has been increased by means of separate reinforcement of X.

In a CER study with rats, Suiter and LoLordo (1971) demonstrated blocking of inhibitory conditioning. When electric shock was negatively contingent upon an AX compound, CS A became a conditioned inhibitor of suppression, unless shock had previously been negatively contingent upon CS X alone, in which case inhibitory conditioning to A was blocked. This result suggests that the effect of non-reinforcement of a simultaneous compound of CSs A and X in conditioning inhibition to A is diminished when the inhibitory strength of X has been increased by separate non-reinforcement of X.

All the data which have been described in this section can be summed up in the statements: (1) Reinforcement of a compound CS has a smaller incremental, associative effect upon an element of the compound when the other element has been reinforced separately, and a larger effect when the other element has been non-reinforced separately; (2) Non-reinforcement of a compound CS has a larger decremental, associative effect upon one element of the compound when the other element has been reinforced separately, and a smaller effect when the other element has been non-reinforced separately.

The Rescorla-Wagner model's explanation of these results will be presented later in this chapter. First, the basic elements of the model will be introduced.

THE RESCORLA-WAGNER MODEL

The Rescorla-Wagner model (Rescorla & Wagner, 1972; Wagner & Rescorla, 1972) was designed to predict changes in the associative strengths of various CSs present on reinforced or non-reinforced conditioning trials. The net associative strength of CS A at the start of a trial is called V_A, and the change in the associative strength of A as a result of reinforcement or non-reinforcement on that trial is ΔV_A. Since we have been distinguishing learning (associative strength) from performance (the overt response), we need some rule which translates associative strength into measures of performance, e.g., percentage CRs in blocks of trials. Rescorla and Wagner simply stated that measures of performance should preserve the ordering of associative strengths across CSs. As we shall see, the restriction to ordinal predictions does not constrain the model unduly; a diversity of novel predictions, many of which will be considered in this chapter, can still be made.

The model predicts only the associative effects of classical conditioning procedures. Consequently, any experimental tests of predictions from the model should include control groups which permit the assessment of associative and non-associative contributions to the outcomes. Furthermore, the occurrence of non-associative effects which are not predicted by the model should not be viewed as disconfirmations, although the frequent occurrence of such effects would point to the need for rules specifying the contribution of associative and non-associative determinants of performance.

The Wagner-Rescorla model calculates changes in associative strengths on the basis of a traditional assumption (e.g., Hull, 1943), that the increment in V_A as a result of a reinforced presentation of CS A depends on the discrepancy between V_A and the asymptotic associative strength which the particular UCS can support. The sort of equation used to calculate ΔV_A when A is reinforced is $\Delta V_A = K(\lambda - V_A)$, where λ is the asymptotic associative strength supportable by the reinforcer. The value of λ is specific to a given UCS, and thus may change during an experiment if the intensity or duration of the UCS is changed. According to the equation above, if CS A is repeatedly followed by the UCS, V_A should increase in negatively accelerated fashion over trials.

To add some necessary detail, the equation describing changes in the associative strength of CS A as a result of a reinforcement of A really looks like this:

$$\Delta V_A = \alpha_A \beta (\lambda - V_A) \qquad (1)$$

The magnitude of the increment in V_A is a function of two learning rate parameters, α_A and β, each of which can assume values between zero and one. The term α_A represents the learning rate parameter for CS A, and is called its salience. The term β represents the learning rate parameter for the particular UCS, and may depend upon its intensity. In any particular case, the values of these parameters have to be estimated on the basis of data, e.g., three groups of subjects would receive repeated reinforced presentations of CSs A, B, and C, and the ordinal values of α_A, α_B, and α_C would be estimated on the basis of the relative rates of conditioning of the three groups.

The unique feature of the Rescorla-Wagner model arises when we consider the case of compound CSs, as when CS A and CS B are presented simultaneously and reinforced. How do we calculate ΔV_A and ΔV_B in this case? One option would be to use equation (1) above, so that:

$$\Delta V_A = \alpha_A \beta (\lambda - V_A) \qquad (2)$$

$$\Delta V_B = \alpha_B \beta (\lambda - V_B) \qquad (3)$$

These equations are based on the assumption that each element of the compound gains associative strength as though the other element were not present, i.e., that changes in the associative strengths of the two elements are independent. This assumption is implausible in light of the data presented in the preceding section. Consider Kamin's blocking effect, the failure of excitatory conditioning to CS A in an $AX+$ compound as a result of prior reinforcement of X alone. CS A was certainly conditionable when presented alone, but not when accompanied by an already conditioned X. The Wagner-Rescorla model is able to account for these results by maintaining that the critical determinant of ΔV_A and ΔV_X as a result of an $AX+$ trial is $(\lambda - V_{AX})$, or the difference between the maximum associative strength supportable by the UCS and the combined associative strengths of CSs A and X. Thus the equations describing the changes in the associative strengths of CSs A and X when the AX compound is reinforced are:

$$\Delta V_A = \alpha_A \beta (\lambda - V_{AX}) \qquad (4)$$

$$\Delta V_X = \alpha_X \beta (\lambda - V_{AX}) \qquad (5)$$

The model further assumes that the associative strength of the compound is the algebraic sum of the associative strengths of the elements, e.g., $V_{AX} = V_A + V_X$. To summarize, the Rescorla-Wagner model stipulates that the necessary and sufficient condition for the conditioning of an increase in the associative strengths of all cues present on a trial is a positive discrepancy between the asymptotic associative strength supportable by the reinforcer which occurs on that trial and the net associative strength of all cues present on that trial. Neither contiguity nor a positive contingency between CS and UCS is a necessary or sufficient condition for excitatory conditioning to the CS, according to the explicit statement of the model. Implicitly, however, it relies on contiguity of the CS trace with the UCS in defining a trial, just as the earlier contingency view did.

Given this description of the model, we can account for the results of the studies on variations in the effectiveness of reinforcement by Kamin (1969) and Wagner and Saavedra (in Wagner, 1969). If 16 pairings of CS X and electric shock resulted in asymptotic conditioning in Kamin's experiment, then $V_X = \lambda$, according to the model. If we assume that the associative strength of the added novel CS A was zero on the first (compound) trial, then on that trial $(\lambda - V_{AX}) = (\lambda - V_X)$. Thus, since $\lambda = V_X$ it follows that $(\lambda - V_X) = 0$, and neither V_A nor V_X should change, regardless of the values assigned to the αs and to β. Nor should further reinforcements of AX alter V_A or V_X, because $(\lambda - V_{AX})$ will remain zero. Thus in accord with Kamin's finding, the number of $AX+$ trials should not affect blocking. Kamin did find that the number of prior reinforcements of X alone affected blocking; blocking was only partial if prior conditioning to X was preasymptotic. The Rescorla-Wagner model accounts for that result as well, because in that case $V_X < \lambda$, thus $(\lambda - V_{AX})$ is positive, and both A and X will gain associative strength (according to Equations 4 and 5) in a negatively accelerated fashion over trials until $V_{AX} = \lambda$, and the process stops. Conditioning to the added element A will be partially blocked and will be less than in a group that received no prior $X+$ trials, because V_X and thus V_{AX} are smaller in the naive group on the first compound trial.

Turning to the experiment by Wagner and Saavedra, only Groups AX+/X+ and AX+ will be considered, since the model's predictions about non-reinforcement have not been described yet. To explain the effects of the added $X+$ trials upon conditioning to A, the model needs three equations. On $AX+$ trials,

$$\Delta V_A = \alpha_A \beta (\lambda - V_{AX}) \qquad (6)$$

$$\Delta V_X = \alpha_X \beta (\lambda - V_{AX}) \qquad (7)$$

and on X+ trials in Group AX+/X+

$$\Delta V_X = \alpha_X \beta (\lambda - V_X) \qquad (8)$$

In Group AX+, V_A will grow steadily, and to some moderate level so long as α_X is not much greater than α_A. Parenthetically, V_A will grow more slowly than it would if only A+ trials were presented, because when AX is reinforced, both V_A and V_X are increased, and their sum, V_{AX}, increases rapidly, thereby reaching λ and stopping the process sooner. This phenomenon is called overshadowing. Returning to the comparison of Groups AX+/X+ and AX+, the critical point is that in the former group, X gains associative strength on both AX+ and X+ trials. Since the two kinds of trials occur in a mixed order, the value of V_X, and hence of V_{AX} on compound trials, will be greater on the average in Group AX+/X+ than in Group AX+. Thus $(\lambda - V_{AX})$ will be smaller in Group AX+/X+, and ΔV_A will also be smaller. Consequently V_{AX} will reach λ, ending the growth of associative strength, with V_A at a smaller value in Group AX+/X+ than in Group AX+, implying that the rabbits in the former group should respond less frequently on test presentations of CS A. Wagner and Saavedra obtained this result.

Earlier it was argued that a satisfactory model of classical conditioning must be able to account for inhibition, and for variations in the effectiveness of non-reinforcement. How does the model treat non-reinforcement? First, extending the assumption that the value of λ is directly related to the intensity of the UCS, it assumes that the asymptotic associative strength supportable by non-reinforcement, λ_0, is zero. Given this assumption, the model employs the same equations to calculate changes in associative strengths resulting from reinforcement and non-reinforcement. Non-reinforcement will yield decrements in associative strength only when $(\lambda_0 - V_{AX})$ is negative, i.e., only when V_{AX} is positive. Intuitively, the model asserts that non-reinforcement results in decrements in associative strength only when it occurs against a background of reinforcement. That is, it stipulates that the necessary and sufficient condition for the conditioning of a decrease in the associative strengths of all cues present on a trial is a negative discrepancy between the asymptotic associative strength supportable by the reinforcer which occurs on that trial and the net associative strength of all cues present on that trial. Thus, a negative contingency between CS and UCS is neither necessary

nor sufficient for the conditioning of decreases in associative
strengths of CSs.

The model assumes no special inhibitory process, but identi-
fies inhibition with a net negative value of V. A CS is described
as a conditioned inhibitor only when it has negative associative
strength. A CS with a negative value of V has, according to the
model, the effects that one would expect a conditioned inhibitor
to have in summation and retardation-of-acquisition tests. For
example, if CS A evokes a conditioned eyeblink, and CS X is a conditioned inhibitor of the eyeblink, then $V_{AX} = V_A + V_X$, and since
$V_X < 0$, $V_{AX} < V_A$. Thus there should be fewer eyeblink responses
to the AX compound than to A alone. Further, in a retardation-of-
acquisition test, for any value of $\alpha_A < 1.0$, a larger number of
reinforced presentations of A will be required to bring an initially
negative V_A to some positive value than will be required to bring
an initially zero V_A to the same value. Thus, it is reasonable to
identify an inhibitory CS with one having negative V. The experimental conditions which the model deems necessary for the pro-
duction of a negative value of V will be considered later in this
chapter. First, the model will be applied to those experiments
described in the preceding section which included non-reinforced
trials.

Let's consider the other half of the study by Wagner and
Saavedra (in Wagner, 1969) which involved the relative amount of
conditioning to CS A in Groups AX+/X- and AX+. The model requires
three equations to explain the data; on $AX+$ trials:

$$\Delta V_A = \alpha_A \beta_R (\lambda_R - V_{AX}) \qquad (9)$$

$$\Delta V_X = \alpha_X \beta_R (\lambda_R - V_{AX}) \qquad (10)$$

And on X- trials in Group AX+/X-:

$$\Delta V_X = \alpha_X \beta_0 (\lambda_0 - V_X) = \alpha_X \beta_0 (-V_X) = -\alpha_X \beta_0 V_X \qquad (11)$$

The subscripts "R" and "0" in these equations represent reinforcement and non-reinforcement, respectively. In Group AX+/X-, CS X
loses associative strength on $X-$ trials (Equation 11), offsetting
its gain on $AX+$ trials (Equation 10). Thus on $AX+$ trials V_X, and
hence V_{AX}, will be smaller on the average in Group AX+/X- than in
Group AX+. Consequently $(\lambda - V_{AX})$ will be larger in Group AX+/X-,
and ΔV_A will be larger in this group. Thus the model predicts a
greater final V_A in Group AX+/X- than in Group AX+, and a greater
percent CRs to A in the former, the result obtained by Wagner and
Saavedra.

Next consider the study by Wagner, Saavedra, and Lehman (in Wagner, 1969), in which CS X lost more associative strength as a result of repeated non-reinforcements when it was compounded with a strong excitor A than when it was compounded with a weak excitor, B. The model assumes that $V_A > V_B$ at the end of the first phase of conditioning during which there had been 224 $A+$, 224 $X+$, and only 28 $B+$ trials, and the data were in accord with this prediction. Then in phase two, Group AX- received 32 $AX-$ trials, and Group BX-, 32 $BX-$ trials. The model calculates:

$$\Delta V_X = \alpha_X \beta_0 (0-V_{AX}) \text{ for Group AX-} \qquad (12)$$

$$\Delta V_X = \alpha_X \beta_0 (0-V_{BX}) \text{ for Group BX-} \qquad (13)$$

Since $V_{BX} < V_{AX}$ at the start of phase 2, then ΔV_X should be a larger negative number for Group AX-, which implies X should lose more associative strength as a result of non-reinforced presentations of AX than as a result of non-reinforced presentations of BX. This prediction was confirmed; Group AX- made fewer eyeblinks in the presence of CS X during reconditioning (see Figure 2).

Finally, let's apply the Rescorla-Wagner model to the study by Wagner and Saavedra (in Rescorla & Wagner, 1972) in which CS X became a stronger conditioned inhibitor of the eyeblink response when it was non-reinforced in the presence of a strong excitor A than when it was non-reinforced in the presence of a weak excitor B. As a result of 548 $C+$ trials, 240 $A+$ trials, and only 8 $B+$ trials in the first phase of the study, V_A was much greater than V_B. In the second phase, Group AX- received non-reinforced compounds of a novel CS X with CS A intermixed with reinforced presentations of A. Group BX- received $BX-$ and $B+$ trials. The effects of $AX-$ and $BX-$ trials upon V_X are calculated as follows:

$$\text{for Group AX-:} \quad V_X = \alpha_X \beta_0 (0-V_{AX}) = \alpha_X \beta_0 (-V_{AX}) \qquad (14)$$

$$\text{for Group BX-:} \quad V_X = \alpha_X \beta_0 (0-V_{BX}) = \alpha_X \beta_0 (-V_{BX}) \qquad (15)$$

V_A is much greater than V_B at the start of phase 2, and will continue to be so because $A+$ and $B+$ trials will keep V_A and V_B close to their initial levels. Consequently, CS X should lose more associative strength on $AX-$ trials than on $BX-$ trials. Since V_X began this phase at zero, the model predicts that it should reach a larger negative value in Group AX- than in Group BX-, before the negative values of V_X offset the positive values of V_A and V_B in the two groups, the quantities in parentheses become zero, and the associative strengths of all the cues become fixed. In any case,

a summation test should reveal the differential inhibitory strengths
of CS X in the two groups, and it did (see Figure 3). The Wagner-
Rescorla model thus correctly predicts the conditioning of inhibition to CS X in an AX-/A+ design, and also correctly predicts that
the strength of the inhibitory effect should be directly related
to the excitatory strength of CS A.

According to the Wagner-Rescorla model the standard extinction
procedure, i.e., simple non-reinforcement of an excitatory CS,
should cause the CS to lose its excitatory strength, but not to
become a net inhibitor. This can be seen if we assume that V_A is
positive, because of prior reinforcement of CS A. Then ΔV_A on extinction trials is given by equation (16):

$$\Delta V_A = \alpha_A \beta_0 (0-V_A) \qquad\qquad (16)$$

For the first extinction trial ΔV_A will be negative, reducing the
net V_A somewhat. Thus $(0-V_A)$ will be a smaller negative number on
the next trial, and V_A will again be reduced, but by somewhat less.
Thus, over extinction trials, V_A should decay to zero, at which
point $(0-V_A)$ will be zero, and the process should stop. Reberg's
(1972) finding that an extinguished CS fails to become a net inhibitor is thus consistent with the model.

When a novel CS B is introduced at the start of the extinction
treatment, and accompanies CS A on every extinction trial, then CS
B should become a net inhibitor, i.e., $(0-V_{AB})$ should be negative
on early extinction trials, making V_B negative. When $V_{AB} = 0$, the
process should stop, leaving V_B negative. According to the model,
the same treatment that extinguishes the excitatory strength of a
former CS+, but does not make it an inhibitor, should cause a novel
cue to become inhibitory. Rescorla (1978a) has obtained evidence
which supports this prediction.

Simple non-reinforcement of inhibitory CSs, which are those
with negative values of V, should cause them to gain associative
strength and approach zero because $\Delta V_A = \alpha_A \beta_0 \ 0-(-V_A) = \alpha_A \beta_0 (V_A)$.
Thus V_A should grow over extinction trials until it reaches zero,
and the process stops. That is, simple extinction of an inhibitory
CS should be the mirror image of simple extinction of an excitatory
CS. As we shall see, there are data which cause one to doubt this
symmetry.

A noteworthy feature of the model is that it assigns no unique
role to non-reinforcement or to a negative contingency between CS
and UCS in producing decrements in associative strength. A decrement in associative strength will occur whenever the quantity
$(\lambda - V_{AB...X})$ is negative, that is, whenever the algebraic sum of
the associative strengths of all the cues present on a trial is

greater than the associative strength that the UCS can support. This can happen when λ suddenly decreases, for example because of a decrease in UCS intensity. The model predicts that a novel cue added to a CS+ just before UCS intensity is decreased should become inhibitory. Decrements in the associative strengths of CS elements should also occur if the associative strength of a compound is made larger than λ, and reinforcement of the compound is continued. This effect could be produced by conditioning each element to asymptote before compounding them. Then continued reinforcement of the compound should cause each element to lose associative strength, with the magnitude of the decrement directly proportional to the salience (α) of a given element. Data supporting this prediction have been obtained in the CER procedure (e.g., Kamin & Gaioni, 1974; Rescorla, 1970; but see St. Claire-Smith & Mackintosh, 1974 for results contrary to the prediction). The model also predicts that when a novel cue is added to the supra-asymptotic compound, and reinforcement is continued, the novel cue should become inhibitory, even though it is always followed by the UCS.

Variations in the Contingency between CS and UCS

To account for the effects of variations in the contingency between CS and UCS, the Rescorla-Wagner model makes one additional assumption. It is that background or static cues can be treated like any other stimuli i.e. that they play the same role in compounds that explicit CSs do.

Given this assumption, suppose we want to predict the effects of the limiting positive contingency, i.e., the case in which p(UCS/CS) = 1.0 and p(UCS/$\overline{CS}$) = 0. The CS will be called A, and background X. The duration of the CS will provide the unit of time, and intertrial intervals will be divided into units that long. Three equations are required to describe changes in the associative strengths of A and X. First, on reinforced trials both the explicit CS A and background stimuli, X, are present, so:

$$\Delta V_A = \alpha_A \beta_R (\lambda_R - V_{AX}) \qquad (17)$$

$$\Delta V_X = \alpha_X \beta_R (\lambda_R - V_{AX}) \qquad (18)$$

No UCSs are presented between trials, so:

$$\Delta V_X = \alpha_X \beta_0 (0 - V_X) \qquad (19)$$

Intuitively, although V_A can only grow (Equation 17), V_X will increase on reinforced trials (Equation 18), but it will decrease on non-reinforced trials, i.e., during the intertrial period (Equation 19). Consequently V_X will remain near zero, while V_A grows steadily until it reaches λ_R. Asymptotically, then, $V_A = \lambda_R$, and $V_X = \lambda_0 = 0$. That is, X will be neutral.

If p(UCS/CS A) is held constant across groups but p(UCS/$\overline{\text{CS } A}$) is varied from zero to p(UCS/CS A), V_A will change over trials as shown in the upper half of Figure 4, which is derived from some simulations by Rescorla and Wagner (1972). In these simulations, A was presumed to be present one-fifth of the time in an irregular sequence, A was assumed to be five times as salient as X ($\alpha_A = 0.5$, $\alpha_X = 0.1$), and the learning rate parameter for reinforcement twice as large as the learning rate parameter for non-reinforcement ($\beta_R = 0.1$, $\beta_0 = 0.05$). The upper part of Figure 4 illustrates changes in V_A over trials for groups given the pairs of conditional probabilities p(UCS/CS A) and p(UCS/$\overline{\text{CS } A}$) equal to 0.8-0, 0.8-0.2, 0.8-0.4, and 0.8-0.8. Holding p(UCS/CS A) = 0.8, we see that the asymptotic V_A decreases as p(UCS/$\overline{\text{CS } A}$) increases. Intuitively, this is because X is now being reinforced, at least some of the time, alone, so that $V_X > 0$, and is proportional to p(UCS/$\overline{\text{CS } A}$). Consequently, ($\lambda - V_{AX}$) will be inversely proportional to p(UCS/$\overline{\text{CS } A}$), and so will ΔV_A. Asymptotic conditioning occurs when V_{AX} reaches λ, stopping the process. At asymptote, V_A can be represented as $1/VX$, showing that the asymptotic value of V_A decreases as p(UCS/$\overline{\text{CS }}A$) increases. The results of several experiments using the CER procedure (Keller, Ayres & Mahoney, 1977; Rescorla, 1968) which were presented in the preceding chapter, confirm this prediction.

The curve labeled 0.8-0.8 shows that a zero-contingency between CS A and a UCS should yield an asymptotic V_A of zero, according to the model. The model makes this prediction for all shock densities, given a zero-contingency. However, early in a zero-contingency procedure V_A may be positive, as shown in Figure 4. This should tend to happen when AX is reinforced early, before V_X is very large. Thus, the model predicts that early pairings of A and the UCS which precede presentations of the unsignaled UCS should be most likely to yield transient excitatory conditioning to A. Ayres, Benedict, and Witcher (1975) have obtained precisely this result in a CER experiment with rats (see Chapter 3). Rescorla and Wagner (1972) have noted that the strength of initial excitatory conditioning to the CS in a zero-contingency procedure should increase with the overall probability of the UCS, the proportion of session time during which the CS is present, the relative salience of the CS (α_A/α_X), and the relative effectiveness of reinforcement (β_R/β_0).

Asymptotically, the zero-contingency procedure should result in a positive value of V_X, thus, there should be excitatory condi-

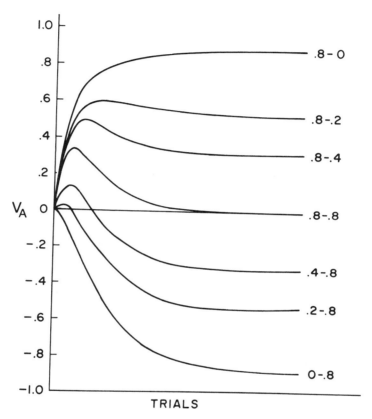

Figure 4.4 Predicted associative strength of CS *A* across trials as a function of p(UCS/CS A), represented by the first number in each pair, and p(UCS/$\overline{\text{CS A}}$), represented by the second number in each pair (after Rescorla & Wagner, 1972).

tioning to background cues. The strength of such excitatory conditioning should be directly related to the average frequency of
UCS presentations and to the relative effectiveness of reinforcement (Rescorla and Wagner, 1972; p. 89). According to the model,
excitatory conditioning to X is responsible for the ultimate loss
of excitatory strength by the explicit CS, A, under conditions in
which A had been excitatory initially. The argument goes something
like this. As X becomes strongly excitatory in a zero-contingency
procedure, then V_{AX} equals (or exceeds) the asymptote, λ_R. When
this happens, reinforcement of the AX compound produces no further
increment (or a decrement) in V_A. Further, AX is sometimes non-
reinforced, also producing decrements in V_A. Thus, CS A should
lose the excitatory strength it had acquired earlier, whereas X
should remain excitatory by virtue of reinforcements of X alone,
that is, the presentations of the UCS during the intertrial intervals. Recently, Keller, Ayres, and Mahoney (1977) found that the
amount of conditioned suppression evoked by a CS progressively diminished to zero when rats received additional hours of exposure
to a zero-contingency procedure that initially resulted in strong
excitatory conditioning to that CS, confirming the model's prediction.

The Rescorla-Wagner model's treatment of background cues differs from the treatment of such cues that was implicit in Rescorla's
earlier (1967) contingency model of conditioning, described in
the previous chapter. The contingency model treated the stimulus
conditions called "CS" and "$\overline{CS}$" symmetrically, thereby implying
that there was no conditioning to either CS or $\overline{CS}$. As described
earlier, the Rescorla-Wagner model redefines the two stimulus situations as "CS plus background" and "background alone." and predicts that excitatory conditioning does occur to background cues
in the zero-contingency procedure, implying that discrimination of
reinforced versus non-reinforced stimuli within an experimental
session is not required for excitatory conditioning.

The lower half of Figure 4 illustrates the predicted changes
in the associative strength of a CS A upon which the UCS is negatively contingent. Since UCSs occur during intertrial intervals,
the background cues, X, will become excitatory. In the simplest
case in which no UCSs occur on AX trials, then A should become inhibitory, that is, V_A should become negative. This is because V_X
and hence V_{AX} will be positive early in conditioning, so that
$\Delta V_A = \alpha_A \beta_0 (0 - V_{AX})$ will be negative. Asymptotically, since V_{AX}
should approximate zero, the asymptote for non-reinforcement, V_A,
should equal $-V_X$. If a novel CS B is then added to CS A, and the
negative contingency between the CSs and the UCS is maintained,
then $(\lambda - V_{ABX}) = 0$, and there should be no inhibitory conditioning

to B. As was noted earlier, Suiter and LoLordo (1971) obtained
this result, using rats and the CER procedure.

According to the model, the negative contingency case, which
can be represented $AX-/X+$, is formally equivalent to the Pavlovian
conditioned inhibition paradigm. Contiguity of background cues and
the UCS is the basis for the inhibition that is subsequently conditioned to a discrete CS which is less frequently reinforced than
background cues alone.

Figure 4 also illustrates that the magnitude of the inhibitory
CR to CS A should decrease as p(UCS/CS A) increases from zero,
thereby reducing the magnitude of the negative contingency between
CS and UCS. Intuitively, inhibition is conditioned to A insofar
as there is an "overexpectation" of the UCS on AX trials.

Thus far the discussion has been designed to impress the
reader with the scope of the Wagner-Rescorla model. It correctly
predicts the results of a variety of manipulations of the contingency between CS and UCS, including at least some of the arrangements of an overall zero-contingency which result in some excitatory conditioning. Moreover, as we have seen, it correctly predicts
several variations in the effectiveness of reinforcement or non-
reinforcement in conditioning a target element of a compound CS
when another element is reinforced or non-reinforced separately.

The next section of this chapter takes a closer look at two
provocative phenomena of compound conditioning, overshadowing and
blocking. The Rescorla-Wagner model's accounts of these phenomena
will be evaluated in light of some recent data.

Overshadowing and Blocking

In his 1927 book, Pavlov discussed several experiments in which
a compound CS was repeatedly paired with food. In several cases,
one element of the compound elicited no salivation when tested alone
following compound conditioning even though it would have elicited
considerable salivation had it been conditioned alone from the
start. Pavlov called this phenomenon, weaker conditioning to A in
an $AB+$ treatment than in an $A+$ treatment, overshadowing, and suggested that the "strong" (i.e., more conditionable) CS somehow prevented the "weak" CS from forming connections with the UCS.

Kamin (1969) systematically studied overshadowing in the CER
paradigm with rats. When the compound of light and a 50 dB noise
was followed by shock, there was virtually no conditioning to the

noise; the suppression ratio for noise was 0.42 on a test trial. However, when the light and a more intense, 80 dB, noise were followed by shock, the noise subsequently elicited moderate suppression of food-getting behavior. The suppression ratio for noise was 0.25 on a test trial. Next, Kamin determined the relative rates of acquisition of conditioned suppression to the light, the 50 dB noise, and the 80 dB noise, in three separate groups of rats, in order to relate the conditionability of the various elements treated separately to the outcomes of the compound conditioning procedures. The visual CS yielded much faster acquisition than the 50 dB noise, but only slightly faster acquisition than the 80 dB noise. This outcome prompted Kamin to suggest an account of overshadowing based on the phenomenon of blocking described earlier in this chapter. He maintained that when light was compounded with the 50 dB noise, the former conditioned very rapidly, and reached asymptote after a few trials, at which point there had been virtually no conditioning to the 50 dB noise. Once conditioning was asymptotic Kamin maintained that the perfectly predicted shock UCS was no longer surprising, and since surprise was necessary for conditioning, there was little conditioning to the noise. When the 80 dB noise was compounded with the light, there should have been a moderate amount of conditioning to noise by the trial of asymptotic conditioning to the light. Consequently the noise should have elicited moderate conditioned suppression when tested alone.

Kamin presented further evidence that the relative conditionability of the various elements critically determined the results of compound conditioning procedures. He increased the intensity of the shock UCS to 4 mA, a level which produces very rapid conditioning to each of the CSs taken separately. After the compound of light and 50 dB noise was repeatedly paired with the 4-mA shock, a test trial revealed considerable conditioning to noise. This outcome supported Kamin's contention that overshadowing could be predicted from the relative conditionability of the separate elements.

The Wagner-Rescorla model's explanation of overshadowing is similar to Kamin's, insofar as it is based upon blocking. When a compound of light and noise (L and N) is repeatedly paired with a shock UCS, then the changes in the associative strengths of the elements over reinforced trials are given by Equations 20 and 21:

$$\Delta V_N = \alpha_N \beta (\lambda - V_{LN}) \qquad (20)$$

$$\Delta V_L = \alpha_L \beta (\lambda - V_{LN}) \qquad (21)$$

On the other hand, when light and noise are individually paired with shock in two additional groups, then ΔV_N and ΔV_L are given by Equations 22 and 23:

$$\Delta V_N = \alpha_N \beta (\lambda - V_N) \qquad (22)$$

$$\Delta V_L = \alpha_L \beta (\lambda - V_L) \qquad (23)$$

All these equations ignore background cues, but this omission creates no problems in this case.

The model predicts no overshadowing as a result of the first reinforced trial, because V_N, V_L, and hence V_{LN} are all zero, so that the values of ΔV_L and ΔV_N are the same for single-cue and compound-cue groups. As a result of trial 1, both V_L and V_N will increase, in proportion to α_L and α_N. Then, on the next trial V_{LN} will be greater than either V_L or V_N, thus $(\lambda - V_{LN})$ will be smaller than either $(\lambda - V_L)$ or $(\lambda - V_N)$. Consequently both elements should gain associative strength more slowly in the compound cue group (Equations 20 and 21) than in the single cue groups (Equations 22 and 23). In fact, at asymptote, when $V_{LN} = \lambda$ for the compound group, and conditioning stops, V_L and V_N should equal λ for the single cue groups, but $V_N = \lambda(\alpha_N/\alpha_N + \alpha_L)$ and $V_L = \lambda(\alpha_L/\alpha_N + \alpha_L)$ for the compound cue group. Thus if one element is much more salient than the other, i.e., has a much larger α, the stimulus with the smaller α will be strongly overshadowed. Furthermore, if the two αs are roughly equal, there should be reciprocal overshadowing, with both asymptotic associative strengths equal to $\lambda/2$.

To account for Kamin's data, the model would have to assume that $\alpha_L > \alpha_{N80} > \alpha_{N50}$, and further, that the last was very small. On this assumption, there should have been detectable overshadowing of light by the 80 dB noise, i.e., reciprocal overshadowing should have occurred when the compound of light and 80 dB noise was paired with a 1-mA shock. Kamin (1969) did not present the data required for assessment of reciprocal overshadowing, but Mackintosh (1976) has reported an experiment on just that point, using conditioned suppression of licking by rats. Preliminary research resulted in the identification of equally salient noise and light CSs at three levels of salience, low, moderate, and high. At each level of salience, four groups were conditioned. Groups conditioned to noise alone or to the compound were later tested on noise alone, in order to reveal any overshadowing of noise by light. Conversely, groups conditioned to light alone or to the compound were tested on light alone, to reveal any overshadowing of light by noise.

Substantial overshadowing occurred only with the least salient CSs, and in that case overshadowing was reciprocal. Mackintosh noted that the failure of reciprocal overshadowing with equally and highly salient CSs contradicts the prediction of the Wagner-Rescorla model. He offered an alternative account (Mackintosh, 1975b, 1976), in which the learning rate parameter for a CS, α, may change with experience. Specifically, he assumed that α_A will increase over trials if A is the best predictor of the outcome of the trial. On the other hand, if A is not the best predictor of the trial outcome, α_A will decrease over trials. This formulation predicts reciprocal overshadowing with the least salient CSs, because α_N and α_L will increase over trials in the single-CS groups, but will remain constant and small in the compound-CS groups, because N and L are equally good predictors of reinforcement. On the other hand, when the CSs are initially quite salient, i.e., α_L and α_N are initially large, there will be little room for increases in the αs in the single-CS groups, and thus little or no overshadowing will be observed.

Allowing α to change with experience is more than a minor modification of the Wagner-Rescorla model, which attempts to explain the phenomena of selective stimulus control without recourse to the concept of attention. Although many of the implications of allowing α to change with experience remain to be worked out, this modification permits explanation of the phenomena of latent inhibition and learned irrelevance, two phenomena which were not explained by the Wagner-Rescorla model.

Kamin's blocking procedure, which was described earlier, has recently become a popular vehicle for theory testing. It has already been shown that the Wagner-Rescorla model predicts the complete blocking observed in Kamin's experiment. The outcome of another study by Kamin (1969) is also in accord with the model. Kamin showed that it was a necessary condition for blocking that the UCS be unchanged on the transition trial, which is the first $AB+$ trial. When the shock intensity was increased from 1 mA to 4 mA on the transition trial there was considerable suppression conditioned to the added B element. Kamin argued that conditioning occurred to the added CS because it was followed by a surprising UCS. The Rescorla-Wagner model predicts conditioning to the added element by making the reasonable assumption that the asymptotic associative strength supportable by a 4-mA shock is greater than that supportable by a 1-mA shock. An increase in λ on the transition trial would permit increases in associative strength to be conditioned to both A and B. If shock were omitted on the transition trial, the model predicts that the added element should become inhibitory, and this, too, seems to happen. Thus far, the model successfully predicts the outcomes of several blocking procedures. However, it has difficulty accounting for the results of other experiments on blocking.

First, Kamin's basic result was oversimplified in the earlier discussion of blocking. Although he did observe complete blocking on a single test presentation of B following $A+$, then $AB+$ trials, in a more sensitive reconditioning test he was able to detect that there had been some conditioning to the added element. Further, all the conditioning to the added element occurred on the transition trial. This sort of discontinuity is hard for the Wagner-Rescorla model to explain, but one could suppose that conditioning to A was slightly less than asymptotic at the end of the first phase, so that there were small increments in both V_A and V_B on the transition trial, after which $V_{AB} = \lambda$ and the process stopped. Perhaps because of this counter-argument, the problem posed by Kamin's result has received relatively little attention. Recently, however, Mackintosh (1975a) has asked whether any blocking occurs on the first compound trial. The Rescorla-Wagner model demands at least some. In a series of studies of conditioned suppression of licking by rats, Mackintosh observed evidence of blocking in a group that received eight $AB+$ trials following $A+$ trials, but no blocking in a group that received a single $AB+$ trial following $A+$ trials. Mackintosh's own account of his finding was based on the assumption, described earlier, that changes in the salience of a CS depend upon its relative validity as a predictor of the trial outcome. CS B is not redundant until after the first $AB+$ trial, thus α_B will not differ from the control value on that trial, and no blocking will occur. Once CS B is shown to be redundant, α_B will decline precipitously, and blocking will occur.

Dickinson, Hall, and Mackintosh (1976) have recently conducted an experiment on blocking which poses great problems for the Rescorla-Wagner analysis. In a CER experiment, four groups of rats received a series of A trials followed by a series of AB trials. The four groups differed in the number of shocks presented in each phase of the experiment. In a 2 X 2 design, groups received either a single shock or two shocks, 8-sec apart, at the end of each A trial in phase one, and the same alternatives at the end of each AB trial in phase two. Figure 5 reveals that either adding an unexpected shock or deleting an expected (second) shock reduced blocking, relative to the performance of a control group which received the same number of shocks on all trials.

Dickinson, et al., asserted that the reduction in blocking which results from deletion of the second shock is difficult to attribute to changes in λ on the transition trial, for if λ changed at all,

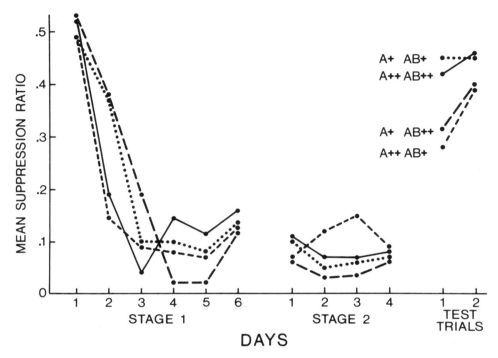

Figure 4.5   Mean suppression ratios to CS *A* in Stage 1, to the AB
compound in Stage 2, and to CS *B* on test trials.
Group A+, AB+ received a single shock on all conditioning trials, Group A++, AB++, a double shock on
all trials, Group A+, AB++, a single shock during
Stage 1 and a double shock during Stage 2, and Group
A++, AB+, a double shock during Stage 1 and a single
shock during Stage 2.   Data are from Dickinson, Hall,
and Mackintosh, 1976.

it should have decreased, making $(\lambda-V_{AB})$ negative, and thus conditioning inhibition, rather than the observed excitation, to B. They argued instead that the occurrence of a surprising event, omission of an expected shock, on the transition trial effectively made the added element B informative, thereby preventing a decline in α_B and producing more conditioning to B than in the control group which received two shocks on every trial. On this account, a surprising event need not be one which is itself a reinforcer in a given situation (e.g., Gray & Appignanesi, 1973). Wagner (1977) has recently proposed an information-processing model of reinforcement in classical conditioning that attempts to describe the conditions under which surprising posttrial events will facilitate or retard conditioning. This model appears to predict the outcomes of the aforementioned experiments on blocking, however, it is beyond the scope of this chapter.

Conclusions

The Rescorla-Wagner model was a significant advance in theory-building in classical conditioning primarily because of its broad scope and testability. It predicted the outcomes of the sorts of experiments which arose from earlier contiguity- and contingency-based positions on the basis of the discrepancy between the maximum associative strength which a given UCS will support and the combined associative strengths of all cues present on a trial. Moreover, its central concept of discrepancy has generated a host of new experiments.

The model has had progressively increasing influence on research in classical conditioning since its publication in 1972. Now, in 1978, its influence may be beginning to wane, in part because it does not very well predict the outcomes of several classes of experiments. First, as we have seen, it seems to predict the results of several experiments on blocking and overshadowing less well than does a competing account (Mackintosh, 1975b) which is based on the notion of surprise and allows the saliences of cues to change with experience.

Second, the model fails to predict the phenomena of latent inhibition and learned irrelevance, since it asserts that the associative strength of a novel CS should be unchanged by either CS-alone or zero-contingency procedures. Neither latent inhibition nor learned irrelevance is an associative effect in the sense used throughout these chapters; thus it can be argued that their occurrence does not disconfirm the model. Nonetheless, they are provocative phenomena, and point to the need for a further increase in the scope of our models.

The Rescorla-Wagner model predicts that conditioned inhibition will be extinguished in a CS-alone procedure. In several experiments with the CER procedure, Zimmer-Hart and Rescorla (1974) disconfirmed this prediction; non-reinforced presentations of a conditioned inhibitor of fear failed to make it less inhibitory. This outcome suggests that the model's symmetrical treatment of excitation and inhibition may be inappropriate. Alternatively, Rescorla (Rescorla, 1978b; Zimmer-Hart & Rescorla, 1974) has suggested that it may be useful to view an inhibitor as a stimulus which only has an effect when excitatory cues are present, in which case it raises the threshold for action of those cues.

These and other problems for the Rescorla-Wagner model will have to be resolved by any new, general theory of classical conditioning, whether it be a variant of the Rescorla-Wagner model, a more process-oriented model such as Wagner's (e.g., Wagner, 1977), or a theory like Mackintosh's (Dickinson, Hall, & Mackintosh, 1976; Mackintosh, 1975b), which incorporates both sorts of approach.

REFERENCES

Ayres, J. J. B., Benedict, J. O., & Witcher, E. S. Systematic manipulation of individual events in a truly random control in rats. JOURNAL OF COMPARATIVE AND PHYSIOLGOICAL PSYCHOLOGY, 1975, 88, 97-103.

Dickinson, A., Hall, G., & Mackintosh, N. J. Surprise and the attenuation of blocking. JOURNAL OF EXPERIMENTAL PSYCHOLOGY: ANIMAL BEHAVIOR PROCESSES, 1976, 2, 313-322.

Gray, T., & Appignanesi, A. A. Compound conditioning: Elimination of the blocking effect. LEARNING AND MOTIVATION, 1973, 4, 374-380.

Hull, C. L. PRINCIPLES OF BEHAVIOR. New York: Appleton-Century-Crofts, 1943.

Kamin, L. J. Predictability, surprise, attention, and conditioning. In B. A. Campbell & R. M. Church (Eds.), PUNISHMENT AND AVERSIVE BEHAVIOR. New York: Appleton-Century-Crofts, 1969.

Kamin, L. J., & Gaioni, S. J. Compound conditioned emotional response conditioning with differentially salient elements in rats. JOURNAL OF COMPARATIVE AND PHYSIOLOGICAL PSYCHOLOGY, 1974, 87, 591-597.

Keller, R. J., Ayres, J. J. B., & Mahoney, W. J. Brief versus extended exposure to truly random control procedures. JOURNAL OF EXPERIMENTAL PSYCHOLOGY: ANIMAL BEHAVIOR PROCESSES, 1977, 3, 53-65.

Mackintosh, N. J. Blocking of conditioned suppression: Role of the first compound trial. JOURNAL OF EXPERIMENTAL PSYCHOLOGY: ANIMAL BEHAVIOR PROCESSES, 1975, 1, 335-345. (a)

Mackintosh, N. J. A theory of attention: Variations in the associability of stimuli with reinforcement. PSYCHOLOGICAL REVIEW, 1975, 82, 276-298. (b)

Mackintosh, N. J. Overshadowing and stimulus intensity. ANIMAL LEARNING & BEHAVIOR, 1976, 4, 186-192.

Pavlov, I. P. CONDITIONED REFLEXES. Translated by G. V. Anrep. Oxford: Oxford University Press, 1927.

Reberg, D. Compound tests for excitation in early acquisition and after prolonged extinction of conditioned suppression. LEARNING AND MOTIVATION, 1972, 3, 246-258.

Rescorla, R. A. Pavlovian conditioning and its proper control procedures. PSYCHOLOGICAL REVIEW, 1967, 74, 71-80.

Rescorla, R. A. Probability of shock in the presence and absence of shock in fear conditioning. JOURNAL OF COMPARATIVE AND PHYSIOLOGICAL PSYCHOLOGY, 1968, 66, 1-5.

Rescorla, R. A. Reduction in the effectiveness of reinforcement after prior excitatory conditioning. LEARNING AND MOTIVATION, 1970, 1, 372-381.

Rescorla, R. A. Conditioned inhibition and excitation. In A. Dickinson & R. A. Boakes (Eds.), MECHANISMS OF LEARNING AND MOTIVATION: A MEMORIAL VOLUME FOR JERZY KONORSKI. Hillsdale, N. J.: Lawrence Erlbaum Associates, 1978. (a)

Rescorla, R. A. Some comments on a model of conditioning. In J. Baerwaldt & G. McCain (Eds.), RECENT DEVELOPMENTS IN LEARNING THEORY: THE ARLINGTON SYMPOSIUM. Stamford, Conn.: The Greylock Press, 1978. (b)

Rescorla, R. A., & Wagner, A. R. A theory of Pavlovian conditioning: Variations in the effectiveness of reinforcement and non-reinforcement. In A. H. Black & W. F. Prokasy (Eds.), CLASSICAL CONDITIONING II: THEORY AND RESEARCH. New York: Appleton-Century-Crofts, 1972.

St. Claire-Smith, R., & Mackintosh, N. J. Complete suppression to a compound does not block further conditioning to each element. CANADIAN JOURNAL OF PSYCHOLOGY, 1974, 28, 92-101.

Suiter, R. D., & LoLordo, V. M. Blocking of inhibitory Pavlovian conditioning in the conditioned emotional response procedure. JOURNAL OF COMPARATIVE AND PHYSIOLOGICAL PSYCHOLOGY, 1971, 76, 137-144.

Wagner, A. R.   Stimulus-selection and a "modified continuity theory."
    In G. H. Bower & J. T. Spence (Eds.), THE PSYCHOLOGY OF
    LEARNING AND MOTIVATION, Vol. 3.   New York: Academic Press,
    1969.

Wagner, A. R.   Priming in STM: An information processing mechanism
    for self-generated or retrieval-generated depression in per-
    formance.   In T. J. Tighe & R. N. Leaton (Eds.), HABITUATION:
    PERSPECTIVES FROM CHILD DEVELOPMENT, ANIMAL BEHAVIOR AND NEURO-
    PHYSIOLOGY.   Hillsdale, N. J.: Lawrence Erlbaum Associates,
    1977.

Wagner, A. R., & Rescorla, R. A.   Inhibition in Pavlovian condi-
    tioning: Application of a theory.   In R. A. Boakes & M. S.
    Halliday (Eds.), INHIBITION AND LEARNING.   London: Academic
    Press, 1972.

Zimmer-Hart, C. L., & Rescorla, R. A.   Extinction of Pavlovian
    conditioned inhibition.   JOURNAL OF COMPARATIVE AND PHYSIOLOGI-
    CAL PSYCHOLOGY, 1974, 88, 837-845.

5. REWARD TRAINING: METHODS AND DATA

M. E. Rashotte

Florida State University
Tallahassee, Florida U.S.A.

Reward training may be summarized by the equation

$$S:R \rightarrow S^V$$

which designates that a specified response (R) performed in a given stimulus situation (S) results in presentation of a stimulus that is currently valued by the animal (S^V). Thorndike's (1898) experiment in which hungry cats pulled a loop in a box to gain access to food illustrates one variety of reward training in which S^V is an appetitive stimulus. Non-appetitive forms of S^V include periods of freedom from an aversive stimulus (i.e., "escape training"), certain classes of exteroceptive stimuli (e.g., a change in ambient illumination level for rats), stimulus conditions which allow certain responses to be performed (e.g., presentation of a manipulable puzzle to monkeys), direct stimulation of certain brain areas (e.g., electrical stimulation of the hypothalamus), and classically conditioned stimuli previously paired with S^V (i.e., "conditioned reinforcement" training). When the parameters of training are properly arranged, an occurrence of $S:R \rightarrow S^V$ results in increased likelihood that R will occur when the animal subsequently encounters S. It is this result which is summarized in the Empirical Law of Effect (Chapter 1).

Stimulus conditions which are effective as S^V in a given training situation must be determined empirically on a species-by-species and even individual-by-individual basis. Thorndike (1911, p. 245) set the precedent: "... for any animal in any given condition [S^V] cannot be determined with precision and surety save by observation," and he proposed that a stimulus "which the animal does nothing to avoid, often doing such things as attain and pre-serve it" would be classified as an instance of S^V. A modern dis-cussion of this approach can be found in Premack (1965). It was once a central problem to specify the properties common to all instances of S^V (e.g., Hull, 1943). In recent years, however, a relatively limited subset of S^Vs have been employed in a few train-ing situations to gain understanding of the (assumed) general pro-cess by which an occurrence of S:R → S^V changes the likelihood of R. The most studied S^Vs are food or water, offset of painful elec-tric shock, and conditioned stimuli associated with these events.

The Origin of the First Response. Reward training requires that a designated response be performed before S^V is presented. How, then, does the first response come about which allows reward training to begin? The answer lies in interactions of the animal's biology, its past experiences, and the nature of the experimental situation in which the animal is trained. When these variables are "set" appropriately the experimenter can insure that the instrumental response occurs "spontaneously," and reward training can begin with-out further manipulations. Two examples of this principle are shown for rats in Figure 1.

The left panel shows averaged running speed for a group of hungry rats run down a runway once a day for 57 days (Logan, 1960). No food reward was given in the runway; the rats were fed later each day in their home cages. Locomotion occurred from the outset and running speed measured at the beginning of the runway ("Start" measure) even showed a modest increase across trials. The right panel shows that bar pressing also occurs "spontaneously" when a bar is inserted into the living cages of hungry or thirsty rats for 1 hr a day (Schoenfeld, Antonitis, & Bersh, 1950). In this case, the response declined in strength with repeated testing and, so, seemed determined by a predisposition to explore novel stimuli. In the initial session the food-deprived rats pressed more often than rats deprived of water, perhaps because reaching and grasping move-ments made with their paws while eating food once a day transferred to the bar.

Experimenters sometimes achieve the first response by inten-tionally manipulating an animal's experience with S^V. For example, in "response shaping by the method of successive approximation" the experimenter first uses the reward training procedure to strengthen some spontaneously occurring fraction of the desired response (e.g., orienting towards a stimulus). Then the criterion for S^V is freq-uently changed, each time demanding a response which approximates the desired response more closely. A second method, "autoshaping," involves the classical conditioning of motor responses (Hearst & Jenkins, 1974). In a third method, "putting through," the experi-

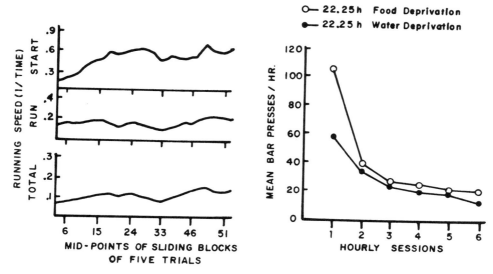

Figure 5.1   Examples of the "spontaneous" occurrence of two responses studied in reward training. The left panel
shows running speed of a group of rats in a runway on
57 successive days when S^v was not presented. The
"Start" measure shows speed as the rats left the startbox; the "Run" measure shows speed in the rest of the
runway; the "Total" measure is the sum of the two
(after Logan, 1960, p. 41). The right panel shows
rate of bar pressing by groups of hungry or thirsty
rats when a bar was presented in their home cages.
Each data point is the average of responding in a 1-hr
session immediately before the daily ration was given
(after Schoenfeld, Antonitis, & Bersh, 1950).

menter forces the response to occur (e.g., by moving the animal's
limb), and then presents S^v. A few repetitions of the passive-
$R \rightarrow S^v$ sequence results in the spontaneous occurrence of R in many
cases.

Finally, the first instrumental response might arise from imitation. Some evidence indicates that simply watching an animal
perform the designated instrumental response increases the likelihood that the observer will perform that response (e.g., Hayes &
Hayes, 1952; Kohn & Dennis, 1972; Zentall & Levine, 1972).

Parameters of Reward Training

   Trials and free-responding procedures. Reward training is
carried out with a trials or free-responding procedure. In the
trials procedure successive occurrences of $S:R \rightarrow S^v$ are separated
by time and often by a radical change in stimulus conditions, as
when a rat is kept in its home cage for 24 hours between successive
trials on which it runs to food in a runway. In the free-responding
procedure the concept of a trial is meaningless because the animal
is allowed to obtain S^v repeatedly by responding at its own pace,
without interruption by the experimenter. For example, a rat may
be placed in a box with a lever and allowed to press as it will to
obtain food.

   The trials procedure originated with Thorndike, free-responding
(sometimes "free operant") with Skinner, and each procedure has been
associated with a distinctive approach in studying reward training.
The trials procedure has been favored for analyzing the associative
determinants of performance. This is so because the associative
possibilities inherent in reward training (S-S and S-R connections)
are clearly operationalized and readily manipulated experimentally
when successive occurrences of $S:R \rightarrow S^v$ are controlled by the experimenter. The free-responding procedure does not encourage associative analysis, however, because the stimulus conditions preceding each response are difficult to specify and because the experimenter does not control the pacing of successive responses. Many
advocates of this procedure have favored the amassing of functional
relations between experimental manipulations and performance rather
than the elaboration of an associative theory of reward training
(Schoenfeld & Cole, 1972; Skinner, 1938, 1950). There is a growing
interest in examining the relation between the determinants of performance in the trials and free-responding procedures (e.g., Jenkins, 1970; Platt, 1971; Shimp, 1975).

   Schedule of reward. The simplest reward schedule is summarized
by the statement "Only if R occurs will S^v be presented." In this

case every time the instrumental response occurs in the training situation it will be followed by reward, a schedule known as "continuous reinforcement" or CRF. Other schedules can be arranged by substituting for X in the statement "only if R occurs <u>and condition X is met</u> will S^v be presented." Many of these schedules have powerful effects on performance. Collectively, they reveal much about the processes which determine performance in reward training.

On a <u>probability of reward schedule</u> each response theoretically has the same probability of being followed by S^v. These schedules influence response strength in different ways early and late in training. Goodrich's (1959) experiment in which hungry rats ran in a runway to obtain food provides a good example. The top panel of Figure 2 shows that speed of leaving the startbox was greater in the early trials when the probability of reward (p) was 1.0 than when p = 0.5. However, by the end of training, the group with the lower probability was running faster. This finding has been obtained in similar experiments where the values of the lower probability studied fell between 0.17 and 0.83 (e.g., Weinstock, 1958). Free-responding experiments also show that asymptotic performance is stronger when p < 1.0 than when p = 1.0 (e.g., Sidley & Schoenfeld, 1964).

These results pose the interesting theoretical question of why asymptotic response strength is facilitated when p < 1.0 relative to the case in which p = 1.0. The possibility that the group with p = 1.0 becomes satiated after a few trials and therefore is less motivated to respond can be rejected because similar results are obtained when trials are separated by 24-hour intervals, long enough for any satiation effects from reward to dissipate (e.g., Wagner, 1961; Weinstock, 1958). A more likely possibility is that the function of non-reward trials in probability schedules changes as training proceeds. Initially, non-reward trials seem to be neutral, or possibly to detract from performance; later, non-reward trials potentiate responding. A theoretical account of this apparent change in the function of non-reward has been developed by Amsel (1958). He proposed that non-reward is a relatively neutral event until a number of rewards have been experienced whereupon it becomes frustrating (i.e., aversive) and provides an additional source of motivation for groups trained with p < 1.0. When trials are highly massed or when a free-responding procedure is used, a part of this motivation can come from lingering aftereffects of frustrative non-rewards. Non-reward must provide motivation in another way as well, however, because the data show the same results when trials are widely spaced. Amsel has proposed that stimuli in the early parts of the runway which precede frustrative non-reward

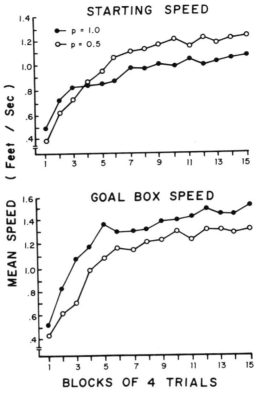

Figure 5.2   Effects of probability of reward on rats' running
speed in two sections of the runway. The top panel
shows the speed of two groups as they left the startbox, the bottom panel shows speed as the groups entered the goalbox. The probability of being fed on
each trial was 1.0 for one group and 0.5 for the other.
Each data point is the average of a group's speeds on
the four trials of each training session (after Goodrich, 1959, Exp. 1).

in the goalbox come to evoke a classically conditioned form of the frustration reaction (and thereby generate a mild aversive state) which motivates stronger asymptotic performance when p < 1.0. Of course, such a classically conditioned source of motivation would survive long intertrial intervals.

It may be noted that higher asymptotic performance when p < 1.0 is not always obtained (e.g., Bacon, 1962; Rashotte, Adelman, & Dove, 1972). There are several reasons why this might happen. First, Goodrich's data indicate that speed is greater when p < 1.0 provided that training is continued for a sufficient length of time. Initial experiments with probability schedules were not run long enough and it was mistakenly concluded that response strength is lower when p < 1.0 (Jenkins & Stanley, 1950; Spence, 1960, p. 103). Second, even after extensive training response strength might not be greater when p < 1.0 if the experimental conditions impose a "ceiling" on responding so that no group can respond more strongly than the group for which p = 1.0 (e.g., Rashotte, et al., 1972). Third, there is evidence that the effect occurs more strongly when the magnitude of reward is large (Wagner, 1961); failure to produce the effect may be related to this factor. Finally, there is some evidence that probability schedules influence running speed differentially in different sections of the runway so that the effect may depend on where the response measure is taken. For example, the data in the top panel of Figure 2 which showed higher asymptotic speed when p = 0.5 were obtained when speed was measured as the animal emerged from the startbox. The data in the bottom panel of Figure 2 were obtained from the same animals and on the same trials but when speed was measured in the last few centimeters of the runway as the animal neared the food hopper. In this measure response speed of the group receiving the lower probability of reward never exceeded the speed of the group for which p = 1.0. Obviously, different conclusions would be drawn about the effect of reward probability on response strength depending on the portion of the locomotor response measured.

The particularly challenging problem of explaining how reward probability exerts one effect on asymptotic responding near the site of reward but a different effect on responding earlier in the runway seems handled by Amsel's theory. It proposes that conditioned frustration evoked by stimuli near the goalbox should be strong and therefore the animal should be reluctant to enter that part of the runway. In the early parts of the runway conditioned frustration should be only mildly aversive and should motivate the running response (e.g., Amsel, 1967; Wagner, 1961).

A patterned schedule of reward arranges that reward and nonreward occur in a predictable sequence across successive responses.

The most studied pattern is "single alternation" in which only
every other response is rewarded. Tyler, Wortz and Bitterman
(1953) showed that with sufficient training hungry rats run quickly on reward trials and slowly on the alternate non-reward trials
in a runway. Their finding has been widely replicated and comparable results are obtained when other responses are studied (e.g.,
lever pressing: Gonzalez, Bainbridge, & Bitterman, 1966). Patterns
more complex than single alternation have important effects on performance in extinction (Chapter 8) but only subtle effects during
training.

A single alternation schedule is useful for analyzing the
stimulus conditions which become associated with responding during
reward training. The fact that rats come to respond differentially
on alternate trials is widely regarded as evidence that the rat has
learned a discrimination between different stimuli present on the
two types of trials (i.e., S:R → S^V vs. S^1:R → S^O). There are
several potential sources of different stimuli. One is the experimenter who might inadvertently signal the two types of trials. For
example, in most runway experiments the experimenter manually baits
the food cup only before reward trials and then manually places the
rat in the apparatus, making it possible for differential auditory
and/or olfactory stimuli to be correlated with the two types of
trials. Another possibility is that rats leave distinctive chemical
stimuli in the runway on reward and non-reward trials (e.g., different odors) and these stimuli differentially signal the availability of reward to rats run later in the session (Morrison &
Ludvigson, 1970). Experiments designed to eliminate these two
sources of stimuli show that rats can still learn to respond differentially on a single alternation schedule (e.g., Flaherty &
Davenport, 1972; Gonzalez, et al., 1966; Ludvigson & Sytsma, 1967).
Consequently, an appeal to other sources of stimuli is required in
these cases.

One important possibility is that the two types of trials
have distinctive aftereffects which carry over to the next trial
as stimuli. Because an animal's experience in the goalbox is different on reward and non-reward trials the animal might carry sensory traces of these different experiences to its next trial and
learn the discrimination on the basis of those stimuli. This proposal implies that if all other sources of differential stimuli
were eliminated the animals could learn to respond differentially
only when the intertrial interval is sufficiently short that traces
of one trial persist until the next. A considerable body of evidence indicates that this is the case. Differential responding is
found readily when alternating trials are separated by only a few
seconds (e.g., Capaldi, 1958; Tyler, et al., 1953), less readily

when minutes intervene between trials (e.g., Flaherty & Davenport, 1972; Katz, Woods, & Carrithers, 1966), and almost never when the intertrial interval is 24 hours (Surridge & Amsel, 1966).

Another potential source of differential stimuli has been emphasized in the theorizing of E. J. Capaldi (1966, 1967) He has proposed that the events of one trial can be carried forward to the next in the form of "memories." The idea is that the animal "stores" the events of a trial in memory and that these events are "retrieved" when it next encounters the trial stimulus. "Memories" are regarded as more permanent than sensory traces of a previous trial and therefore, could provide differential stimuli when trials are widely spaced in time. In fact, a recent experiment claims to show that previous failures to find differential responding at long intertrial intervals with the single alternation schedule resulted from weak "retrieveal cues" being present at the start of each trial (Jobe, Mellgren, Feinberg, Littlejohn, & Rigby, 1977, Exp. 4). That experiment showed that when the same stimulus was present in the startbox and goalbox of the runway the rats readily learned to run quickly on reward trials and slowly on non-reward trials even though the intertrial interval was 24 hours. The stimulus condition in the startbox was conceptualized as a cue which allowed the previous day's experience in the goalbox to be "retrieved" (i.e., a "memory" of reward or non-reward). Rats with different stimuli in the start- and goalboxes did not learn to run differentially. The effect was also shown to depend upon large reward magnitude and long confinement time in the empty goalbox on non-reward trials, the latter manipulations presumably making the reward and non-reward events more salient in memory. Emphasis on "memories" as a part of the stimulus condition which becomes associated with performance is in tune with growing interest in animal memory processes (e.g., Honig & James, 1971; Medin, Roberts, & Davis, 1976). A satisfactory account of how a "memory" of a previous trial's events is formed and later reinstated in single alternation schedules remains to be provided, however. It is possible, for example, that in these cases "memories" refer to a kind of associative linkage rather than to some non-associative memorial process.

The single alternation schedule has implicated a variety of stimuli as potential elements in the associations formed during reward training. It seems most likely that external stimuli (such as experimenter-originated cues or non-specific odors) and internal cues (such as sensory traces and, possibly, "memories" of previous events) combine in complex stimulus compounds which became associated with responding and/or reward (Capaldi & Haggbloom, 1975; Capaldi & Morris, 1976).

Schedules of reward devised for the free-responding procedure are known collectively as operant schedules of reinforcement. Figure 3 illustrates how four basic operant schedules are constructed by arranging fixed or variable applications of temporal or behavioral criteria which must be met before a response can produce S^V. The figure also illustrates in idealized curves the performances established by these schedules.

The ratio schedules require that a designated number of non-rewarded responses be performed before a response can produce reward. In the fixed ratio (FR) schedule that number remains fixed throughout training and this results in a pattern of responding between successive rewards known as "break-run." This pattern is illustrated in the FR cell of Figure 3 and consists in a period of no responding immediately after reward and then a high rate of responding until the next reward. The average length of the pause after reward is directly related to the number of responses required by the schedule (Felton & Lyon, 1966). In the variable ratio (VR) schedule the criterion number of responses fluctuates unpredictably around some average from one occurrence of reward to the next. The resulting performance is characterized by a high steady rate of responding between rewards, as illustrated in the figure. In the fixed-interval (FI) schedule a response is followed by reward only if a specified amount of time has elapsed since some event (usually the occurrence of the immediately preceding reward). After an intermediate amount of training responding between successive rewards takes on the distinctive "scallop" pattern shown in the figure: there is little or no responding immediately after a reward, but as the time interval draws to a close responding becomes increasingly strong. The number of responses in successive intervals is extremely variable, even after extended training (Dews, 1970), but on the average the absolute duration of the pause after reward is directly related to the length of the interval and occupies approximately 66% of it (Schneider, 1969). After lengthy training on FI schedules the "scallop" pattern gives way to a "break-run" pattern (Cumming & Schoenfeld, 1958; Schneider, 1969). In the variable interval (VI) schedule, the temporal criterion fluctuates unpredictably around some average value from one occurrence of reward to the next. When the intervals are properly distributed around the mean, this schedule yields a steady and moderate rate of responding between successive rewards. Responding at any point in time is related to the momentary probability of reward (Catania & Reynolds, 1968; Flesher & Hoffman, 1962). The higher rate of responding on VR than on VI schedules is thought to originate in the greater likelihood that a series of responses will be followed by reward on VR schedules (Skinner, 1938).

Figure 4 shows data from individual pigeons trained to peck a lighted key to obtain reward on two of the schedules discussed above.

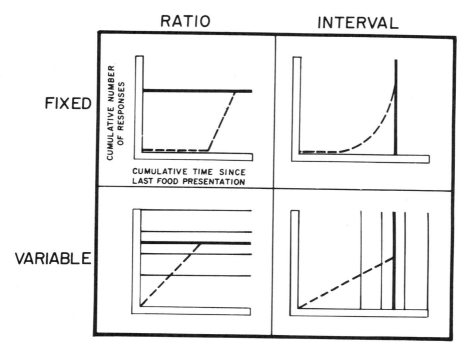

Figure 5.3 Schematic diagram of procedures for arranging four
simple schedules of reinforcement in the free-responding
procedure. The thick solid line in each cell represents
the behavioral or temporal criterion which must be met
before R can produce S^v. In the bottom two cells that
line is the average of variable criteria which are represented by the thin solid lines. The broken line in
each cell represents an idealized distribution of responding established by each schedule in the time interval between successive presentations of S^v.

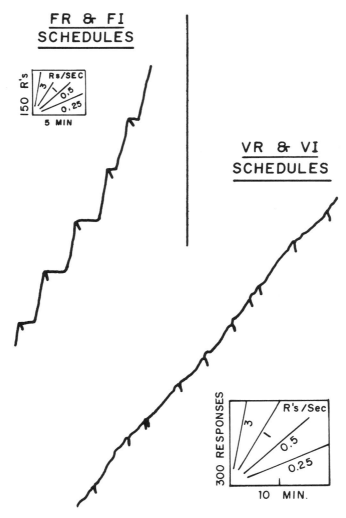

Figure 5.4   Cumulative records of responding by pigeons trained on
             FR 160 (upper tracing) and VI 2-min (lower tracing).
             Reinforcements are shown by diagonally-downward de-
             flections on the records.   The pattern of responding
             between successive reinforcements on FR 160 is the
             "break-run" pattern characteristic of asymptotic per-
             formance on FR and FI schedules.   The pattern of steady
             responding between reinforcements on VI 2-min is char-
             acteristic of asymptotic performance on VI and VR
             schedules (after Ferster & Skinner, 1957, Figure 28
             and Figure 394).

This figure illustrates the patterns of responding established after lengthy training, "break-run" (on FR and FI schedules) and steady responding (on VR and VI schedules). The upper panel shows "break-run" responding by a well trained pigeon on FR 160 (the first key peck after every 159th key peck was followed by grain). Performance is shown while the pigeon earned five successive rewards during a part of one training session. The data are presented in the form of a cumulative record on which the diagonally downward marks indicate presentation of grain. Each key peck moved the recording pen in an upward direction but reduction of the figure prevents individual responses from being seen. A horizontal line indicates a period without responding. Rate of responding (Rs/sec) can be estimated from the slope of the record measured against the inset scale. The bottom panel shows responding by a pigeon trained on VI 2-min (reward was produced by the first response two minutes, on the average, since the last reward). It illustrates the relatively steady rates of responding obtained when pigeons are trained on VI and VR schedules.

Complex schedules of reinforcement can be arranged by combining the basic schedules, by modifying the basic temporal and behavioral criteria for allowing R to produce S^v, and by correlating specific stimulus conditions with different schedules presented serially or concurrently to the same animal. The performances generated by such schedules have many nuances which are discussed in detail elsewhere (Ferster & Skinner, 1957; Honig, 1966; Honig & Staddon, 1977; Schoenfeld & Cole, 1972).

Paradoxically, operant schedules of reinforcement have both hindered and aided study of associative learning processes. On the one hand, some proponents of operant research have argued influentially against the significance of an associationistic theory of learning (e.g., Skinner, 1950, 1969). This argument is made on two levels. On logical grounds it is claimed that the stimulus-response emphasis in associative theory is unsuited for the majority of real-life behaviors which appear not to be evoked by obvious stimuli. A more appropriate analysis, the argument goes, is based on the concept of the "operant," a unit of behavior which is emitted by the animal (rather than evoked by discrete stimuli) and which is particularly influenced by rewarding or punishing consequences. It is a corollary of this reasoning that the free-responding procedure and operant schedules of reinforcement deserve prime experimental and theoretical attention by those who wish to understand the most significant aspects of human and animal behavior. The argument is also advanced on empirical grounds. It is well established that schedules of reinforcement generate highly reproducible performances within individual animals, between animals and

even between species. Some features of these performances have
even led to the assertion that operant schedules are "fundamental
determinants of behavior," determining whether a given event (e.g.,
electric shock) will act as a reward or punisher (Morse & Kelleher,
1970, 1977). In the light of such powerful behavioral control it
can seem frivolous to devote attention to the traditional questions
of associative learning, a point underscored by the many successes
in controlling human behavior achieved by applied psychologists who
employ operant principles (e.g., Reese, 1978). The influence of
the argument against an associative approach can be judged from the
impressive body of empirical and theoretical work produced by those
persuaded by the argument (e.g., Honig, 1966; Honig & Staddon, 1977;
Schoenfeld, 1970; Schoenfeld & Cole, 1972).

On the other hand, in some applications operant schedules of
reinforcement have been particularly useful experimental tools for
advancing associative theory. Several excellent examples are discussed elsewhere in this book. For example, the conditioned suppression experiment in which operant schedules provide a behavioral
baseline on which the effects of classically conditioned stimuli (presented in a trial procedure) are assessed has played a key role in
recent theoretical developments in classical conditioning (Chapter 4).
Also, the free-responding avoidance procedure has stimulated important developments in associative theories of avoidance (Chapter 10).
Furthermore, in recent years there has been increasing awareness
that the performances established by operant schedules are at least
partly influenced by associative processes. For example, classically
conditioned responses evoked by the manipulanda in operant training
devices (i.e., levers, keys, etc.) now appear to be important determinants of schedule-generated performance (e.g., Moore, 1973;
Schwartz & Gamzu, 1977). This line of research has caused the significance of the operant to be reevaluated. A problem for future research and theory is to clarify the various configurations of associative, motivational and other processes which operant reinforcement
schedules bring into play (e.g., Jenkins, 1970; Shimp, 1976).

The final reward schedule to be considered here is the delay
of reward schedule which is employed to study the effect of a delay
between performance of R and presentation of S^V in reward training.
Theorists have distinguished two types of delay by the nature of
responding during the delay interval. In "chaining" delay, responding during the interval is similar in topography to the response
from which the delay is timed. An example is the delay to reward
timed from the first of several lever presses on an FR schedule, or
from locomotor behavior in the initial part of a runway when the
animal continues to run to the goalbox. In "non-chaining" delay,
responding during the delay interval has a different topography
than the response which initiated the delay. An example is the case

in which the lever is withdrawn from the apparatus during the delay interval so that the animal can no longer perform the lever-
press response, or when the animal is confined in a chamber of the
runway to prevent further running.

The experimental evidence indicates that strength of responding is inversely related to the length of delay of reward in both
chaining and non-chaining delay. In the chaining case a comparison of responding in the early portions of short and long FR schedules or in the initial segments of short and long runways indicates
that performance is stronger in the shorter (Felton & Lyon, 1966;
Hull, 1934). The effect of delay in the non-chaining case is illustrated in the left panel of Figure 5 which shows the speed of different groups of rats running to the goalbox of an electrified runway (Fowler & Trapold, 1962). The parameter of the curves is the
length of the delay to shock offset after the rats reached the goalbox. The inverse relation between delay-length and response strength
is very clear.

Delays of reward are detrimental to responding for two reasons.
One is simply that delay reduces temporal contiguity between response and reward. This factor operates in both chaining and non-
chaining delays. The other operates in the non-chaining case and
it involves the strengthening of responses performed during the
delay interval which come to compete with the instrumental response.
Consider, for example, the non-chaining delay data in the right panel
of Figure 5. These data were obtained by Carlton (cited in Spence,
1956, p. 106ff) who trained hungry rats with a trials procedure to
obtain food by leaving a startbox and pressing a lever a short distance away. The lever was withdrawn from the training apparatus as
soon as it was pressed and food was presented either immediately
(0-sec condition) or after a 10-sec delay. The response measure
was the reciprocal of the time to press the lever on each trial. The
0-sec:U and 10-sec:U groups spent the delay period in an unconfined
experimental chamber where the 10-sec delay group could make competing responses such as turning away from the lever and exploring
the chamber. The figure shows that performance was markedly debilitated by the 10-sec delay. Perin (1943) had obtained a similar
finding with this procedure. However, the 0-sec:C and 10-sec:C
groups were confined during the delay period to prevent competing
responses in the 10-sec group. The 10-sec delay was not nearly as
debilitating for the confined animals.

Data from the free-responding procedure also show the effect
of non-chaining delay. For example, on FI schedules when the usual
0-sec delay between the response which produces reward and the presentation of reward is lengthened or made free to vary there is a

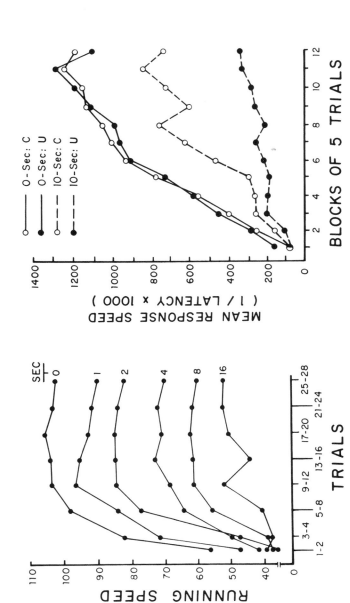

Fig. 5.5 Effects of delay of reward on performance in the runway. The left panel shows the influence of delay in escape training with rats. Each curve represents the speed of running (100/time in sec) by a different group. The parameter of the curves indicates the delay to shock offset after the rats reached the goalbox (after Fowler & Trapold, 1962). The right panel shows latency to barpress for food reward in a trials procedure when reward was delayed for 0 sec or 10 sec. Each curve represents the average speed of a different group of rats. The groups labelled "C" were confined in a small space during the delay interval; groups labelled "U" were unconfined (after Carlton, 1954, cited in Spence, 1956, p. 160ff).

decrement in the overall response rate (Dews, 1969; Shull, 1970; Staddon & Frank, 1975). The importance of responses during the delay interval in non-chaining delays on FI schedules is suggested by some data. For example, overall response rate is decreased when pigeons are required <u>not</u> to peck the response key during a delay after the FI period has elapsed, but response rate is increased when a high rate of responding occurs during the delay (Ferster & Skinner, 1957, p. 464ff, p. 416ff).

While the inverse relation between response strength and delay of reward is clear, a given delay is not expected to have a comparable effect in all training procedures and in all species. For one thing, certain aspects of a training procedure might reduce or enhance the decremental effect of a delay. Carlton's experiment discussed above provides one example (see also Grice, 1948). Also, there may be species differences in the ability to tolerate delays. In early experiments where a delay was placed between presentation of a <u>stimulus</u> and the opportunity to respond for reward, it was noted that rats and dogs could respond successfully with delays of a few seconds to a few minutes if they maintained bodily orientation to the site where the response was to be made. In raccoons, monkeys, and children, however, behavior during the delay period seemed relatively unimportant, possibly indicating the involvement of "representative" processes (Hunter, 1913; Tinklepaugh, 1928). In some recent studies of <u>response</u>-reward delay, it has been claimed that rats can tolerate exceptionally long delays without performance decrement if the experimental procedure properly activates memory processes (Lett, 1973, 1974, 1975, 1977). This claim should be treated with caution, however, since these findings have not been replicated (Roberts, 1976, 1977).

At the present, it seems that a theoretical analysis of the performance decrements engendered by delays between R and S^v in reward training must take into account the absolute duration of the delay and the nature of the responses performed during the delay.

<u>Reward magnitude</u>. Figure 6 illustrates that both rate of acquisition and asymptotic level of responding are positively related to reward magnitude when it is operationalized in different ways. The top panel (Crespi, 1942) shows running speed of three groups of rats which ran in a runway to different numbers of food pellets (16, 64, or 256 pellets; approximately 0.3, 1.3, or 5 g of food, respectively). The bottom panel (Kraeling, 1961) shows running speed when groups ran to a fixed volume of liquid reward which differed in concentration of sucrose (2.5, 5, or 10% concentration). Comparable functions are obtained with a wide range of reward events, including different amounts of shock-intensity reduction in escape

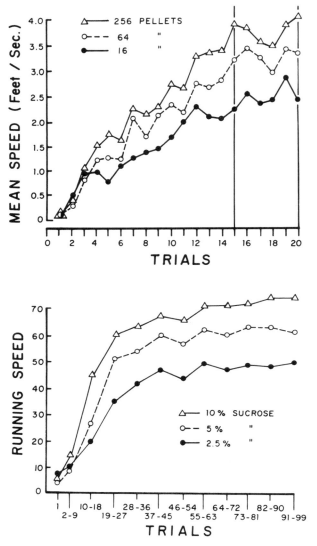

Figure 5.6   Effects of reward magnitude on performance in the runway.
The top panel shows the running speed of three groups of
rats which received different numbers of food pellets on
each trial (after Crespi, 1942).   The bottom panel shows
running speed (100/time in sec) for three groups receiving different concentrations of sucrose solution (after
Kraeling, 1961).   In both experiments, the intertrial
interval was 24 hours.

training (Bower, Fowler, & Trapold, 1959; Campbell & Kraeling , 1953). There is some evidence that asymptotic differences in performance between large and small food reward may disappear with extended training (e.g., McHose & Moore, 1976).

Studies of reward magnitude with operant schedules yield results comparable to those obtained with the trials procedure. For example, when rats are trained to lever press on FI schedules overall response rate is positively related to number of food pellets per reward (Meltzer & Brahlek, 1968), and to sucrose concentration (Guttman, 1953, 1954; Stebbins, Mead, & Martin, 1959). On VI schedules monkeys' rate of lever pressing is an increasing function of sucrose concentration (Conrad & Sidman, 1956), and pigeons' key pecking rate is greater when reward is 10 food pellets than when it is 1 pellet (Gonzalez & Champlin, 1974). Rate of key pecking is not so clearly affected in pigeons when reward magnitude is varied by altering the duration of grain-hopper presentations (Catania, 1963; Jenkins & Clayton, 1949; Keesey & Kling, 1961). This is understandable because different hopper durations do not insure that the pigeons will eat different numbers of seeds, or different weights of food, from the seed mixture presented in the hopper.

One complication of studying magnitude of reward with the free-responding procedure is that animals trained with high magnitudes may become satiated and show performance decrements for that reason. This problem also arises in the massed trials procedure where many rewards are presented in each training session. For example, Guttman (1953) found that rats pressed a lever at a lower rate on a continuous reinforcement schedule when they obtained 32% sucrose solution than when 8% and 16% concentrations were employed. Findings such as this appear related to satiation (Conrad & Sidman, 1956; Hodos & Kalman, 1963), a problem largely avoided when only one trial a day is given, as in the Crespi and Kraeling experiments shown in Figure 6.

The generally positive relation found between response strength and reward magnitude poses a number of interesting theoretical questions. Does reward magnitude affect the strength of the association formed in instrumental training? Does it affect some motivational process which interacts with associative strength? Does it affect the characteristics of some "representational" process through which the animal retains an "image" of the reward event which, in turn, influences performance? These questions will be addressed in the next chapters where it will become clear that magnitude of reward plays a central role in theories of reward training.

   Rate of reinforcement. An influential parameter of training
in free-responding procedures is the rate at which reinforcers are
obtained. Figure 7 shows the effects of varying rate of reinforcement by manipulating the average inter-reinforcement interval on
VI schedules (Catania & Reynolds, 1968). Each data point is the
average of six pigeons' asymptotic key peck rates on VI schedules
whose average inter-reinforcement interval was set to provide S^v
at the rates shown. The data show a negatively accelerated monotonically increasing function relating response rate and reinforcement rate.

   The function shown in Figure 7 relates two molar variables,
rate of responding and rate of reinforcement computed over the
entire training session. However, rate of reinforcement seems to
exert its effect at a more molecular level. For example, Catania
and Reynolds (1968) showed that the momentary probability of reinforcement is an important determinant of free responding. When the
VI schedule arranged a constant momentary probability of reinforcement response rate was steady throughout the training session,
whereas if the momentary probability fluctuated responding was more
variable. An extreme comparison would be between the behavioral
effects of, say, a FI 60-sec schedule and a VI 60-sec. Both arrange
an average of one reinforcement per minute but the momentary probability of reinforcement in the inter-reinforcement interval is
vastly different. On the FI schedule the probability is zero for
60 sec after each food presentation and it is possible for the
animal to learn a discrimination on this basis. Consequently the
pronounced pause after reinforcement on FI schedules but not on VI
schedules is understandable (see Figure 4).

   Quality of reward. Experiments using a variety of events as
S^v were cited above more or less interchangeably to illustrate how
the parameters of reward training influence performance. This was
done to emphasize that different reward events have comparable
effects in many cases. However, some findings indicate that qualitatively different rewards have distinctive effects on performance.
Some of these cases are instructive because they reveal an associative determinant of performance which may be operative in many
reward training situations.

   One finding is illustrated in Figure 8 which shows cumulative
records of a pigeon key pecking on a VI 3-min schedule for water or
food reward. The water-reward panel shows responding in a session
when the highest rate of key pecking for water was obtained (Ferster & Skinner, 1957). The food-reward panel shows about average
performance when the same pigeon was trained with food. The slopes
of the records for these sessions indicate that the highest rate of
key pecking maintained by water reward was lower than the average
rate of key pecking maintained by food.

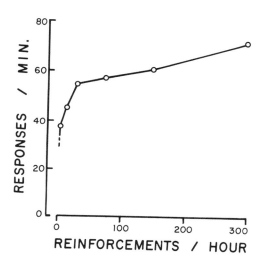

Figure 5.7   Effects of rate of reinforcement on keypecking by
pigeons in the free-responding procedure (after
Catania & Reynolds, 1968).

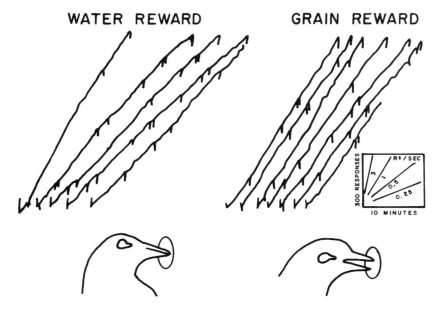

Figure 5.8 Effects of quality of reward on keypecking by pigeons.
            The upper panels show cumulative records by a pigeon a
            keypecking on a VI 3-min schedule for water or food
            reward. Each panel shows responding in an entire ses-
            sion. The recording pen reset when it reached the top
            of the panel, and the records have been cropped to-
            gether to save space. Reward presentations are marked
            by downward deflections of the pen. The inset scale
            allows the rate of responding to be estimated (after
            Ferster & Skinner, 1957, p. 374). The bottom panels
            show the topography of the pigeon's autoshaped key
            peck when water or food is the reward event (after
            Jenkins & Moore, 1973).

What factors might produce these differences in performance? One possibility is that the amounts of water and food given as reward were not functionally equivalent. Another is that the level of water deprivation was not equivalent to the level of food deprivation. Either possibility poses a hazard when the effects of different rewards are compared (Bitterman, 1960). A more interesting associative possibility is that a classically conditioned form of the consummatory response influenced performance of the pecking response. The bottom panels of Figure 8 show tracings of high-speed photographs made at the moment a pigeon contacted the response key in an autoshaping experiment when different keys signalled water and grain presentations (Jenkins & Moore, 1973). The pigeon made drinking movements on the key that signalled water, each contact occurring with closed beak, for a relatively long duration and with weak force. When the food-key was illuminated, the pigeon made forceful, short-duration contact with an open beak as if it were picking up grain. Related key peck topographies have been observed when key pecking on operant schedules is reinforced with food or water (Smith, 1974; Wolin, 1948). Thus, a reasonable supposition is that different rates of key pecking are found with food and water reward because the key peck topography is determined by the nature of the reward event. Water reward leads to low rates <u>via</u> long-duration, low-force key contacts; food reward leads to relatively high rates <u>via</u> contacts which are short in duration and high in force.

Other indications that classically conditioned forms of the response to reward influence performance have recently come to light. For example, the Brelands reported numerous cases in which attempts to train animals for commercial purposes were aborted because reward-related responses, rather than the desired instrumental response, came to predominate in the training situation (Breland & Breland, 1966). Boakes, Poli, and Lockwood (1975) have described related findings in laboratory experiments. Sevenster (1973) found it easier to train three-spined sticklebacks to bite a plastic rod when the response evoked by the reward event included biting than when it didn't. Hull (1977) has reported that rats trained to lever press for food or water press the lever in distinctive ways which can be identified by observers as relating to the type of reward. The implication of all these findings is that reward events of different quality may influence performance through an associative process which allows classically conditioned consummatory responses to be evoked by stimuli in the training situation.

The possibility that a classically conditioned form of the response made to the reward event interacts with instrumental responding has been an influential theoretical idea in Hullian theory (e.g., Amsel, 1958; Hull, 1931; Miller, 1935; Spence, 1956). Reasoning from the premise that instrumental behavior is adaptive, Hull concluded that, when fully elaborated, the classically conditioned "goal" reaction must be compatible with performance of the instrumental response. If early in training it interfered with performance, Hull expected the competing portions of the classically conditioned response to drop out. Only non-competitional forms of the goal response should occur concurrently with the instrumental response. While subsequent findings indicate that Hull's conclusion was wrong for at least some training arrangements (e.g., Williams & Williams, 1969), his emphasis on classically conditioned "goal" responses as determinants of performance in reward training seems to have been well-placed (Hearst & Jenkins, 1974; Moore, 1973).

Spatial contiguity of stimulus, response and reward. There has been little explicit concern with the spatial arrangements of stimulus, response and reward in reward training. However, recent experiments indicate that manipulation of this variable can provide evidence about the associative determinants of performance. One such experiment is considered here.

Grastyán and Vereczkei (1974) trained cats in a runway to wait on the start-platform until an auditory stimulus signalled that food was available in the goalbox. Between trials the cats were free to move in the runway, but only those runs to the goalbox were rewarded which began from the start-platform in the presence of the auditory stimulus. In one condition the stimulus was presented from a loudspeaker behind the start-platform, and so it was spatially discontiguous to the location of reward and to the place towards which the instrumental running response was directed. In another condition the loudspeaker was in the goalbox so that stimulus, response and reward were spatially contiguous.

The upper panel of Figure 9 shows latency to leave the start-platform by a representative cat on successive days of the experiment. When the loudspeaker was in the startbox, performance improved the first 3 days (average latency fell from about 30 sec to about 10 sec) but subsequently deteriorated until, on the final day of the discontiguous condition, average latency was about 60 sec. When the loudspeaker was moved to the goalbox, performance improved substantially in one or two sessions. The bottom panel of Figure 9 shows performance on individual trials of single sessions with the stimulus in the two locations. In session 4 (when the stimulus was in the startbox) latencies generally became longer as more rewarded trials were given; in session 15 (when the stimulus was in the goalbox) performance was strong on all trials in the session.

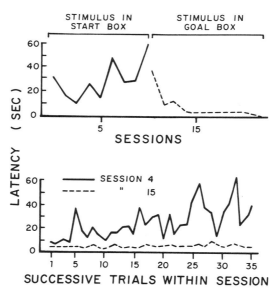

Figure 5.9 Effects of locating the trial stimulus in opposite ends of a runway. The top panel shows the average latency per session when a cat left the startbox to run for food. The trial stimulus was in the startbox for the initial sessions and then was switched to the goalbox. The bottom panel shows latency on the individual trials of session 4 (stimulus in startbox) and session 15 (stimulus in goalbox) for the same cat (after Grastyán & Vereczkei, 1974).

Grastyán and Vereczkei's experiment provides some important leads about why these two arrangements of stimulus, response and reward had such large effects on performance. The deterioration in performance when the loudspeaker was in the startbox was related to the emergence of stimulus-directed responses which competed with running to the goalbox. Early in training the stimulus evoked only a brief turning of the head towards the loudspeaker. However, after a few trials this reaction became more pronounced and the cats remained "frozen," gazing at the stimulus source for a period on each trial. After more trials the cats approached and sniffed the loudspeaker when the stimulus was first presented, and many cats showed the competing nature of the two responses very clearly. For example, on some trials they alternately ran part way to the goalbox and then back to the loudspeaker; some refused to eat when they finally reached the goalbox, and even retched and vomited as soon

as the stimulus was presented. When training was continued sufficiently long the instrumental response came to predominate.

Taken together, these observations imply that as training progressed the trial stimulus became classically conditioned through
repeated pairings with food and eventually evoked strong approach
and contact responses. When the stimulus was in the startbox these
responses competed with the instrumental response for a substantial
portion of training. When the stimulus was in the goalbox the classically conditioned responses were compatible with the instrumental
response and performance was not debilitated. The implication is
that classically conditioned responses may be an important but
hidden influence in many reward training procedures.

One instructive aspect of the Grastyán and Vereczkei (1974)
finding, and its proposed classical conditioning explanation, is
that an apparent failure to strengthen an instrumental response in
reward training need not imply a failure of associative principles.
On the contrary, poor or worsening performance may originate in
associative processes which combine to have a deleterious effect on
performance. The major effects obtained here when the trial stimulus in a more-or-less standard laboratory apparatus was relocated
indicate the limitations on our current understanding of the associative basis of reward training (see also LoLordo, McMillan, & Riley,
1974). In view of these limitations we should be cautious in concluding that the associative approach is inadequate, or that associative learning is "constrained," simply because in some novel training situation the instrumental response fails to change in a manner
consistent with changes observed in traditional laboratory experiments. A more appropriate conclusion may be that the training conditions have caused familiar associative processes to combine in
ways that produce unexpected performances. Such a conclusion is
susceptible to experimental test and should be considered as a
serious possibility when an anomalous result is obtained.

Motivation. Operationally, an animal's motivational state is
manipulated by depriving it of some commodity (e.g., food or water)
in appetitive training or by varying the intensity of the painful
stimulus applied to the animal in escape learning. In both cases,
motivational level has a profound effect on performance.

Figure 10 provides one of the many possible illustrations of
this fact. Kintsch (1962) trained rats to run to water reward and
arranged a 24-hr interval between trials. In this experiment,
all rats were deprived of water for 23 hr before the daily
training trial. Motivational level was manipulated by prewatering
the rats 5 min before the trial was given. The low drive group

received 6 ml of water, the medium drive group 3 ml and the high
drive group was given no pre-watering. The figure clearly shows
that motivational level affected the rate of acquisition and the
asymptote reached. The results could not be attributed to differences in the amount of reward consumed in the runway because each
group rapidly drank all the water presented there.

The theoretical problem posed by motivational manipulations
is to determine how they interact with associative processes to
determine performance.

Significance of the Response-Reward Relation

Theories and data surveyed above suggest that performance in
reward training is at least partly determined by stimulus-reward
(i.e., classical conditioning) relations embedded in the training

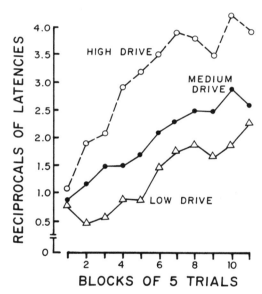

Figure 5.10 Effects of motivational level on performance in a runway. Each curve shows the average speed of a group of
thirsty rats leaving the startbox to run to water reward. The parameter of the curves is the amount of
water given immediately prior to the trial: 6 ml for
the low drive group; 3 ml for the medium drive group;
0 ml for the high drive group (after Kintsch, 1962).

procedure. Is it possible that the paradigmatic emphasis on the response-reward relation (i.e., S:R → S^v) is <u>entirely</u> misplaced and that the varied performances characteristic of reward training result from the stimulus-reward relations these procedures arrange? This possibility has been seriously examined in recent years, prompted mostly by the discovery of autoshaping (Brown & Jenkins, 1968).

The various demonstrations that motor responses such as key pecking and lever pressing are susceptible to strong control by stimulus-reward relations has been unsettling. Much of our information about reward learning has come from experiments in which a response-reward relation was imposed on these responses or on locomotion, which is also subject to stimulus-reward influence (e.g., Hearst & Jenkins, 1974). There are decided cases in which performance maintained with only a stimulus-reward relation imposed has many of the features of performance in similar situations when a response-reward relation is imposed. For example, Staddon and Simmelhag (1971) found that the behavior of pigeons trained to key peck for food on a FI 12-sec schedule was very similar to that observed when pigeons simply received food every 12 sec irrespective of responding. In both cases the probability of orienting and pecking towards the wall where the food hopper was located increased as the time of feeding approached. The principal difference in the behaviors established by the two procedures was that on FI 12-sec virtually all pecks were made towards the key (as required) while in the response-independent case pecks were dispersed along the hopper wall.

While an emphasis on the similarity of behaviors established with and without a response-reward relation imposed has been instructive, the differences are equally important. Moore (1973) has pointed out that the performances of pigeons in many reward training procedures can be accounted for by stimulus-reward pairings but that it is unreasonable to expect all behaviors in instrumental training to originate from those pairings. It would seem, for example, that the variety of skilled movements which can be established in different species when a response-reward requirement is imposed do not readily submit to a classical conditioning interpretation. On the other hand, classical conditioning cannot be rejected as the source of performance simply because the instrumental response does not resemble the response evoked by S^v. For example, Holland (1977) and others have shown that the form of the classically conditioned response is determined by the properties of the CS as well as by the nature of the US. At the present, the most reasonable conclusion seems to be that the response- <u>and</u> the stimulus-reward relations both contribute to the performances we observe.

The problem is to determine how the two relations combine in a given training situation to influence responding.

A new procedure developed by Woodruff, Conner, Gamzu, and Williams (1977) is of considerable interest because it allows the response-reward (R-S^V) and the stimulus-reward (S-S^V) relations to be varied independently while reward training is in progress. They trained pigeons to key peck for food reward on what they termed a "T*" schedule. During each session the response key was alternately lighted red or green. When red (the trial stimulus) was present, key pecks produced food twice a minute on the average. When the key was green (the intertrial stimulus) food was never presented. The logic of the experiment requires the reasonable assumption that pecks at the red key were potentially controlled by a response-reward relation (i.e., pecks → food) and by a stimulus-reward relation (i.e., red → food). The experiment manipulated the strength of both relationships.

The R-S^V relation was made strong, moderate, or weak by imposing delays between R and S^V. In the strong R-S^V group, S^V immediately followed each R that produced it. In the moderate R-S^V group, a random half of the S^{V}s were delayed for variable times, averaging 4 sec from the response that produced them; the remaining S^{V}s were presented immediately for this group. The weak R-S^V group had all S^{V}s delayed an average of 4 sec. The question posed in the experiment was whether the debilitating effects of weakening the response-reward relation would depend on the prevailing strength of the stimulus-reward relation.

The left panel of Figure 11 shows that the rate of pecking the red key was high for all three groups and statistical analyses indicated no reliable differences among them. The data shown here were obtained when the red trial stimulus was present for 12 sec alternately with the green intertrial stimulus which was present for 48 sec. These stimulus durations were intended to make the S-S^V relation (between red key and food) strong. When the trial stimulus is short relative to the duration of the intertrial stimulus, stimulus-reward control of key pecking seems to be enhanced (e.g., Terrace, Gibbon, Farrell, & Baldock, 1975). Thus, in this condition the debilitating effects of weak R-S^V control of key pecking were overcome by the strong control over key pecking exerted by the S-S^V relation.

The right-hand panel of Figure 11 shows a different picture. Here, when S-S^V control was relaxed a main effect of the different strengths of R-S^V relation was clearly evident. The stronger the R-S^V relation the stronger the pecking on the red key. In this

case, the S-S^V influence was weakened by making the trial duration
long (48 sec) relative to the intertrial duration (12 sec). A
unique feature of the procedure was that through computer control
of the experiment the mean rate of S^V occurrences during the trial
stimulus was held constant across the groups despite differences in
the R-S^V relation. This was achieved by arranging response-independent presentations of S^V whenever responding failed to maintain
an average of two reinforcers per minute (see Woodruff, et al., 1977,
for details). Thus, this procedure insured that in both phases of
the experiment the strength of the S-S^V relation was equally strong
for all groups while the strength of the R-S^V relation varied between groups. Clearly, the data show that the effectiveness of the
R-S^V relation as a factor controlling the instrumental key-peck
response depended heavily on the prevailing strength of the S-S^V
relation controlling that response.

The experimental evidence discussed here implicates both R-S^V
and S-S^V relations as influential factors in reward training. The
important problem for future study is to determine more fully how
these two relations articulate in the various reward training procedures. On the basis of their results, Woodruff, et al., suggested

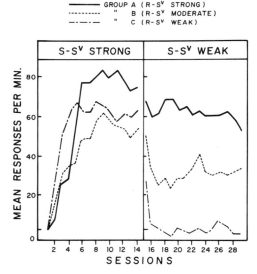

Figure 5.11 Effects of jointly manipulating the R-S^V and S-S^V re-
             lations in reward training. Each curve shows the av-
             eraged keypecking rate of a group of pigeons. See text
             for additional details (after Woodruff, Conner, Gamzu,
             & Williams, 1977).

that the stimulus-reward relation may be particularly influential in discrete-trial procedures where the ratio of trial to intertrial durations is typically small (they found the S-S^v relation predominant when that ratio was small); conversely, the R-S^v relation may predominate in the free-responding procedure where there is no obvious trial period (they found R-S^v particularly influential when the trial/intertrial duration ratio was large). These suggestions are very provocative and suggest some new directions for research and theorizing in reward training. When we finally arrive at a proper weighting of S-S^v and R-S^v influences in theories of reward training we will have made major progress toward the goal of understanding the associative basis of instrumental performance.

The Learning-Performance Distinction

In closing this chapter it may be well to restate the learning-performance distinction. It is basic to make this distinction because behavior is determined by a variety of factors, including age, sex, species, motivational level, etc. Experience is only one of the determinants of an animal's performance in any situation, and in the present context we are most interested in how the associative residue of that experience affects performance. Accordingly, to understand the associative basis of performance we attempt to hold the other factors constant in our experiments and manipulate only those variables which our theories indicate will alter associative processes.

The learning-performance distinction is fundamental to the traditional distinction between "learning theory" and "behavior theory." Learning theories are concerned with the factors influencing associative processes. Behavior theories are concerned with the determinants of performance.

Conclusion

Reward training has been the center of much experimental attention in the past 80 years. In this chapter it has been possible to survey only some of the principal methods and variables which have been investigated. It is clear from the chapter that performance in reward training is greatly influenced by the schedule of reward, reward magnitude, rate of reinforcement, quality of reward, the spatial contiguity of stimulus-response-reward, and the motivational condition of the animal. The discussion of these variables was confined here to the case where a single response is studied in the experimental situation. Most of these variables have also been investigated when multiple responses are available simultaneously so that the animal can choose among different reward outcomes. The interested reader is referred to a more comprehensive source for a review of that literature (e.g., Honig & Staddon, 1977; Mackintosh, 1974). In all cases, the data pose the important problem of separating the influences of response-reward and stimulus-reward relations on the performances we observe. We have made some important progress toward solving that problem in the past few years. The next three chapters will review some of the important theoretical and empirical developments in specific topics in reward training. In these chapters many of the parameters of reward training surveyed here will appear in key roles, and the questions posed here about the role of classical conditioning as a determinant of performance in reward training will be greatly embellished.

REFERENCES

Amsel, A. The role of frustrative nonreward in noncontinuous reward situations. PSYCHOLOGICAL BULLETIN, 1958, 55, 102-119.

Amsel, A. Partial reinforcement effects on vigor and persistence. In K. W. Spence & J. T. Spence (Eds.), THE PSYCHOLOGY OF LEARNING AND MOTIVATION, Vol. 1. New York: Academic Press, 1967.

Bacon, W. E. Partial reinforcement extinction effect following different amounts of training. JOURNAL OF COMPARATIVE AND PHYSIOLOGICAL PSYCHOLOGY, 1962, 55, 998-1003.

Bitterman, M. E. Toward a comparative psychology of learning. AMERICAN PSYCHOLOGIST, 1960, 15, 704-712.

Boakes, R. A., Poli, M., & Lockwood, M. J. A study of misbehavior: Token reinforcement with rats. Paper delivered at the annual meeting of the Psychonomic Society, Colorado, 1975.

Bower, G. H., Fowler, H., & Trapold, M. A. Escape learning as a function of amount of shock reduction. JOURNAL OF EXPERIMENTAL PSYCHOLOGY, 1959, 58, 482-484.

Breland, K., & Breland, M. ANIMAL BEHAVIOR. New York: The Macmillan Co., 1966.

Brown, P. L., & Jenkins, H. M. Auto-shaping of the pigeon's keypeck. JOURNAL OF THE EXPERIMENTAL ANALYSIS OF BEHAVIOR, 1968, 11, 1-8.

Campbell, B. A., & Kraeling, D. Response strength as a function of drive level and amount of drive reduction. JOURNAL OF EXPERIMENTAL PSYCHOLOGY, 1953, 45, 97-101.

Capaldi, E. J. The effect of different amounts of training on the resistance to extinction of different patterns of partially reinforced responses. JOURNAL OF COMPARATIVE AND PHYSIOLOGICAL PSYCHOLOGY, 1958, 51, 367-371.

Capaldi, E. J. Partial reinforcement: A hypothesis of sequential effects. PSYCHOLOGICAL REVIEW, 1966, 73, 459-477.

Capaldi, E. J. A sequential hypothesis of instrumental learning. In K. W. Spence & J. T. Spence (Eds.), THE PSYCHOLOGY OF LEARNING AND MOTIVATION, Vol. 1. New York: Academic Press, 1967.

Capaldi, E. J., & Haggbloom, S. J. Response events as well as goal events as sources of animal memory. ANIMAL LEARNING & BEHAVIOR, 1975, 3, 1-10.

Capaldi, E. J., & Morris, M. D. A role of stimulus compounds in eliciting responses: Relatively spaced extinction following massed acquisition. ANIMAL LEARNING & BEHAVIOR, 1976, 4, 113-117.

Catania, A. C. Concurrent performances: A baseline for the study of reinforcement magnitude. JOURNAL OF THE EXPERIMENTAL ANALYSIS OF BEHAVIOR, 1963, 6, 299-300.

Catania, A. C., & Reynolds, G. S. A quantitative analysis of the responding maintained by interval schedules of reinforcement. JOURNAL OF THE EXPERIMENTAL ANALYSIS OF BEHAVIOR, 1968, 11, 327-383.

Conrad, D. G., & Sidman, M. Sucrose concentration as a reinforcement for lever pressing by monkeys. PSYCHOLOGICAL REPORTS, 1956, 2, 381-384.

Crespi, L. P. Quantitative variation of incentive and performance in the white rat. AMERICAN JOURNAL OF PSYCHOLOGY, 1942, 55, 467-517.

Cumming, W. W., & Schoenfeld, W. N. Behavior under extended exposure to a high-value fixed-interval schedule. JOURNAL OF THE EXPERIMENTAL ANALYSIS OF BEHAVIOR, 1958, 1, 245-263.

Dews, P. B. Studies on responding under fixed-interval schedules of reinforcement: The effects on the pattern of responding of changes in the requirements at reinforcement. JOURNAL OF THE EXPERIMENTAL ANALYSIS OF BEHAVIOR, 1969, 12, 191-199.

Dews, P. B. The theory of fixed-interval responding. In W. N. Schoenfeld (Ed.), THE THEORY OF REINFORCEMENT SCHEDULES. New York: Appleton-Century-Crofts, 1970.

Felton, M., & Lyon, D. O. The post-reinforcement pause. JOURNAL OF THE EXPERIMENTAL ANALYSIS OF BEHAVIOR, 1966, 9, 131-134.

Ferster, C. B., & Skinner, B. F. SCHEDULES OF REINFORCEMENT. New York: Appleton-Century-Crofts, 1957.

Flaherty, C. F., & Davenport, J. W. Successive brightness discrimination in rats following regular versus random intermittant reinforcement. JOURNAL OF EXPERIMENTAL PSYCHOLOGY, 1972, 96, 1-9.

Fleshler, M., & Hoffman, H. S. A progression for generating variable-interval schedules. JOURNAL OF THE EXPERIMENTAL ANALYSIS OF BEHAVIOR, 1962, 5, 529-530.

Fowler, H., & Trapold, M. A. Escape performance as a function of delay of reinforcement. JOURNAL OF EXPERIMENTAL PSYCHOLOGY, 1962, 63, 464-467.

Gonzalez, R. C., Bainbridge, P., & Bitterman, M. E. Discrete-trials lever pressing in the rat as a function of pattern of reinforcement, effortfulness of response and amount of reward. JOURNAL OF COMPARATIVE AND PHYSIOLOGICAL PSYCHOLOGY, 1966, 61, 110-122.

Gonzalez, R. C., & Champlin, G. Positive behavioral contrast, negative simultaneous contrast and their relation to frustration in pigeons. JOURNAL OF COMPARATIVE AND PHYSIOLOGICAL PSYCHOLOGY, 1974, 87, 173-187.

Goodrich, K. P. Performance in different segments of an instrumental response chain as a function of reinforcement schedule. JOURNAL OF EXPERIMENTAL PSYCHOLOGY, 1959, 57, 57-63.

Grastyán, E., & Vereczkei, L. Effects of spatial separation of the conditioned signal from the reinforcement: A demonstration of the conditioned character of the orienting response or the orientational character of conditioning. BEHAVIORAL BIOLOGY, 1974, 10, 121-146.

Grice, G. R. The relation of secondary reinforcement to delayed reward in visual discrimination learning. JOURNAL OF EXPERIMENTAL PSYCHOLOGY, 1948, 38, 1-16.

Guttman, N. Operant conditioning, extinction, and periodic reinforcement in relation to concentration of sucrose used as reinforcing agent. JOURNAL OF EXPERIMENTAL PSYCHOLOGY, 1953, 46, 213-224.

Guttman, N. Equal-reinforcement values for sucrose and glucose solutions compared with equal-sweetness values. JOURNAL OF COMPARATIVE AND PHYSIOLOGICAL PSYCHOLOGY, 1954, 47, 358-361.

Hayes, K. J., & Hayes, C. Imitation in a home-reared chimpanzee. JOURNAL OF COMPARATIVE AND PHYSIOLOGICAL PSYCHOLOGY, 1952, 45, 450-459.

Hearst, E., & Jenkins, H. M. SIGN-TRACKING: THE STIMULUS-REINFORCER RELATION AND DIRECTED ACTION. Austin, Texas: The Psychonomic Society, 1974.

Hodos, W., & Kalman, G. Effects of increment size and reinforcer volume on progressive ratio performance. JOURNAL OF THE EXPERIMENTAL ANALYSIS OF BEHAVIOR, 1963, 6, 387-392.

Holland, P. C. Conditioned stimulus as a determinant of the form of the Pavlovian conditioned response. JOURNAL OF EXPERIMENTAL PSYCHOLOGY: ANIMAL BEHAVIOR PROCESSES, 1977, 3, 77-104.

Honig, W. K. (Ed.), OPERANT BEHAVIOR: AREAS OF RESEARCH AND APPLICATION. New York: Appleton-Century-Crofts, 1966.

Honig, W. K., & James, P. H. R. (Eds.), ANIMAL MEMORY. New York: Academic Press, 1971.

Honig, W. K., & Staddon, J. E. R. (Eds.), HANDBOOK OF OPERANT BEHAVIOR. Englewood Cliffs, N. J.: Prentice-Hall, 1977.

Hull, C. L. Goal attraction and directing ideas conceived as habit phenomena. PSYCHOLOGICAL REVIEW, 1931, 38, 487-506.

Hull, C. L. The rat's speed-of-locomotion gradient in the approach to food. JOURNAL OF COMPARATIVE PSYCHOLOGY, 1934, 17, 393-422.

Hull, C. L. PRINCIPLES OF BEHAVIOR. New York: Appleton-Century-Crofts, 1943.

Hull, J. H. Instrumental response topographies of rats. ANIMAL LEARNING & BEHAVIOR, 1977, 5, 207-212.

Hunter, W. S. The delayed reaction in animals and children. BEHAVIOR MONOGRAPHS, 1913, 2, 21-30.

Jenkins, H. M. Sequential organization in schedules of reinforcement. In W. N. Schoenfeld (Ed.), THE THEORY OF REINFORCEMENT SCHEDULES. New York: Appleton-Century-Crofts, 1970.

Jenkins, H. M., & Moore, B. R. The form of the auto-shaped response with food and water reinforcements. JOURNAL OF THE EXPERIMENTAL ANALYSIS OF BEHAVIOR, 1973, 20, 163-181.

Jenkins, W. O., & Clayton, F. L. Rate of responding and amount of reinforcement. JOURNAL OF COMPARATIVE AND PHYSIOLOGICAL PSYCHOLOGY, 1949, 42, 174-181.

Jenkins, W. O., & Stanley, J. C. Jr. Partial reinforcement: A review and critique. PSYCHOLOGICAL BULLETIN, 1950, 47, 193-234.

Jobe, J. B., Mellgren, R. L., Feinberg, R. A., Littlejohn, R. L., & Rigby, R. L. Patterning, partial reinforcement, and N-length effects at spaced trials as a function of reinstatement of retrieval cues. LEARNING AND MOTIVATION, 1977, 8, 77-97.

Katz, S., Woods, G. T., & Carrithers, J. H. Reinforcement after-
    effects and intertrial interval. JOURNAL OF EXPERIMENTAL
    PSYCHOLOGY, 1966, 72, 624-626.

Keesey, R. E., & Kling, J. W. Amount of reinforcement and free-
    operant responding. JOURNAL OF THE EXPERIMENTAL ANALYSIS OF
    BEHAVIOR, 1961, 4, 125-132.

Kintsch, W. Runway performance as a function of drive strength and
    magnitude of reinforcement. JOURNAL OF COMPARATIVE AND PHYSIO-
    LOGICAL PSYCHOLOGY, 1962, 55, 882-887.

Kohn, B., & Dennis, M. Observation and discrimination learning
    in the rat: Specific and nonspecific effects. JOURNAL OF
    COMPARATIVE AND PHYSIOLOGICAL PSYCHOLOGY, 1972, 78, 292-296.

Kraeling, D. Analysis of amount of reward as a variable in learning.
    JOURNAL OF COMPARATIVE AND PHYSIOLOGICAL PSYCHOLOGY, 1961, 54,
    560-565.

Lett, B. T. Delayed reward learning: Disproof of the traditional
    theory. LEARNING AND MOTIVATION, 1973, 4, 237-246.

Lett, B. T. Visual discrimination learning with a 1-min delay of
    reward. LEARNING AND MOTIVATION, 1974, 5, 174-181.

Lett, B. T. Long delay learning in the T-maze. LEARNING AND MOTI-
    VATION, 1975, 6, 80-90.

Lett, B. T. Regarding Roberts's reported failure to obtain visual
    discrimination learning with delayed reward. LEARNING AND
    MOTIVATION, 1977, 8, 136-139.

Logan, F. A. INCENTIVE: HOW THE CONDITIONS OF REINFORCEMENT AFFECT
    THE PERFORMANCE OF RATS. New Haven: Yale University Press,
    1960.

LoLordo, V. M., McMillan, J. C., & Riley, A. L. The effects upon
    food-reinforced pecking and treadle-pressing of auditory and
    visual signals for response-independent food. LEARNING AND
    MOTIVATION, 1974, 5, 24-41.

Ludvigson, H. W. & Sytsma, D. The sweet smell of success: Apparent
    double alternation in the rat. PSYCHONOMIC SCIENCE, 1967, 9,
    283-284.

Mackintosh, N. J. THE PSYCHOLOGY OF ANIMAL LEARNING. London:
    Academic Press, 1974.

McHose, J. H., & Moore, J. N. Reinforcer magnitude and instrumental
    performance in the rat. BULLETIN OF THE PSYCHONOMIC SOCIETY,
    1976, 8, 416-418.

Medin, D. L., Roberts, W. A., & Davis, R. T. PROCESSES OF ANIMAL
    MEMORY. Hillsdale, N. J.: Lawrence Erlbaum Associates, 1976.

Meltzer, D., & Brahlek, J. A. Quantity of reinforcement and fixed-interval performance. PSYCHONOMIC SCIENCE, 1968, 12, 207-208.

Miller, N. E. A reply to "Sign-Gestalt or Conditioned Reflex?" PSYCHOLOGICAL REVIEW, 1935, 42, 280-292.

Moore, B. R. The role of directed Pavlovian reactions in simple instrumental learning in the pigeon. In R. A. Hinde & J. Stevenson-Hinde (Eds.), CONSTRAINTS ON LEARNING. London: Academic Press, 1973.

Morrison, R. R., & Ludvigson, H. W. Discrimination by rats of con-specific odors of reward and nonreward. SCIENCE, 1970, 167, 904-905.

Morse, W. H., & Kelleher, R. T. Schedules as fundamental determi-nants of behavior. In W. N. Schoenfeld (Ed.), THE THEORY OF REINFORCEMENT SCHEDULES. New York: Appleton-Century-Crofts, 1970.

Morse, W. H., & Kelleher, R. T. Determinants of reinforcement and punishment. In W. K. Honig & J. E. R. Staddon (Eds.), HAND-BOOK OF OPERANT BEHAVIOR. Englewood Cliffs, N. J.: Prentice-Hall, 1977.

Perin, C. T. A quantitative investigation of the delay-of-reinforcement gradient. JOURNAL OF EXPERIMENTAL PSYCHOLOGY, 1943, 32, 37-51.

Platt, J. R. Discrete trials and their relation to free-behavior situations. In H. H. Kendler & J. T. Spence (Eds.), ESSAYS IN NEOBEHAVIORISM: A MEMORIAL VOLUME TO KENNETH W. SPENCE. New York: Appleton-Century-Crofts, 1971.

Premack, D. Reinforcement theory. In D. Levine (Ed.), NEBRASKA SYMPOSIUM ON MOTIVATION. Lincoln: University of Nebraska Press, 1965.

Rashotte, M. E., Adelman, L., & Dove, L. D. Influence of percentage-reinforcement on runway running of rats. LEARNING AND MOTIVA-TION, 1972, 3, 194-208.

Reese, E. P. HUMAN BEHAVIOR: ANALYSIS AND APPLICATION. 2nd Edition. Dubuque, Iowa: William C. Brown, 1978.

Roberts, W. A. Failure to replicate visual discrimination learning with a 1-min delay of reward. LEARNING AND MOTIVATION, 1976, 7, 313-325.

Roberts, W. A. Still no evidence for visual discrimination learning: A reply to Lett. LEARNING AND MOTIVATION, 1977, 8, 140-144.

Schneider, B. A. A two-state analysis of fixed-interval responding in the pigeon. JOURNAL OF THE EXPERIMENTAL ANALYSIS OF BEHA-VIOR, 1969, 12, 677-687.

Schoenfeld, W. N. (Ed.), THE THEORY OF REINFORCEMENT SCHEDULES. New York: Appleton-Century-Crofts, 1970.

Schoenfeld, W. N., Antonitis, J. J., & Bersh, P. J. Unconditioned response rate of the white rat in a bar pressing apparatus. JOURNAL OF COMPARATIVE AND PHYSIOLOGICAL PSYCHOLOGY, 1950, 43, 41-48.

Schoenfeld, W. N., & Cole, B. K. STIMULUS SCHEDULES: THE t-τ SYSTEMS. New York: Harper & Row, 1972.

Schwartz, B., & Gamzu, E. Pavlovian control of operant behavior: An analysis of autoshaping and its implications for operant conditioning. In W. K. Honig & J. E. R. Staddon (Eds.) HANDBOOK OF OPERANT BEHAVIOR. Englewood Cliffs, N. J.: Prentice-Hall, 1977.

Sevenster, P. Incompatability of response and rewards. In R. A. Hinde & J. Stevenson-Hinde (Eds.), CONSTRAINTS ON LEARNING. London: Academic Press, 1973.

Shimp, C. P. Perspectives on the behavioral unit. Choice behavior in animals. In W. K. Estes (Ed.), HANDBOOK OF LEARNING AND COGNITIVE PROCESS. Vol. 2. Hillsdale, N. J.: Lawrence Erlbaum Associates, 1975.

Shimp, C. P. Organization in memory and behavior. JOURNAL OF THE EXPERIMENTAL ANALYSIS OF BEHAVIOR, 1976, 26, 113-130.

Shull, R. L. The response-reinforcement dependency in fixed-interval schedules of reinforcement. JOURNAL OF THE EXPERIMENTAL ANALYSIS OF BEHAVIOR, 1970, 14, 55-60.

Sidley, N. A., & Schoenfeld, W. N. Behavior stability and response rate as functions of reinforcement probability on "random ratio" schedules. JOURNAL OF THE EXPERIMENTAL ANALYSIS OF BEHAVIOR, 1964, 7, 281-283.

Skinner, B. F. THE BEHAVIOR OF ORGANISMS. New York: Appleton-Century-Crofts, 1938.

Skinner, B. F. Are theories of learning necessary? PSYCHOLOGICAL REVIEW, 1950, 57, 193-216.

Skinner, B. F. CONTINGENCIES OF REINFORCEMENT: A THEORETICAL ANALYSIS. New York: Appleton-Century-Crofts, 1969.

Smith, R. F. Topography of the food-reinforced key peck and the source of 30-millisecond interresponse times. JOURNAL OF THE EXPERIMENTAL ANALYSIS OF BEHAVIOR, 1974, 21, 541-551.

Spence, K. W. BEHAVIOR THEORY AND CONDITIONING. New Haven: Yale University Press, 1956.

Spence, K. W. The roles of reinforcement and non-reinforcement in simple learning. In K. W. Spence (Ed.), BEHAVIOR THEORY AND LEARNING: SELECTED PAPERS. Englewood Cliffs, N. J.: Prentice-Hall, 1960.

Staddon, J. E. R., & Frank, J. A. The role of the peck-food contingency on fixed-interval schedules. JOURNAL OF THE EXPERIMENTAL ANALYSIS OF BEHAVIOR, 1975, 23, 17-23.

Staddon, J. E. R., & Simmelhag, V. L. The "superstition" experiment: A reexamination of its implications for the principles of adaptive behavior. PSYCHOLOGICAL REVIEW, 1971, 78, 3-43.

Stebbins, W. C., Mead, P. B., & Martin, J. M. The relation of amount of reinforcement to performance under a fixed-interval schedule. JOURNAL OF THE EXPERIMENTAL ANALYSIS OF BEHAVIOR, 1959, 2, 351-356.

Surridge, C. T., & Amsel, A. Acquisition and extinction under single alternation and random partial-reinforcement conditions with a 24-hour intertrial interval. JOURNAL OF EXPERIMENTAL PSYCHOLOGY, 1966, 72, 361-368.

Terrace, H. S., Gibbon, J., Farrell, L., & Baldock, M. D. Temporal factors influencing the acquisition and maintenance of an autoshaped keypeck. ANIMAL LEARNING AND BEHAVIOR, 1975, 3, 53-62.

Thorndike, E. L. Animal intelligence: An experimental study of the associative processes in animals. PSYCHOLOGICAL MONOGRAPHS, 1898, 2, (4, Whole No. 8).

Thorndike, E. L. ANIMAL INTELLIGENCE: EXPERIMENTAL STUDIES. New York: Macmillan, 1911.

Tinklepaugh, O. L. An experimental study of representative factors in monkeys. JOURNAL OF COMPARATIVE PSYCHOLOGY, 1928, 8, 197-236.

Tyler, D. W., Wortz, E. C., & Bitterman, M. E. The effect of random and alternating partial reinforcement on resistance to extinction in the rat. AMERICAN JOURNAL OF PSYCHOLOGY, 1953, 66, 57-65.

Wagner, A. R. Effects of amount and percentage of reinforcement and number of acquisition trials on conditioning and extinction. JOURNAL OF EXPERIMENTAL PSYCHOLOGY, 1961, 62, 234-242.

Weinstock, S. Acquisition and extinction of a partially reinforced running response at a 24-hour intertrial interval. JOURNAL OF EXPERIMENTAL PSYCHOLOGY, 1958, 46, 151-158.

Williams, D. R., & Williams, H. Auto-maintenance in the pigeon: Sustained pecking despite contingent non-reinforcement. JOURNAL OF THE EXPERIMENTAL ANALYSIS OF BEHAVIOR, 1969, 12, 511-520.

Wolin, B. R.   Difference in manner of pecking a key between pigeons
    reinforced with food and with water.   CONFERENCE ON THE EXPERI-
    MENTAL ANALYSIS OF BEHAVIOR, Note #4 (mimeographed), April 5,
    1948.   (Reprinted in A. C. Catania [Ed.], CONTEMPORARY RESEARCH
    IN OPERANT BEHAVIOR.   Glenview, IL: Scott, Foresman & Co., 1968)

Woodruff, G., Conner, N., Gamzu, E., & Williams, D. R.   Associative
    interaction: Joint control of key pecking by stimulus-
    reinforcer and response-reinforcer relationships.   JOURNAL OF
    THE EXPERIMENTAL ANALYSIS OF BEHAVIOR, 1977, 28, 133-144.

Zentall, T. R., & Levine, J. M.   Observational learning and social
    facilitation in the rat.   SCIENCE, 1972, 178, 1220-1221.

6. REWARD TRAINING: LATENT LEARNING

M. E. Rashotte

Florida State University
Tallahassee, Florida, USA

Reward training has provided a forum for many developments in associative learning theory. The present chapter is the first of three which summarize the major empirical phenomena and theoretical advances derived from the study of latent learning, contrast effects and extinction. These developments have helped our thinking evolve to its current status about the functions of the reward event (S^V) and about the nature of associative processes in reward training.

At the time of the first latent learning experiments, S^V was understood to strengthen S-R associations in reward training. This view had been embodied in Thorndike's (1911) positive Law of Effect, and in later years came to be known as "S-R reinforcement theory." According to that theory, each occurrence of $S:R \rightarrow S^V$ yielded an increment in strength of the association between S and R, and performance indexed associative strength because the response became directly connected to the stimulus (Chapter 1).

Simple temporal contiguity of S and R without S^V was rejected as sufficient for substantive associative learning. In an experiment, Thorndike asked blindfolded humans to repeatedly draw a line 4 inches long and gave them no indication of the accuracy of their performance. There was no tendency for the length of the line drawn to converge on those lengths drawn most frequently on the initial trials. When the subjects were informed about their accuracy, responding quickly converged on the correct length. This type of result seemed indicative of the power of "effects" over "frequency" (Thorndike, 1932). In animal research, Gengerelli (1928) reasoned that if S-R contiguity were important, hungry rats forced to make repeated errors in a maze during non-reinforced runs should build up associative strength to perform those incorrect responses. After many such runs the rats were <u>forced</u> to run the correct pathway once or twice and were rewarded. On subsequent trials when the rats were free to choose the correct or incorrect paths they rapidly learned to choose the rewarded path, indicating a major role for "effects" and little role for S-R contiguity as a determinant of associative strength.

In this theoretical context, the latent learning experiments (and the "contrast" and "extinction" experiments to be discussed in the next chapters) stimulated theoretical controversy and experiment. This resulted in modifications to existing S-R reinforcement theory, proposals of new versions of S-R theory which no longer emphasized the strengthening effects of S^v, and the formation of an entirely new type of associative theory in which animals were said to learn that one stimulus event follows another.

Early Latent Learning Experiments

<u>Non-rewarded trials procedure</u>. Blodgett (1929) systematically studied how a series of non-rewarded trials in a complex maze influence performance when reward is subsequently introduced. These experiments led him to coin the term "latent learning". One of Blodgett's experiments was conducted in the multiple T-maze whose floor-plan is shown in Figure 1. A trial began when a rat was placed in the startbox (lower left on the floor-plan) and ended shortly after the rat reached the goalbox (upper right). One trial was given each day and the rats were always trained hungry.

Figure 1 shows that for a group fed in the goalbox on every trial (Group I) there was a gradual reduction in errors as training progressed. This result is consistent with the general expectation of S-R reinforcement thoery that associative strength grows as a negatively accelerated function of the number of rewarded trials. Two other groups ran from startbox to goalbox 3 or 7 times before rewarded trials began. On these non-rewarded trials the rats were simply detained for two minutes in the empty goalbox and removed to a holding cage where they were fed 1-hour later. Figure 1 shows that there was little improvement in performance on non-reward trials, a result consistent with the S-R reinforcement view. However, after only one rewarded trial the performance of those groups improved to approximately the current level of Group I which had been rewarded on all trials. S-R reinforcement theory predicted a gradual reduction in errors once rewarded trials began, as found in Group I, not the sudden improvement in performance shown by

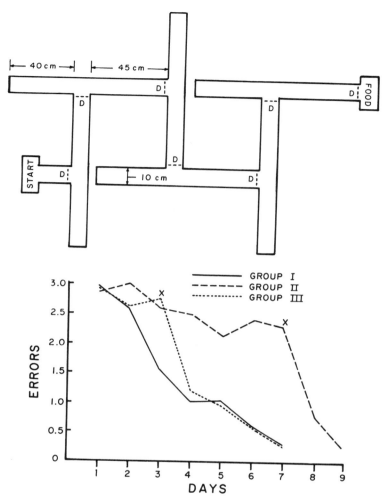

Figure 6.1   The top panel shows one of the mazes used in Blodgett's
study of latent learning. "D" indicates the location
of a door which closed behind the animal after it passed
that point in the maze.   The bottom panel shows data
obtained from three groups given one trial a day in
that maze.   Group I was rewarded with food when it
reached the goalbox on every trial.   Rewarded trials
began for Groups II and III on the trial marked with
an X.   The dependent variable is the number of choice
points in which the rats made at least one entrance of
as much as a body's length (not counting tail) into a
blind alley.   (After Blodgett, 1929.)

these groups. Comparable results have been reported by others using approximately the same procedure (e.g. Tolman & Honzik, 1930).

On the face of it, Blodgett's data indicate that rats learn about the correct pathway to the goal whether they receive S^V when they reach the goalbox or not. S^V seems to allow learning to be expressed in performance (hence, the need for a learning-performance distinction). This, in fact, was the interpretation proposed by Blodgett and by Tolman as the "latent learning" hypothesis which appears to contradict the assumption of S-R learning theory that reinforcement is necessary for the formation of associations. There are, however, two alternative hypotheses which are more compatible with S-R reinforcement theory and which need to be rejected before "latent learning" is accepted.

Once is that the relatively sudden improvement in performance found after only one rewarded trial in Blodgett's non-rewarded groups may not reflect prior learning at all. The argument is that before reward training began, non-rewarded groups were handled more, and also had more opportunities to habituate emotional responses to the maze, than did the group rewarded from the outset. Thus, performance on rewarded trials may have been facilitated in some way by these non-associative aspects of pre-training rather than by learning which was supposed to have occurred on non-rewarded trials. It seems safe to reject this possibility. For one thing, Blodgett (1929) obtained strong evidence against it by showing that a group run backwards through the maze (from the goalbox to startbox) on non-rewarded trials failed to show a sudden improvement in performance when running forward through the maze was rewarded. For another, even when it was later demonstrated that handling and simple exposure to the maze do contribute to the sudden drop in errors when reward is first introduced, it was also demonstrated that these factors are not sufficient to account for the entire latent learning effect (Karn & Porter, 1946).

The second possibility is that Blodgett's procedure may have allowed some unrecognized sources of S^V to operate on the otherwise non-rewarded trials. For example, simply removing the animals from the empty goalbox, or placing them in the holding cage where they are later fed, may have rewarding properties which allow at least a weak association to be formed between the stimuli of the maze and the responses which eventuate in an animal's reaching the goalbox. In an effort to test this possibility Haney (1931) devised a second type of latent learning procedure.

Continuous non-rewarded exposure procedure. Haney (1931) arranged that non-rewarded exposures to the maze occur in extended blocks of time rather than on trials. The logic is that rats which live in the maze for many hours will have the opportunity to learn about the maze without the possible advantage of unintentional reward after each successful run through the maze in the pre-reward period. Haney allowed rats to live as a group in a complex maze for 18 hours on each of four successive days, providing food and water midway during their daily 6-hour period out of the maze. A control group was treated in the same way except that it lived in a simple rectangular maze. Subsequently, each animal in both groups was trained, one trial a day, to leave the startbox of the complex maze and find food in the goalbox. The results showed that the group pre-exposed to the complex maze made fewer errors than the control group.

Haney's result is consistent with the latent learning hypothesis. Unfortunately, it does not rule out the possibility that reward operated in Blodgett's trial procedure. Neither is it free of alternate interpretations. For example, Haney's design leaves open the possibility that the control group performed more poorly in the complex maze because it did not habituate emotional responses to the maze before rewarded trials began. This seems particularly likely because the control group performed more poorly than the experimentals on the very first rewarded trial in the complex maze, before reward was introduced. Whatever the case, Blodgett's and Haney's experiments were important because they helped stimulate theoretical developments which led to more carefully designed types of latent learning experiments. These, in turn, contributed more to our understanding of the function of S^V in reward training.

Theoretical Developments

Three theoretical developments related to latent learning are discussed below. The first is that a formal statement of S-R reinforcement theory was revised to deal with some of the latent learning data. The second is that a cognitive learning theory was developed which accommodated the latent learning findings and stimulated a variety of experiments intended to challenge S-R theory. The third is that a modified version of S-R theory was proposed in which conditioned responses were invoked to account for some features of performance. These three developments provide background for latent learning procedures discussed later in this chapter, and for the topics of contrast effects and extinction which are discussed in Chapters 7 and 8.

Formal S-R Reinforcement Theory and Its Revision. Hull incorporated the S-R reinforcement principle into a formal set of theoretical statements designed to account for mammalian behavior (Hull, 1943). In presenting the theory, his strategy was to state in a first volume the principles from which complex behavior could be derived in a rigorous fashion. He intended to provide detailed derivations of complex phenomena in later volumes but managed to complete only one volume which was published posthumously (Hull, 1952). To derive phenomena such as latent learning, Hull was forced to modify the original principles and, by reviewing the details of his revision, we can illustrate one way in which the latent learning experiments had an impact on S-R reinforcement theory.

The associative element in Hull's theory was symbolized by $_SH_R$ or "habit strength," a term which incorporated the view that associations form between stimulus and response. Postulate 4 of the 1943 theory restated the Law of Effect in a more detailed form. It asserted that increments in habit strength occur when an S-R sequence is followed by S^V, and that successive increments summate to yield a simple positive growth function relating "strength of $_SH_R$" to the "number of times the S-R event has been followed by S^V." The asymptote of the function was positively related to magnitude of reinforcement and negatively related to delay of reinforcement. Thus, in the original set of principles S^V was entirely a "learning" variable, influencing only associative strength.

In Hull's theory, habit strength combined with other theoretical variables to yield a variable known as "excitatory potential" ($_SE_R$) which was directly related to performance. An abbreviated version of Hull's formulation is given by the equation,

$$_SE_R = {_SH_R} \times D$$

in which the associative variable ($_SH_R$) combines in a multiplicative fashion with the motivational or "drive" variable (D) to yield $_SE_R$, whose value determines the strength of performance of response. It may be noted that this formulation incorporates the learning-performance distinction.

In a situation such as Blodgett's where reinforcement is absent (or minimal) on non-rewarded trials Hull's theory demanded that no (or, at best, weak) $_SH_R$ develop prior to the introduction of reward. The theory also demanded that once rewarded trials began, habit strength should grow gradually to the limit specified by the magnitude and delay of S^V. Clearly, the theory was unable to handle Blodgett's data without some revision. Although several avenues were open, Hull (1952), chose to revise the theory by making S^V a performance variable as well as a learning variable.

In its new form, the relevant portions of the theory were stated,

$$_SE_R = {}_SH_R \times D \times K$$

which embodied two major changes from the original. First the strength of $_SH_R$ was determined by the number of reinforced trials but not by the parameters of S^V. Second, a new variable, K ("incentive motivation"), was added as a multiplier of $_SH_R$, and the value of K was determined by the magnitude of S^V. The other terms in the formulation were left unchanged. The significance of this revision is that it maintains the S-R reinforcement principle by continuing to insist that S^V is necessary for learning, but provides an additional role for S^V as a performance variable which, like drive (D), multiplies $_SH_R$ to determine the value of $_SE_R$ and, thereby, the strength of behavior. In this formulation a sudden change in the value of K would produce an abrupt shift in performance so that the theory could now encompass sudden shifts in performance following a change in the magnitude of reinforcement. In fact, this revision of Hull's theory was largely prompted by experiments on the effects of shifts in reward-magnitude to be discussed in the next chapter.

As revised in 1952, Hull's theory attempted to account for the Blodgett-type result in the following way. It assumed that some minimal reinforcement occurred on non-rewarded trials (e.g., through the rat's being removed from the maze). Thus, there should be an increment in $_SH_R$ on each trial, just as in the rewarded group, and all groups should be equal in $_SH_R$ and D on every trial. The theory predicts differences in performance because of different magnitudes of reinforcement and, therefore, different values of incentive motivation, K; performance should be poor on non-rewarded trials relative to rewarded trials because K is so much smaller when the animal is simply removed from the goalbox than when it is fed there. Finally, by assuming that K can change abruptly from a low to a high value after only one rewarded trial, Hull was able to "derive" from this theory the sudden improvement in performance found in Blodgett's non-rewarded group when S^V was first introduced.

Although Hull's derivation was plausible it shed little light on the nature of latent learning and, in fact, was not particularly influential in guiding additional research on the phenomenon. However, his attempt to deal with latent learning was a clear sign that the simple version of S-R reinforcement theory encompassed by the Law of Effect was not adequate to account for some forms of learning. In particular, it seems incorrect to conceptualize S^V solely as a source of associative strength without, in some way, acknowledging its motivational role.

S-S contiguity theory. The latent learning experiments were
among several lines of evidence which led Tolman to formulate a
theory that contrasted sharply with the S-R reinforcment position
(Tolman, 1932). Key features of the theory are illustrated in its
treatment of learning in a multiple-unit T-maze, such as that used
by Blodgett (Figure 1). The basic idea is that the animals learn
"what-leads-to-what" as they run through the maze ("sign-Gestalt
expectations" in Tolman's terminology). On the very first exposure
to the maze the stimulus conditions at each choice point arouse a
set of varied expectations about the events that might occur if
the animal turned one way or the other. For example, a hungry
animal might expect to find food, no food, stimuli to explore,
no stimuli to explore, and so on. These original expectancies
were thought to come from various sources, including the animal's
current deprivation state and its past experiences in similar en-
vironments. The theory states that as a consequence of running
through the maze these original expectations are "refined" to conform to the actual situations the animal encounters.

Figure 2 shows, in a schematic diagram, the assumed learning
about the correct pathway in the maze early and late in training.
The learned expectancies are symbolized by the ellipses. After a
few training trials, the stimulus conditions at the first choice
point, S_1, are shown to evoke an expectation that certain features
of that stimulus condition (O_1) when reacted to in a certain way
will result in ($\rightarrow$) a second stimulus condition (O_2). That is, the
animal has learned that a right turn at S1 results in its being
exposed to S2. (For simplicity, the diagram does not show that the
animal also learns that turning left results in a cul de sac). The
theory asserts that the "S1-leads-to-S2" expectation will increase
towards asymptotic strength every time the animal turns right at
S1 and encounters S2 (which "confirms" the correctness of the expectancy). Reinforcement, in the sense of an hedonic event, is not
necessary for the formation of the expectancy. The same process is
shown as occurring at each choice point of the maze and at the
end of the maze where the sight of the goalbox (S4) evokes an expectancy that food will be found there. After more trials, the
figure shows, the individual expectancies evoked in various parts
of the maze should fuse so that S1 will evoke an expectancy about
the entire correct pathway and food. In different words, the net
result of repeatedly running through the maze is that the animal
is left with a so-called "cognitive map" of the maze environment.

In Tolman's theory, S^V was a performance variable which func-
tioned to change the "demand" of the stimuli in the goalbox. At
the choice point of a maze, the expectancy which includes the more

1. EARLY IN TRAINING

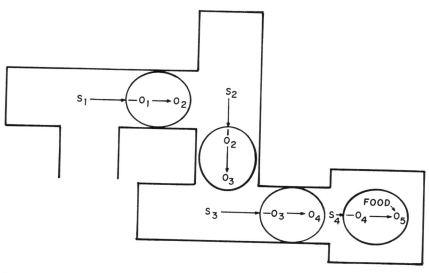

2. LATER IN TRAINING

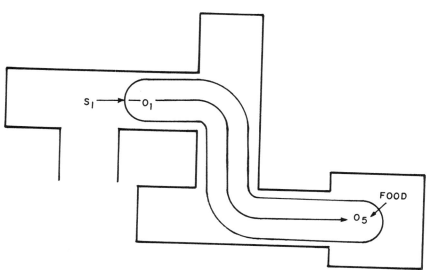

Figure 6.2   Schematic representation of sign-Gestalt expectations
formed for the correct pathway of a maze early and
later in training.   (After Tolman, 1932, p. 147.)

demanded stimulus will control behavior. For example, an expectancy that food will be found following a right turn will have
greater demand than an expectancy that a cul de sac will be found
following a left turn. Consequently, the animal will turn right.

Tolman's account of Blodgett's latent learning experiment
was as follows. On non-rewarded trials the animals should form
expectancies about the correct pathway of the maze, but should
not show much improvement in goal-directed performance because
the goalbox stimuli at the end of that pathway have little more
"demand" than do other stimuli in the maze. On the first rewarded
trial, however, the goalbox stimulus is associated with food and
the theoretical consequence is that the goalbox stimulus, which
is already incorporated in the animal's cognitive representation
of the maze, becomes the most demanded stimulus. Accordingly,
on the very next trial the cognitive representation of the correct
pathway will control behavior more so than representations of incorrect pathways, and the animal will show a sudden reduction in
errors.

Tolman's view of maze learning can be distinguished from the
S-R reinforcement view in two principal ways. It asserts that
expectancies (i.e., stimulus-stimulus associations) form rather
than stimulus-response associations; and, it asserts that these
expectancies form through temporal contiguity rather than through
the action of a hedonic reinforcing event. Hence, Tolman's position is often termed "S-S contiguity" theory. The history of
learning theory between 1930 and about 1960 shows that the influential positions of Tolman and Hull guided most of the theoretically motivated research in reward training (Hilgard & Bower, 1975).
The ability of these theories to make unique predictions narrowed
over the years, principally because Tolman's theory stimulated
research which forced revisions in Hullian theory.

One development in S-R theory which provided considerable
leverage was Hull's concept of the classically conditioned goal
response. This concept played an important role in conceptualizing
new types of latent learning experiments and its details are summarized below.

S-R theory: The classically conditioned goal response. In
early papers, Hull outlined ways in which complex behavioral phenomena which we label "knowledge," "foresight," "purpose" and "expectancy" might be reduced to stimulus-response terms (Hull, 1930,
1931). An important idea in these analyses was that the response
which terminates a given behavior sequence becomes anticipatory,

intruding into earlier parts of the sequence through a process of classical conditioning. In this view, the stimuli which regularly precede the goal are analogous to Pavlovian CSs, and the goal-stimulus and goal-response are analogous to the US and UR. Hull symbolized the goal stimulus-goal response sequence as S_G-R_G, and the conditioned goal response as r_G.

Broadly speaking, the principle of stimulus-substitution led Hull to expect that the form of r_G would be highly similar to the form of R_G. However, the similarity may not be great in some cases. For one thing, only those portions of the goal response which can "split off" and occur independently of the goal stimulus would be evoked by the pre-goal stimuli. For example, in a maze where the goal response requires the animal to turn right and eat food, the animal might run along the right side of a straight pre-goal pathway which has no right-branching turns, and might make chewing movements or salivate as it runs. For another thing, Hull reasoned that since the goal stimulus resulted in satisfaction of the animal's need, it would be maladaptive for r_G to interfere with the animal reaching the goal. Thus, he believed that should an incompatible form of r_G emerge through stimulus-substitution, that form would be modified until it is compatible with performance of the instrumental response. However, even in cases where r_G receded to a covert level Hull's reasoning required that r_G would be distinctively related to the goal event and would obey the principles of Pavlovian conditioning. The importance of this point becomes evident in Hull's utilization of r_G in formulating an S-R analogue of the cognitive theorist's "expectancy".

Hull's logic was as follows. In a maze where the animal turns right and eats in the goalbox, the stimuli in the initial sections of the maze should, through conditioning, come to evoke right-turning and consummatory responses as r_G. As with any other response, an occurrence of r_G will result in proprioceptive feedback stimuli which are acknowledged in the complete designation of r_G as "r_G-s_G." When r_G-s_G occurs in the initial segment of a maze it may be said that the animal experiences a representation of the goal response performed on previous runs through the maze (within the limits noted above). That is, the classically conditioned r_G-s_G is the S-R theorist's analogue of "expectancy" of a goal, and one way in which this expectancy-like process might come to control the instrumental response is shown in Figure 3.

Early in training the stimulus in the initial portion of a maze (S1) evokes ($\rightarrow$) an instrumental response (R_1) which results in ($\leadsto$) exposure of the animal to the next part of the maze (S2), and so on, until the animal encounters the goal stimulus (S_G) which

evokes the goal response (R_G). Because of this arrangement the
pre-goal stimuli (S1, S2, and S3) may stand in a temporal relationship to S_G-R_G which is conducive to the formation of associations
(---→) between these stimuli and the goal response through classical conditioning. Once the pre-goal stimuli evoke a conditioned
goal response (panel 1B of Figure 3), the feedback stimulus of r_G-
s_G should become conditioned to the instrumental response, as would
other stimuli present during reward training. When this has occurred, the associative determinants of instrumental performance include a representation of the goal response as shown in the bottom
panel of Figure 3. In this way, it is possible to argue that S-R
processes can account for behavioral phenomena which seem to impli-
cate animals' "expectancy" or "anticipation" of the goal as a factor which directs instrumental behavior.

Hull's analysis was very clever and proved to be an important
theoretical step which allowed S-R theory to compete with cognitive
theory in accounting for many phenomena, including some latent
learning data to be discussed later in this chapter. However, despite its apparent objective footing in principles of classical conditioning there are several fundamental uncertainties about r_G-s_G.
One is that the nature of the presumed CSs which evoke r_G-s_G is
obscure. The stimuli which accompany instrumental responding are
undoubtedly stimulus compounds comprised of drive-related stimuli
(e.g., hunger cramps), proprioceptive feedback stimuli from the
instrumental response, and exteroceptive stimuli from the training
apparatus (Hull, 1931). Thus, it is uncertain that discrete stimulus compounds can be identified as neatly as implied by the terms
S1, S2 and S3 in the schematic diagrams shown in Figure 3. Even
so, at the time Hull outlined this analysis there was little ap-
preciation of overshadowing, blocking and related processes as factors in conditioning with stimulus compounds (Chapter 4), factors
which greatly increase the complexity of the analysis. A second
problem is that assertions about the conditions under which stimuli
distant from the goal evoke r_G-s_G have little empirical content
since the nature of classical conditioning to stimuli in a series
of CSs preceding the US is not well understood (Baker, 1968; Razran,
1965). This is especially true when the individual CSs are complex
stimulus compounds whose duration and temporal relation to the US
may be variable from trial to trial (because the CS parameters depend entirely on the animal's behavior). Third, it is not certain
what components of the goal response should be included as conditionable in the form of r_G-s_G. Hull emphasized both the motor and
the consummatory behavior in the goalbox as conditionable to pre-
goal stimuli, an emphasis which Miller (1935) used to predictive
advantage (see below). However, Spence (1956) principally emphasized the consummatory component of the goal response as condition-

1. EARLY IN TRAINING

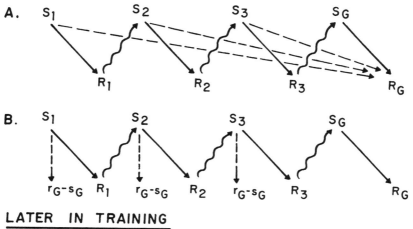

2. LATER IN TRAINING

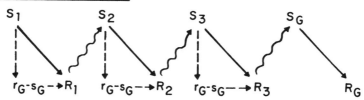

Figure 6.3   Schematic representation of the development of classi-
              cally conditioned goal responses during instrumental
              training.

able in his theory. Fourth, the manner in which r_G-s_G influences
instrumental behavior remains to be clarified. Figure 3 shows
Hull's (1931) emphasis in which the feedback stimulus of r_G-s_G was
presumed to become directly conditioned to the instrumental response. Hull (1952) attributed conditioned reinforcement to the
action of r_G-s_G. Later, Spence (1956) proposed that r_G-s_G also
influences instrumental responding in a motivational way when he
tied the strength of the incentive motivation variable, K, in
Hullian theory to the strength of r_G-s_G. More recently, the autoshaping literature has shown that classically conditioned stimuli
can directly evoke motor responses (Brown & Jenkins, 1968; Moore,
1973; Rashotte, Griffin, & Sisk, 1977) and, so, r_G-s_G may be expressed in instrumental responding in this way.

These uncertainties about the classically conditioned goal
response are major ones and leave the concept open to the charge
of "vagueness" which is often made about the concept of "expectancy" in Tolman's theory (e.g., MacCorquodale & Meehl, 1954).
Nevertheless, the concept proved to be influential for several
reasons. It allowed S-R theory to compete more effectively with
Tolman's cognitive theory, it proved to be an important idea in
later experiments on latent learning (and "contrast" and "extinction" discussed in Chapters 7 & 8), and it explicitly recognized
the potential for classically conditioned responses to influence
performance in reward training, an idea that is currently fashion-
able (Chapter 5).

Later Experiments on the Latent Learning Hypothesis

One series of experiments examined the latent learning hypothesis with what have been termed "irrelevant-incentive" procedures.
The question posed was whether rats can learn the location of a
reward object (e.g., food) in a maze when they encounter that object while satiated for it. Tolman's theory implied that learning
should take place under these conditions. A number of experiments
on this problem by Spence's group, and by others, brought new theoretical emphases to the topic of latent learning and to reward
training in general.

A second type of experiment was pioneered by Tolman (1933) who
studied the effects on instrumental responding of directly placing
the animal in the goalbox and pairing the goalbox stimuli with a
new reinforcement condition. This so-called "placement" procedure
also produced important information about the way in which S^V influences instrumental performance.

As a rule, these later experiments on latent learning were conducted in relatively simple training apparatuses because the mazes of Blodgett, Haney, and others proved to be too complex to permit isolation of all the factors thought pertinent. Hence, most later experiments were conducted in single-unit T-and Y-mazes, and even in straight runways.

Irrelevant-incentive procedure: Balanced drives. In an experiment completed in 1940 (Spence & Lippitt, 1940) and reported in full some time later (Spence, Bergmann, & Lippitt, 1950), rats were trained in the Y-maze shown in Figure 4 while satiated for food and water. (To motivate the rats to run it proved necessary to place them briefly in a cage with other rats when they were removed from the goalboxes). The goalbox in one arm of the maze had food pieces scattered on the floor, the other goalbox had a water spout. Each animal received free-choice and forced-run trials designed to insure that they ran 14 times to each goalbox _and_ that on each run they noticed the food or water objects there (i.e., the animals always sniffed at the food pieces, and if they did not spontaneously explore the water spout they were placed in front of it). During these trials the animals were never observed to eat the food or lick the spout in the goalbox of the mazes. On the reasonable assumption that the rats noticed these "irrelevant" reward objects in the goalboxes, Tolman's expectancy theory would assert that the rats learned that one arm leads to food and the other to a water spout. To test for this learning the rats were subsequently made hungry or thirsty and allowed to choose between the arms. Care was taken to insure that side-preferences could not account for the outcome so that the experiment could clearly determine whether the animals use their presumed learning about the locations of food and water when these objects are later made "relevant" incentives through deprivation.

Figure 4 shows that on the seven pre-test days when the animals were satiated for food and water, less that 50% of them chose the side whose goal-object was to be made relevant through deprivation on the first test day. On the first trial of the test day, there was a marked shift in preference towards the side leading to the now-relevant incentive (61.5% of the rats chose the relevant arm), a change in preference that was statistically significant. Subsequently, the animals received a second test with reversed deprivation conditions and, while the preference reversed, it was not significantly different from the pre-test baseline. The findings of this experiment were replicated in all essentials by Meehl and MacCorquodale (1948).

The first test-day's data shown in Figure 4 demonstrate that

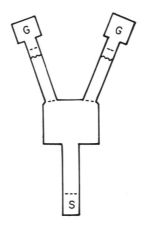

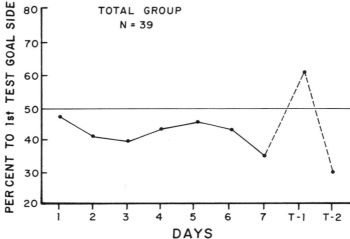

Figure 6.4   The top panel shows the floor plan of the maze used to
              test "irrelevant incentive" learning by animals sati-
              ated for food and water. The start box is marked by
              S, the goalboxes by G. (After Spence & Lippett, 1946.)
              The bottom/ panel shows the percentage of animals
              choosing the side of the maze which was appropriate
              for the deprivation condition of the first test. The
              points plotted are the percentage choosing that side
              on the first choice trial of each day. The animals
              were satiated for food and water on days 1-7. They
              were deprived of food (or water) on test-day 1 (T-1).
              The deprivation conditions were reversed on T-2.
              (After Spence, Bergmann, & Lippitt, 1950.)

rats can learn the location of food and water in a maze while those objects are irrelevant to their present need states. This finding is in agreement with Tolman's (1932) theory which asserts that animals form expectancies about "what-leads-to-what" through simple contiguity of events. Spence et al. (1950) showed how these findings could also be interpreted in S-R terms by appealing to the classically conditioned goal response. They argued that when the rat encounters food or water in the goalboxes when satiated, the sight of these stimuli evokes a weak conditioned consummatory response of eating (r_E) or drinking (r_D) and, therefore, the stimuli of each arm of the maze should become conditioned to evoke r_E or r_D, as well. Following the logic of r_G-s_G theory, Spence et al. argued that the distinctive proprioceptive feedback stimuli from each type of conditioned goal response should become conditioned to responses in the left and right arms of the maze (e.g., r_E-s_E----→ R_{left}; r_D-s_D-----→ R_{right}). Given this possibility, they argued that when a drive state was introduced for the test, it should make the conditioned goal response appropriate to that drive occur more vigorously, thereby increasing the strength of its feedback stimulus (e.g., when deprived of food, r_E-s_E > r_D-s_D). Then, the choice of the test trial was accounted for by assuming that the stronger feedback stimulus excites its maze-arm response (e.g., R_{left}) more strongly than does the other feedback stimulus and, therefore, the animal should choose the appropriate side of the maze.

While the S-R account of these findings is complicated and makes many assumptions, it is probably as reasonable as the "expectancy" account which appears less complex but, in fact, leaves many issues to be resolved (e.g., how "expectancy" translates into behavior). In any event, this type of experiment encouraged use of the r_G-s_G construct in theoretical analyses of reward learning and, ultimately, led Spence (1956) to a revised version of Hullian theory in which r_G-s_G played an important role.

Irrelevant-incentive procedure: Unbalanced drives. Spence and Lippitt (1946) tested the rat's ability to learn the location of an irrelevant incentive in a maze while highly motivated to obtain a different goal object. They pointed out that Tolman's theory implies that animals will learn "what-leads-to-what" irrespective of their prevailing motivational state, contiguity of events being the principle through which expectancies form. Accordingly, they evaluated Tolman's theory by training two groups of thirsty rats in the Y-maze shown in Figure 4, with water in the right arm and food (Group F), or nothing (Group O) in the left. Both groups were satiated on food when they ran in the maze. The procedure mixed free-choices with forced trials to insure that the animals had approximately equal experience with both arms of the maze. Group F was never observed to eat the food when it entered the left goalbox. Of course, the "correct" response for the rats on free-choice trials was to choose the right maze-arm where they could partially relieve their thirst. The rats readily learned this discrimination.

The test for expectancy learning came subsequently when the rats were made hungry and both groups received food in the left goalbox while water continued to be available in the right. According to Tolman's theory, Group F, which had encountered food in the left goalbox while thirsty, should have learned the location of food and therefore should choose the left arm on the very first trial when tested under hunger motivation. Figure 5 shows that none of the 10 animals in that group performed as the theory expected; on the first trial all animals continued to choose the right arm leading to water. Group O rats performed the same way on the first trial. Worse for expectancy theory, Figure 5 shows that in learning the new discrimination (go-left-for-food), Group F showed no advantage over Group O even though, theoretically, the former group should already have a well developed expectancy that food is in the left arm.

Spence and Lippitt noted that while their finding challenged the adequacy of Tolman's theory, it is consistent with S-R reinforcement theory, such as that originally outlined by Hull (1943). According to that theory, the response of turning right should have been strongly reinforced by water in both groups during initial training because both were thirsty; neither group should have had the left-turning response strengthened because it did not lead to water. Subsequently, when they were required to learn a left-turn response to obtain food while hungry, both groups should have retained an associative tendency to turn right which had to be overcome by reinforced trials in the left arm before correct (i.e., left) choices could predominate. Spence and Lippitt (1946) made no appeal to r_G-s_G in accounting for their results in S-R terms, presumably because the data could be explained without it.

Tolman's response to the Spence and Lippitt (1946) result, and others like it, was to assert that when animals are highly deprived they tend to focus on the relatively narrow problem of finding stimuli related to their need state. Hence, under strong deprivation conditions they learn narrow "strip maps" rather than broad, comprehensive cognitive maps of their environment. Hence, the lack of learning about the location of food in the Spence and Lippitt experiment (Tolman, 1948).

The theoretical problems posed by the two irrelevant-incentive procedures are great. One set of principles does not seem to cover both and, perhaps, it is an error to lump them together as instances of a common process, latent learning (MacCorquodale & Meehl, 1954,

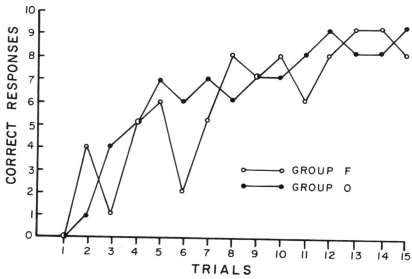

Figure 6.5   Number of animals in each group choosing the left
(food-reward) arm of the maze under food deprivation.
Perfect performance is a score of 10. Both groups had
previously learned to choose the right arm to obtain
water when under water deprivation. During that
training, Group F found food in the left arm when it
ran there, Group O found an empty goalbox. (After
Spence & Lippitt, 1946.)

p. 212). Of the four latent learning procedures reviewed above,
the balanced-drive irrelevant-incentive procedure and the Blodgett
procedure provide the best evidence that animals can learn by responding to stimuli in their environment with only weak (or possibly no) reward present. The large array of experiments conducted
with the Blodgett, Haney, and irrelevant-incentive procedures has
been reviewed in greater detail than is possible here by Thistlethwaite (1951) and MacCorquodale and Meehl (1954).

Goalbox-placement procedure. The final procedure to be reviewed here was first employed by Tolman (1933) in an attempt to
test his theory. Tolman taught rats to choose between two pathways
leading from a choice point. The rats always found food when they
chose the pathway whose entrance was covered by a white curtain;
they encountered a blind alley and electric shock when they ran
through the black-curtained entrance. The rats were overtrained
on the discrimination so that, in Tolman's terms, they should have
had a well formulated expectancy that running through the white

curtain leads to the goalbox and food. Tolman reasoned from his theory that if the rats were placed directly in the goalbox <u>without</u> <u>running</u> <u>the</u> <u>maze</u> and given electric shocks, the goalbox stimuli would acquire a negative connotation for the rat. Consequently, he predicted that when they were returned to the choice point on a subsequent trial the white curtain should activate an expectancy of the goalbox in which shock had just occurred and the animal should refuse to run. Tolman proposed that such a result would confound S-R theory which demands that behavior changes occur as a result of responses being performed.

The results of this experiment were a surprise to Tolman. In his own words,

> Each rat after having been shocked in the food compartment and then carried to the starting point, immediately dashed off gaily and just as usual through the whole discrimination-apparatus and bang whack into the very food compartment in which it had just been shocked.
>   [Tolman, 1933]

To make matters worse for expectancy theory, Tolman found that when he shocked the rats in the goalbox after they had <u>run</u> to it, the rats refused to run on the next trial. Tolman (1933) saw in this result a failure of expectancy theory and a partial victory for S-R theory. However, he suggested that expectancy theory might still be able to account for the result, given suitable (but presently uncertain) changes in its assumptions.

In a subsequent paper, Miller (1935) used the logic of the classically conditioned goal response to show the conditions under which Tolman's placement procedure should work as expectancy theory predicts. Miller's work provides an early and excellent illustration of how r_G-s_G played the same role in S-R theory as "expectancy" played in cognitive theory. Miller reasoned as follows. After extended discrimination training of the sort given by Tolman (1933), the stimulus at the choice point leading to the goalbox should evoke a classically conditioned form of the goal response and the feedback stimulation from r_G-s_G should be conditioned to running down the correct pathway. The general nature of this process was shown in Figure 3. When the rat is placed directly in the goalbox and shocked, feedback stimulation from the goal response should become conditioned to the motor reactions which occur during and following shock (e.g., crouching, withdrawal, etc.). Subsequently, when the animal comes to the choice point and r_G-s_G is evoked by the stimulus leading to the goalbox, the feedback stimulus, s_G, should evoke crouching, withdrawal, etc. (because of its previous conditioning in the goalbox), and the animal should refuse to run. Thus, using r_G-s_G, Miller was able to make the same prediction from S-R theory as Tolman had made from cognitive theory.

However, knowing from Tolman's (1933) work that the prediction is not confirmed, Miller went on to show how this might be understood from the perspective of S-R theory. He noted that a key element in the r_G-s_G account is that the feedback stimulus of r_G-s_G becomes conditioned to the competing responses evoked by shock. If the goal response produced only a weak feedback stimulus, that stimulus might not be sufficiently strong to become conditioned to the competing responses, in which case the animal should run readily to the goalbox when subsequently placed at the choice point (i.e., Tolman's result should occur). According to this logic, the animal should behave as Tolman predicted if the goal response is made highly distinctive, thereby insuring a strong feedback stimulus. Miller tested this reasoning in an experiment whose procedure is shown schematically in Figure 6.

Rats deprived of both food and water were first trained to run down a straight runway to obtain food in a goalbox designed to produce a distinctive goal response (R_G): the rats had to climb an inclined plane, make a sharp right turn and then obtain food. The runway had curtains hung at several points along its length so that an observer could record whether the rats passed through the curtains on the right side, thereby providing an objective measure that the right-turn component of the goal response was evoked by the runway stimuli, as specified by r_G-s_G logic. After extended training to insure that r_G-s_G was fully developed along the length of the runway and, presumably, conditioned to locomotion (Figure 3), the animals were assigned to different groups to receive shocked placements in a goalbox. Figure 6 shows that in this second stage of the experiment half the rats were placed in the original goalbox and shocked, thereby allowing feedback stimulation from the original goal response to be conditioned to the shock-evoked responses. The other rats received shock in a goalbox which required a very different response from that made in the original goalbox: they ran straight in on a flat surface, turned left and drank water. The prediction was that when both groups of rats are subsequently tested in the runway, the group shocked while making the original goal response will show greater interference with running towards the goal than will the group shocked while making the new goal response. (In fact, the experiment counterbalanced the goalbox conditions so that the results did not depend on the specific goal response used in initial training.)

1) Train

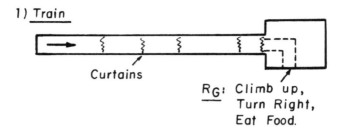

Curtains

R_G: Climb up,
       Turn Right,
       Eat Food.

2) Direct Placement in Goal Box + Shock.

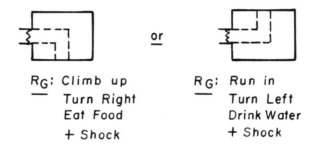

or

R_G: Climb up R_G: Run in
       Turn Right   Turn Left
       Eat Food   Drink Water
       + Shock   + Shock

3) Test

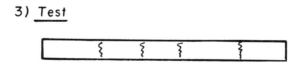

Figure 6.6   Design of Miller's (1935) experiment.

The result of the experiment is shown in Figure 7. The top panel summarizes Miller's evidence from the initial training trials that a conditioned form of the goal response was evoked by the runway stimuli. The percentage of animals passing under the curtain on the side on which the goal-response turn was made increased as the animals neared the goalbox. This result would be expected from conditioning principles. The bottom panel shows the time required to run the alley. On the final trial before shock-placement was given, the animals ran quickly to the goalbox. On the first test trial after shock the rats shocked in the original-training goalbox showed a large increase in time to run the runway, demonstrating clearly the kind of effect Tolman (1933) had expected in his experiment. The rats shocked in the novel goalbox showed some disruption of running, but much less than the experimental group. Thus, Miller (1935) was able to use the terms of an S-R analysis to predict the conditions under which expectancy-like effects should occur.

In view of Miller's success, S-R theory need not be embarrassed by evidence that performance changes can result when goal-events are manipulated independently of the occurrence of the instrumental response which produces them. Thus, what at first appeared to be one major predictive advantage of expectancy theory over S-R theories was eliminated by the r_G-s_G construct. In fact, r_G-s_G was largely responsible for blurring the differences between the two theoretical approaches.

A variety of "placement" procedures have now been employed, all of which indicate that instrumental performance can be modified by direct goalbox placements. For example, Seward (1949) employed a T-maze with distinctive goalboxes on each arm. He showed that after several non-rewarded runs in the maze, rewarded placement in one of the goalboxes resulted in most rats choosing the arm leading to that goalbox. This is a kind of latent learning experiment similar in design to Blodgett's, except that a simple T-maze was employed and reward was introduced independently of the instrumental response. While Seward's result is amenable to an expectancy analysis, it can also be interpreted in terms of S-R theory augmented by the r_G-s_G concept (Spence, 1956, p. 147). Tolman and Gleitman (1949) showed that rats which ran to reward in distinctive goalboxes of a T-maze and were later shocked in one of them, predominantly chose the arm leading to the other goalbox on a test trial. This result is similar to Miller's (1935) finding, except that a choice response was used. Finally, Seward and Levy (1949) showed that a well learned instrumental running response rewarded with food can be weakened by nonrewarded placements in the goalbox. This finding has been termed "latent extinction" and is accounted for both by Tolman's theory and by S-R theory with the aid of r_G-s_G (Moltz, 1957).

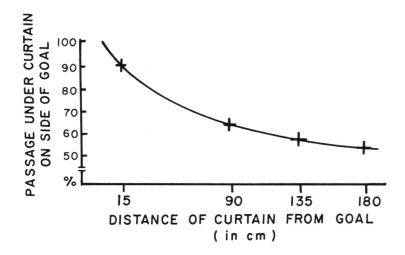

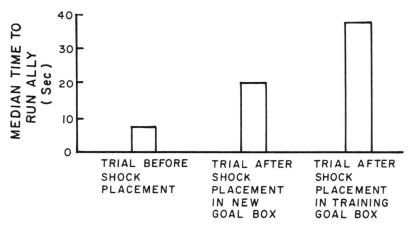

Figure 6.7   The top panel shows the percentage of rats passing
under the curtains in the runway on the side corresponding to the direction of turn in the goalbox.
These data were obtained during initial training
trials.   The bottom panel shows the amount of time
required by the rats to traverse the runway before
and after shocked placements in the goalbox.   Data
are shown separately for groups shocked in the novel
goalbox and in the goalbox used during training.
(After Miller, 1935.)

Conclusion

Interest in latent learning has been largely dormant in the past twenty or so years, a sharp change from the period between 1929 and the mid-'50s when theoretical controversies swirled around it and other topics. One consequence of the theoretical and experimental activity of that period was that multiple functions for the reward event (S^v) were clearly indicated instead of the single function which had been emphasized earlier. As documented in this chapter, the latent learning experiments were instrumental in bringing about this change of view. The main developments were that the Thorndikian principle that S^v acts only to strengthen associations (Hull, 1943; Thorndike, 1911) was abandoned in favor of positions in which S^v was viewed as a performance variable (Tolman, 1932; and later, Spence, 1956), as both a learning and performance variable (Hull, 1952), and as a source of classically conditioned responses which influence instrumental performance (Hull, 1931; Spence, Bergmann, & Lippitt, 1950). Another important consequence of this work was that many fundamental assertions of both S-R and cognitive theory were forced into clearer articulation and, in turn, were shown to have important weaknesses (e.g., Deutsch, 1956; MacCorquodale & Meehl, 1954; Spence & Lippitt, 1946; Thistlethwaite, 1951). Unfortunately, research and theorizing slowed before there was resolution of whether the various latent-learning procedures activate different learning processes. Presumably, these issues will be taken up again in the future.

REFERENCES

Baker, T. W. Properties of compound stimuli and their conditioned components. PSYCHOLOGICAL BULLETIN, 1968, 70, 611-625.

Blodgett, H. C. The effect of the introduction of reward upon the maze performance of rats. UNIVERSITY OF CALIFORNIA PUBLICATIONS IN PSYCHOLOGY, 1929, 4, 113-134.

Brown, P. L., & Jenkins, H. M. Auto-shaping of the pigeon's keypeck. JOURNAL OF THE EXPERIMENTAL ANALYSIS OF BEHAVIOR, 1968, 11, 1-8.

Deutsch, J. A. The inadequacy of the Hullian derivations of reasoning and latent learning. PSYCHOLOGICAL REVIEW, 1956, 63, 389-399.

Gengerelli, J. A. Preliminary experiments on the causal factors in animal learning. JOURNAL OF COMPARATIVE PSYCHOLOGY, 1928, 8, 435-457.

Haney, G. W. The effect of familiarity on maze performance of
    albino rats. UNIVERSITY OF CALIFORNIA PUBLICATIONS IN PSY-
    CHOLOGY, 1931, 4, 319-333.

Hilgard, E. R., & Bower, G. H. THEORIES OF LEARNING, 4th Edition.
    Englewood Cliffs, N.J.: Prentice-Hall, 1975.

Hull, C. L. Knowledge and purpose as habit mechanisms. PSYCHOLOG-
    ICAL REVIEW, 1930, 37, 511-525.

Hull, C. L. Goal attraction and directing ideas conceived as habit
    phenomena. PSYCHOLOGICAL REVIEW, 1931, 38, 487-506.

Hull, C. L. PRINCIPLES OF BEHAVIOR: AN INTRODUCTION TO BEHAVIOR
    THEORY. New York: Appleton-Century-Crofts, 1943.

Hull, C. L. A BEHAVIOR SYSTEM: AN INTRODUCTION TO BEHAVIOR THEORY
    CONCERNING THE INDIVIDUAL ORGANISM. New Haven: Yale Uni-
    versity Press, 1952.

Karn, H. K., & Porter, J. M. The effect of certain pretraining
    procedures upon maze performance. JOURNAL OF EXPERIMENTAL
    PSYCHOLOGY, 1946, 36, 461-469.

MacCorquodale, K., & Meehl, P. E. Edward C. Tolman. In W. K.
    Estes, S. Koch, K. MacCorquodale, P. E. Meehl, C. G. Mueller,
    Jr., W. N. Schoenfeld, & W. S. Verplanck (Eds.), MODERN
    LEARNING THEORY: A CRITICAL ANALYSIS OF FIVE EXAMPLES.
    New York: Appleton-Century-Crofts, 1954.

Meehl, P. E., & MacCorquodale, K. A further study of latent learn-
    ing in the T-maze. JOURNAL OF COMPARATIVE AND PHYSIOLOGICAL
    PSYCHOLOGY, 1948, 41, 372-396.

Miller, N. E. A reply to "sign-gestalt or conditioned reflex?"
    PSYCHOLOGICAL REVIEW, 1935, 42, 280-292.

Moltz, H. Latent extinction and the fractional anticipatory re-
    sponse mechanism. PSYCHOLOGICAL REVIEW, 1957, 64, 229-241.

Moore, B. R. The role of directed Pavlovian reactions in simple in-
    strumental learning in the pigeon. In R. A. Hinde & J.
    Stevenson-Hinde (Eds.), CONSTRAINTS ON LEARNING. London:
    Academic Press, 1973.

Rashotte, M. E., Griffin, R. W., & Sisk, C. L. Second-order condi-
    tioning of the pigeon's keypeck. ANIMAL LEARNING & BEHAVIOR,
    1977, 5, 25-38.

Razran, G. Empirical codifications and specific theoretical impli-
    cations of compound-stimulus conditioning: Perception. In
    W. F. Prokasy (Ed.), CLASSICAL CONDITIONING. New York:
    Appleton-Century-Crofts, 1975.

Seward, J. P. An experimental analysis of latent learning. JOURNAL OF EXPERIMENTAL PSYCHOLOGY, 1949, 34, 177-186.

Seward, J. P., & Levy, H. Latent extinction: Sign learning as a factor in extinction. JOURNAL OF EXPERIMENTAL PSYCHOLOGY, 1949, 39, 660-668.

Spence, K. W. BEHAVIOR THEORY AND CONDITIONING. New Haven: Yale University Press, 1956.

Spence, K. W., Bergmann, G., & Lippitt, R. A study of simple learning under irrelevant motivational-reward conditions. JOURNAL OF EXPERIMENTAL PSYCHOLOGY, 1950, 40, 539-551.

Spence, K. W., & Lippitt, R. O. "Latent" learning of a simple maze problem with relevant needs satiated. PSYCHOLOGICAL BULLETIN, 1940, 37, 429.

Spence, K. W., & Lippitt, R. An experimental test of the sign-Gestalt theory of trial-and-error learning. JOURNAL OF EXPERIMENTAL PSYCHOLOGY, 1946, 36, 491-502.

Thistlethwaite, D. A critical review of latent learning and related experiments. PSYCHOLOGICAL BULLETIN, 1951, 48, 97-129.

Thorndike, E. L. ANIMAL INTELLIGENCE: EXPERIMENTAL STUDIES. New York: Macmillan, 1911.

Thorndike, E. L. FUNDAMENTALS OF LEARNING. New York: Teachers College, 1932.

Tolman, E. C. PURPOSIVE BEHAVIOR IN ANIMALS AND MAN. New York: Century, 1932.

Tolman, E. C. Sign-Gestalt or conditioned reflex? PSYCHOLOGICAL REVIEW, 1933, 40, 246-255.

Tolman, E. C. Cognitive maps in rats and men. PSYCHOLOGICAL REVIEW, 1948, 55, 189-208.

Tolman, E. C., & Gleitman, H. Studies in learning and motivation: I. Equal reinforcements in both end-boxes, followed by shock in one end-box. JOURNAL OF EXPERIMENTAL PSYCHOLOGY, 1949, 39, 810-819.

Tolman, E. C., & Honzik, L. H. Introduction and removal of reward and maze performance in rats. UNIVERSITY OF CALIFORNIA PUBLICATIONS IN PSYCHOLOGY, 1930, 4, 241-256.

7. REWARD TRAINING: CONTRAST EFFECTS

M. E. Rashotte

Florida State University
Tallahassee, Florida, USA

Several important characteristics of reward training have been
revealed by employing procedures in which animals are exposed to
more than one value of the reward event. The present chapter reviews three of those procedures and the contrast effects they produce in instrumental performance. "Contrast effect" is a term bor-
rowed from sensory psychology where it describes the fact that the
perceived difference between two stimuli is exaggerated by the manner of their presentation. A contrast effect in reward training is
characterized by performances which indicate that the influence of
a given reward event is exaggerated by the nature of other reward
events to which the animal is exposed. The procedures to be discussed here yield performances that have been designated successive-,
simultaneous-, and behavioral-contrast effects. They have been an
important stimulant for theoretical development.

Successive Contrast

Negative contrast. The basic design of experiments showing
successive negative contrast involves a minimum of two groups trained
to perform the same response in two phases of the experiment. In
the first phase, the experimental and control groups receive high
valued and low-valued reward, respectively. In the second phase,
both groups receive the low-valued reward. The question of interest
is whether prior experience with the high-valued reward alters the
effectiveness of the low-valued reward in the experimental group.
The control group provides the baseline against which a change in
effectiveness can be assessed.

The classic demonstrations of successive negative contrast in rats are shown in Figure 1. Elliott (1928), working under Tolman's direction, trained two groups of hungry rats to run once a day through a 14-unit T-maze. The rats were allowed to eat for 3 min when they reached the goalbox on each trial. On pre-shift trials, the experimental group ate moistened bran-mash and the control group ate sunflower seeds. Figure 1 shows that on these trials the experimental group made fewer errors, an indication that bran-mash had the greater reward value. After only one trial with the sunflower-seed reward, the experimental group's performance worsened, and subsequently became inferior to that of the control group trained with sunflower seeds from the outset. The second demonstration was made by Crespi (1942). In the pre-shift period, three groups were trained to run one trial a day in a straight runway to 256, 64, or 16 food pellets; in the post-shift period, all groups received 16 pellets. There was a sizable reduction in speed by the downshifted groups after only one trial with the smaller reward, and for several trials speeds were lower for these groups than for the control group always trained with 16 pellets. These experiments demonstrate that the effectiveness of a low-valued reward event is impaired by previous experience in the training situation with a higher-valued reward event, whether "value" is operationalized as reward quality (Elliott) or quantity (Crespi). It is this finding that constitutes a successive negative contrast effect.

The theoretical significance of these results can be illustrated in the context of S-R reinforcement theory. Consider Hull's (1943) expression of that position. He proposed that S^V is necessary for learning, and that the value of S^V sets the limit of associative strength in reward training (see Chapter 6). According to this view, high-reward should yield greater associative strength than low-reward in the pre-shift phase and, other things equal, the high-reward group should respond more strongly. The immediate theoretical consequence of lowering the value of S^V should be to reset the asymptote of associative strength for the experimental group to the lower value set all along for the control group. As a result, the experimental group should enter the post-shift phase with more associative strength than its new S^V can support and, consequently, there must be a loss of associative strength on each trial until the lowered asymptote is reached. While Hull (1943) did not explicitly deal with the details of this case, it is certain that his theory required the associative loss to occur across several trials. For one thing, associative strength changed exponentially in Hull's theory. For another it was thought to be relatively permanent, once established. Thus, the theory must predict that in the post-shift period the experimental group's performance will weaken across several trials until it reaches the level of the control group, whose performance should reflect the asymptotic associative strength appropriate for the lower-valued S^V.

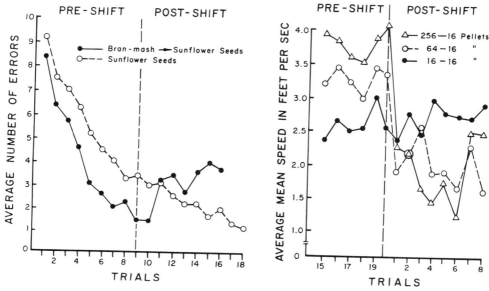

Figure 7.1   Two examples of the successive negative contrast effect
in rats (solid-food reward) when 1 trial was given per
day.   The left panel shows data from an experiment in
which one group of rats received bran-mash and the other
group received sunflower seeds in the pre-shift period
for running through a multiple-unit T-maze.   Beginning
on Trial 10, both groups received sunflower-seed reward.
The dependent variable is the number of errors (entries
into a blind alley) made on each run through the maze
(after Elliott, 1928).   The right panel shows the running speed of three groups of rats in a straight runway.
In the pre-shift period each group received a different
number of pellets in the goalbox (256, 64, or 16).   Beginning on the "shift" trial, marked by the dashed vertical line, all groups received 16-pellet reward (after
Crespi, 1942).

The Elliott-Crespi result posed two obvious problems for Hull's S-R reinforcement theory. One was that responding changed <u>rapidly</u> following the shift in S^V, a sizable performance decrement occurring after a single trial with the lower-valued reward. This problem was handled by his later theoretical revision which assigned S^V the function of a performance variable (Hull, 1952; also see Chapter 6) and was easily incorporated by other theoretical positions that viewed reward as a performance variable (Tolman, 1932). The second problem was that performance of the shifted group <u>undershot</u> the performance level of the control group. This is the defining characteristic of negative contrast and it was not so readily incorporated by Hull nor, indeed, by any theorist.

Important clues to the source of depressed performance in the post-shift period came from qualitative observations of the rat's behavior. Elliot noted that when the new S^V was first encountered the rats did not eat steadily, as before, but seemed to "search" about the goalbox as if looking for the higher-valued S^V. He speculated that they might have entered more blind-alleys in the post-shift period in a search for the higher-valued S^V. Crespi observed "frustration" reactions in the goalbox. When his rats first found the smaller reward, some attempted to jump out of the goalbox, some delayed eating or refused to eat at all, and some reared and looked "frantically" out the top of the goalbox. These behaviors were not observed when the rats received the higher-valued S^V in the pre-shift period, nor were they noted in the control group. Crespi reported that behaviors similar to these occurred in the runway on several subsequent post-shift trials, competed with performance of the instrumental response, and were probably the reason performance of the downshifted groups temporarily undershot the performance level of the control group in his experiment.

These and other observations led to a "learned-expectancy" account of the successive negative contrast effect which emphasizes the following ideas:
    1. In the pre-shift period (and, in reward training in general), rats learn about the specific properties of the reward event. It may be said that they learn to "expect" the specific S^V obtained, or that they form an "internal representation" of the S^V. This idea finds ready expression in Tolman's (1932) expectancy theory and in S-R theory's analogue of expectancy, r_G-s_G (Hull, 1931; Spence, 1956).
    2. The learned expectancy of S^V serves as a standard against which other S^Vs experienced later in training are compared.
    3. When the expected S^V is more valued than the obtained S^V, frustrative and, possibly, searching behaviors are released that cause instrumental performance to be depressed below the level of the

control group. This idea may be conceptualized more generally as competing responses being generated by the negative discrepancy between expected and obtained S^Vs.

4. Post-shift training allows the rats to learn to expect the lower-valued S^V, thereby eliminating the negative discrepency between expected and obtained S^Vs. Once this happens, the source of depressed responding is eliminated and performance will be controlled by the same factors in the experimental and control groups.

These ideas have been incorporated into theoretical accounts of reward learning in various ways, some of which are noted later in this chapter.

It must be emphasized that not all instances of successive negative contrast require the relatively complicated explanatory ideas outlined above. A transient depression of performance following a shift to a lower-valued reward might arise from peripheral sensory interactions or from stimulus generalization decrement. Consider, for example, Bitterman's (1976) demonstration of negative contrast in the honey bee (Apis melifera). The bees initially learned to shuttle between their hive and a table in the laboratory where they could drink from a 40% sucrose solution. The intertrial interval was approximately 4 to 6 min, determined by the bees. After initial training, the solution was downshifted to 20% sucrose. The bees became agitated and were hesitant to drink when they first encountered the lower concentration, but eventually drank without interruption (a successive negative contrast effect). However, this contrast effect could be eliminated by lengthening the interval between the last trial with 40% sucrose and the first trial with 20% sucrose to 24 min and, therefore, Bitterman proposed that it could be accounted for in terms of sensory adaptation. That is, ingestion and regurgitation of the 40% solution allow the taste receptors to adapt which, in turn, reduces the apparent sweetness of the 20% solution and thereby causes the contrast effect. When adaptation is allowed to wane for a few minutes, the bee accepts the 20% solution without distress. Also, generalization decrement could be a source of successive negative contrast in some experiments (Capaldi, 1972; Capaldi & Lynch, 1967; Spear & Spitzner, 1966). This factor can operate when trials are massed so that the lingering aftereffects of one trial become part of the stimulus complex conditioned to the instrumental response on the next trial. In the pre-shift period, a massed-trials experimental group would learn to respond to aftereffect stimuli characteristic of the higher-valued S^V, and when the lower-valued S^V is introduced it might disrupt performance solely by changing these stimulus conditions.

Lloyd Morgan's Canon (see Chapter 1) urges us to distinguish between successive negative contrast effects that require an account in terms of learned expectancies and those for which a simpler explanation will suffice. The distinction can be made by determining whether contrast can be demonstrated under certain temporal conditions. If the effect can be shown when there is a long (e.g., 24-hr) interval between the pre-shift and post-shift trials, an account in terms of the interaction of sensory processes activated by the two S^vs can be ruled out with some certainty. If there is a long interval between successive trials during training, a generalization-decrement account of the sort outlined above can be safely rejected. When a negative contrast effect does occur under these conditions, we must account for the fact that an experience with one S^v is retained by the animal over long temporal delays and that it subsequently influences performance when a different S^v is encountered. The learned-expectancy hypothesis provides a reasonable account of this result.

It should be clear that a learned-expectancy account is required by the Elliott-Crespi data because in both experiments the intertrial interval was 24 hours. Other experiments have strongly implicated learned expectancies as a determinant of successive negative contrast in rats. For example, Figure 2 shows the outcome of an experiment in which the interval between pre-shift and post-shift training was systematically varied (Gonzalez, Fernhoff & David, 1973). The size of the contrast effect became smaller as the interval increased from 1-day through 26, 42, and 68 days where no contrast was found, suggesting that the learned representation of S^v decays over time, probably through a process of forgetting (Gleitman & Steinman, 1964). The fact that contrast occurred reliably in this experiment even when 26 days intervened between pre- and post-shift training is entirely consistent with the learned-expectancy account of successive negative contrast.

There have been numerous studies of the successive negative contrast effect and many outcomes are consistent with the learned-expectancy hypothesis. For example, the size of negative contrast is positively related to the number of pre-shift trials, presumably because the large-reward expectancy develops more strongly with greater training (Vogel, Mikulka & Spear, 1966). Also, negative contrast is more sizable when the discrepancy between the pre- and post-shift reward sizes is larger, a finding that readily fits the learned-expectancy hypothesis (DiLollo & Beez, 1966; Gonzalez, Gleitman & Bitterman, 1962; Peters & McHose, 1974). The size of negative contrast is reduced by conducting the experiment on the baseline of a partial reinforcement schedule (Capaldi & Ziff, 1969; Mikulka, Lehr & Pavlik, 1967), a result which seems to indicate that rats "average" their expectancies across the absolute values of the rewards received (McHose & Peters, 1975; Peters & McHose, 1974).

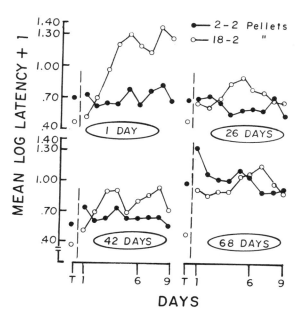

Figure 7.2   The successive negative contrast effect in rats (solid-
food reward) as a function of 1-, 26-, 42- and 68-day intervals between pre-shift and post-shift training.   The
solid circles show the performance of control groups that
received 2 pellets for running in a straight runway in
both the pre- and post-shift phases.   The open circles
show the performance of experimental groups that received
18 pellets in the pre-shift period and 2 pellets in the
post-shift period.   Each panel shows data for experimental and control groups trained with one of the intervals.
The data points at "T" indicate terminal pre-shift performance; the remaining data points show performance on
individual post-shift days.   There were 4 trials per day
at 1-hr intertrial intervals.   The response measure
plotted is a log transformation of the latency to leave
the startbox (in sec).   During the delay intervals, the
animals were maintained at the deprivation level used in
training (after Gonzalez, Fernhoff & David, 1973).

There is also evidence that negative contrast occurs more strongly when rats are highly deprived of food (Ehrenfreund, 1971; Flaherty & Kelly, 1973; cf. Capaldi & Singh, 1973). This result possibly indicates that the strength of expectancies is heightened by more severe deprivation states. Finally, there is evidence that frustrative responses are evoked in rats when the <u>amount</u> of reward is reduced (Bower, 1962; Daly, 1969) or the <u>quality</u> of reward is lessened (Cross & Boyer, 1974), as the learned-expectancy hypothesis requires. Also, successive negative contrast is eliminated by injections of tranquilizing drugs, a finding compatible with the assertion that negative contrast has an emotional basis (Rosen, Glass & Ison, 1967).

The above discussion of successive negative contrast focused almost exclusively on experiments in which rats were trained with solid-food rewards. This body of literature was important in convincing most psychologists that S-R reinforcement theory should be abandoned in favor of theories that emphasize learned expectancies in reward training. However, attempts to demonstrate negative contrast in some other species, or even in rats with certain liquid rewards, have provided important new information which indicates greater complexity than the rat/solid-food data had suggested. In fact, for some of the data, S-R reinforcement theory of the Hullian (1943) variety is perfectly adequate to account for the effects of downshifting the value of reward.

Figure 3 shows the results of an experiment that sought to demonstrate the successive negative contrast effect in rats when the concentration of the sucrose-solution S^V was downshifted (Homzie & Ross, 1962). Two groups of food-deprived rats received 1 ml of sucrose solution in both phases of the experiment for running in a runway. In the pre-shift phase, the experimental group trained with 20% (by weight) sucrose solution ran faster than the control group trained with 1% solution, a finding consistent with the general effects of reward magnitude on instrumental performance (Chapter 5). In the post-shift period, half the 20% group received 1% solution, half received no reward; the control group continued to receive the 1% solution. The figure shows that the shift from 20% to 1% solution produced a <u>gradual</u> (over about 24 trials) reduction in running speed to the level of the unshifted control group, and <u>no</u> undershooting of the control group's performance level. Performance of the group shifted from 20% to no reward fell to a lower level, indicating that the lack of undershooting cannot be attributed to a "floor" effect. This finding is clearly consistent with S-R reinforcement theory's prediction of the effect of a downshift in reward value, and it stands in marked contrast to the results of experiments with solid-food rewards. Furthermore, the experiment was conducted under relatively massed-trials conditions which are

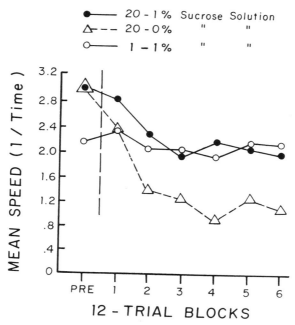

Figure 7.3   Absence of the successive negative contrast effect in rats' locomotion response when the reward is sucrose solution. The data points marked "Pre" show average speed in a runway on the last 12 trials of the pre-shift period for groups receiving 1 ml of 20% or 1% sucrose solution on each trial. In the post-shift period half the 20% group was shifted to 1% sucrose solution (20%-1%), the other half received no reward (20%-0%). The control group continued receiving 1% sucrose solution (1%-1%). The measure shown is running speed in the center portion of the runway (after Homzie & Ross, 1962).

favorable for large negative contrast effects (Capaldi, 1972).
This result has been widely replicated (e.g., Collier, Knarr &
Marx, 1961; Flaherty & Caprio, 1976; Rosen & Ison, 1965).

It is not clear why the negative contrast effect fails to occur
with downshifts in sucrose-solution concentration under the same
general experimental conditions as Crespi employed (i.e., rats;
locomotor response). Some possibilities suggested by the experiments
with solid-food rewards can be largely discounted. For example, the
possibility that the discrepancy between the pre- and post-shift
concentrations of sucrose solution has not been large enough seems
unlikely in view of failure to find the effect when the discrepancy
is sizable (e.g., Flaherty, Riley & Spear, 1973). Another possibility is that not enough pre-shift trials were run, but it, too,
seems remote. The data in Figure 3 were obtained after 84 pre-shift
trials; Rosen (1966) found no evidence of successive negative contrast in an experiment in which the number of pre-shift trials was
12, 42, or 90. These are well beyond the number of trials necessary
to show some evidence of the effect with solid-food S^v. Finally,
the possibility that a downshift in concentration of sucrose solution
is not sufficient to yield a frustrative emotional reaction in rats
is contradicted by the results of experiments by Daly (1974, p. 204-
205), who showed that rats will learn to escape from cues associated
with a downshift in sucrose concentration. This is a traditional
test for the aversiveness of stimuli. It might be noted, however,
that the clearest evidence of frustration in her situation was obtained when concentration shifts were made and the volume of the
liquid was small; in the runway experiments cited above, volume has
been relatively large.

It would be relatively straight forward if the failure of successive negative contrast could be attributed to the use of sweet-
ened solutions as reward. Unfortunately, the matter is not so simple because successive negative contrast is found with sucrose-
solution rewards in rats when the performance measured is consummatory activity (e.g., Flaherty & Caprio, 1976; Vogel, Mikulka &
Spear, 1968), or lever pressing (Weinstein, 1970a, b). Negative
contrast in the consummatory response measure is illustrated in Figure 4. It shows the results of an experiment that varied length of
delay between pre- and post-shift training (Flaherty, Capobianco, &
Hamilton, 1973). In each of the three pairs of groups shown, the
group receiving 32% sucrose solution licked the drinking tube at a
higher rate than the group receiving 4% solution in the pre-shift
period. Negative contrast was clearly evident when 1 or 4 days intervened between pre- and post-shift training, although it was more
durable after 1-day than after 4-days delay. There was no evidence
of contrast when a 17-day delay intervened. It has been demonstrated
in another experiment that negative contrast occurs after a 10-day
delay in this type of experiment (Flaherty & Lombardi, 1977).

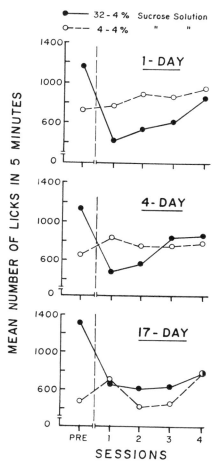

Figure 7.4   Successive negative contrast effect in rats' consummatory
response when the reward is sucrose solution.   Each panel
shows an experimental group and a control group.   The experimental group received 32% sucrose solution for licking
a tube in the pre-shift period, and 4% sucrose solution in
the post-shift period.   The control group always received
the 4% solution.   The panels show the effect of 1-, 4-,
and 17-day intervals between the end of pre-shift training
and the beginning of post-shift training (after Flaherty,
Capobianco, & Hamilton, 1973).

Thus, the available evidence indicates that negative contrast occurs in rats trained with different concentrations of sucrose solution as rewards only when certain responses are measured. For some reason, it does not seem possible to show negative contrast with sucrose-solution rewards in Crespi's runway procedure but, in cases where it is found with this reward, its properties seem similar to those of negative contrast obtained with solid-food reward in the runway. For example, it survives long delays between pre- and post-shift training (see Figure 4) and it becomes larger with larger numbers of pre-shift trials (Vogel, et al., 1968).

An additional complication is posed by some limited evidence that all sweetened-solution rewards do not act in the same manner in the successive negative contrast experiment with rats. Some experiments have shown that, even when the consummatory response is measured, negative contrast does not occur following a downshift in concentration of saccharine solution (Hulse, 1962; Vogel, et al., 1968). In these experiments, the lick-rate of the downshifted group gradually declined until it reached the level of the unshifted control group (the result predicted by S-R reinforcement theory). In another experiment, however, sizable negative contrast was reported when rats' lever pressing was measured and the concentration of saccharine-solution reward was downshifted (Weinstein, 1970b). Clearly, we currently have a limited understanding of the conditions that produce successive negative contrast in rats following a downshift in the value of S^v.

The second line of evidence to be considered in assessing the generality of successive negative contrast comes from comparative studies. Several experiments have attempted to demonstrate successive negative contrast in goldfish and turtles. None has been successful. Instead, these experiments yield results very close to those expected on the basis of S-R reinforcement theory.

Consider experiments conducted with trial spacing equivalent to Crespi's (24-hour intertrial interval). Goldfish trained to swim down a runway or to strike a target to obtain different numbers of Tubifex worms showed the usual effects of magnitude of reward on performance in the pre-shift periods. However, in the post-shift period, the performance of the shifted group did not change following the downshift in reward (Gonzalez, Potts, Pittcoff & Bitterman, 1972; Lowes & Bitterman, 1967). An experiment with painted turtles (Chrysemys picta picta) conducted in a runway with 24-hour intertrial intervals also showed no successive negative contrast effect. In this case, however, post-shift performance gradually weakened to the level of the control group in the post-shift period (Pert & Bitterman, 1970).

With rats, successive negative contrast is larger when trials are massed, probably because of the added influence of stimulus-generalization decrement from trial aftereffects (Capaldi, 1972; see above). Trial-aftereffect stimuli can influence the performance of goldfish under massed-trials conditions (Gonzalez, 1972), so it is reasonable to ask whether successive negative contrast might be demonstrated in goldfish under the more favorable massed-trials procedure. Figure 5 shows the results of such an experiment in which the intertrial interval was 15 sec (Gonzalez Ferry & Powers, 1974). In the pre-shift period, the two groups receiving 40 Tubifex worms on each trial swam faster than the group receiving 1 worm. In the post-shift period, the performance of the downshifted group gradually deteriorated until it reached the level of the unshifted small-reward group. The deterioration of performance following the shift in this experiment, but not in the 1-trial/day experiment with goldfish cited earlier, seems most likely to reflect the influence of stimulus-generalization decrement under highly massed trials. A finding by Mackintosh (1971) is consistent with this interpretation. He reported no significant change in goldfishes' swimming speed following a downshift in reward when trials were spaced at 3- to 5-min intertrial intervals. It may be noted that the result shown in Figure 5 has also been obtained in painted turtles under massed-trials conditions (Pert & Gonzalez, 1974). This result represents yet another failure to find successive negative contrast in turtles, this time under highly favorable conditions for its occurrence in rats.

Overall, the experiments with goldfish and turtles yield evidence consistent with S-R reinforcment theory's predictions about the effect of downshifting reward value. The possibility of phylogenetic differences in learning is raised by these data (e.g., Bitterman, 1975). It may be that the learned behavior of goldfish and turtles is regulated by the S-R reinforcement principle. In rats and other mammals the principle may operate as well, but its influence may be obscured by the involvement of expectancy processes in some situations, at least.

In summary of the entire discussion of successive negative contrast, the available data indicate that this contrast effect is found reliably with some combinations of species, response and reward, but not with others. With rats and solid-food rewards, negative contrast is reliably found in measures of locomotion towards a goal. With rats and certain liquid rewards, negative contrast is not found in locomotion but is found when consummatory activity or lever pressing is measured. In several experiments with goldfish and painted turtles, successive negative contrast has not been reported at all. At the present time, there is no

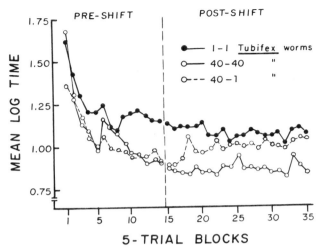

5-TRIAL BLOCKS

Figure 7.5   Absence of the successive negative contrast effect in
              goldfish. Time to swim from startbox to goalbox by
              groups receiving different numbers of Tubifex worms
              as reward is plotted. Unshifted control groups re-
              ceived 1 or 40 worms on every trial of the experi-
              ment. The experimental group received 40 worms per
              trial in the pre-shift period and 1 worm per trial in
              the post-shift period. The intertrial interval was
              15 sec and there were 5 trials per day (after
              Gonzalez, Ferry & Powers, 1974)

good account for this pattern of successes and failures in obtaining successive negative contrast. It is clear, however, that the
relatively straightforward results obtained with solid-food reward in rats is no longer the principal data-base of successive
negative contrast. Thus, the learned-expectancy hypothesis, which
was constructed when only that limited data-base was available,
must be extensively revised or supplemented by as yet undetermined
factors that will allow the failures of successive negative contrast to be understood.

Positive contrast. Hull's (1943) S-R reinforcement theory
asserted that an upwards shift in the value of S^V will increase
the limit of associative strength to a value commensurate with the
higher value of S^V. Thus, if instrumental performance were asymptotic after training with small reward, a shift to a large reward
should result in gradual improvement of performance to the level
of an unshifted control group trained with the large reward from
the outset.

Crespi (1942) presented some influential data which indicated that this prediction is wrong. He had trained a group of rats to run to 16-pellet reward until performance appeared asymptotic, and then had <u>downshifted</u> some of them to 1-pellet, and others to 4-pellet reward. After 14 trials, performance of both groups reached a lower level and training was discontinued for 3 days. Finally, training was resumed with reward <u>upshifted</u> to the 16-pellet level again. Unfortunately, an unshifted 16-pellet control group was not included so that responding for 16-pellet reward following the upshift in reward value had to be compared to performance of the same groups when trained with 16-pellet reward at the beginning of the experiment. Crespi reported that running speed following the upshift <u>exceeded</u> the earlier running speed for 16-pellet reward. He termed this result the <u>elation effect</u>, attributing to it the motivating effects of a positive emotional reaction consequent upon the rats' finding a large reward after experiencing a "frustratingly" small reward (Crespi, 1942, 1944). It is now termed the <u>successive positive contrast effect</u>.

For many years after Crespi's report the replicability of successive positive contrast was viewed as doubtful (Black, 1968; Dunham, 1968; Mackintosh, 1974; Spence, 1956). One possibility was that Crespi's finding was an artifact of the within-group comparison of speeds for 16-pellets reward early and late in the experiment. The suggestion was that the initial performance level for 16-pellets was pre-asymptotic, and that the higher level on 16 pellets following the upshift in S^V simply reflected the true asymptote, not a contrast effect. This possibility received support from failures to find positive contrast in experiments where asymptotic performance on large reward was estimated from a separate control group run for a relatively large number of trials (e.g., Spence, 1956, p. 130ff). However, even these failures to show positive contrast were not entirely conclusive because of the possibility that physical limits of performance prevented the upshift group from responding more strongly than the high-reward control group (i.e., a "ceiling effect," e.g., Bower, 1961).

Recently, the entire issue has been resolved in favor of the reality of successive positive contrast, provided that the procedure employed by Crespi is followed. Crespi attempted to insure that small reward was "frustrating" in his experiment by arranging a downward shift from large to small reward before upshifting to the large reward again. When Crespi's downward-to-upward shift procedure is followed, positive contrast is readily found in the rat's locomotion response (Benefield, Oscós & Ehrenfreund, 1974; Maxwell, Calef, Murray, Shepard & Norville, 1976). In fact, Maxwell, <u>et al</u>. (1976), reported positive contrast with this procedure provided that reward was upshifted while the running response

210

was still showing negative contrast resulting from the prior downshift in reward. Positive contrast was not found if the upshift
occurred after negative contrast had disappeared. This finding is
consistent with Crespi's emphasis on the small reward being frustrative if an upshift in reward value is to result in positive contrast.

Successive positive contrast has also been found in a number of
experiments designed to prevent a possible "ceiling effect" on performance. The logic of these experiments was that performance can
be depressed below its physical maximum by delaying reward on all
trials (see Chapter 5). One example is shown in Figure 6 in which
two groups of rats were trained in a runway with reward delayed for
20 sec on each trial (Mellgren, 1972). The unshifted control group
received 8 pellets throughout the experiment; the experimental group
received 1 pellet in the pre-shift period and 8 pellets following the
shift. Statistical analysis indicated that the shifted group ran
faster than the control group on the third through the eighth block
of post-shift trials, demonstrating that positive contrast occurred
and that it was transitory. Comparable findings have been reported
in other experiments (e.g., Mellgren, 1971; Shanab, Sanders &
Premack, 1969). It may be noted that it is not clear whether the
"delayed-reward" procedure is effective in these experiments because
it depresses performance or because it makes the small reward frustrating prior to the upwards shift in reward, as Crespi thought.

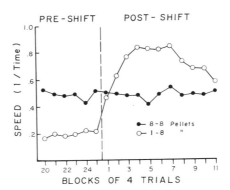

Figure 7.6 Successive positive contrast effect in rats' locomotion response. Reward was delayed 20 sec after the
rats reached the goalbox on every trial. The control
group received eight 0.045 g pellets on every trial;
the experimental group received one 0.045 g pellet on
pre-shift trials and 8 pellets on post-shift trials.
There were 4 trials per day at 3- to 4-min intertrial
intervals (after Mellgren, 1972).

In view of these recent findings, it may now be concluded that successive positive contrast is a genuine effect in rats when solid-food S^Vs are employed. At present, however, there is very little evidence about the generality of positive contrast in various species-response-reward combinations. Some evidence suggests that positive and negative contrast may not be symmetrical in the successive contrast experiment. For one thing, the emotional basis of positive contrast is not certain. The obvious evidence of emotional upset in negative contrast was not paralleled in Crespi's (1942) experiment by behavioral indices of "elation" following an upwards shift in the value of S^V (nor was it found in the monkey which is so expressive in its frustrative/searching reaction following a downshift in reward value; Tinklepaugh, 1928). Also, limited evidence indicates that the size of the positive contrast effect may be <u>inversely</u> related to the number of trials in the pre-shift period (Mellgren, 1971; Nation, Roop & Dickinson, 1976), the opposite of the case with negative contrast. On the other hand, there is evidence that the size of the effect is positively related to size of the discrepancy between the pre- and post-shift amounts (Mellgren, Seybert, Wrather & Dyck, 1973), that the effect occurs following gradual, as well as abrupt, shifts in the value of S^V (Nation, <u>et al.</u>, 1976), and that the size of the effect is lessened when the experiment is conducted on the baseline of a partial reinforcement schedule (Lehr, 1974). These latter findings are consistent with the results of experiments on successive negative contrast that employed rats and solid-food reward. Because of the uncertainty about its occurrence, there has been little theoretical development stimulated by the successive positive contrast effect.

<u>Theoretical implications</u>. It is not possible to describe here the variety of theoretical developments in reward learning stimulated by the study of successive contrast effects. It should be noted, however, that an important consequence of the early work on negative contrast in rats trained with solid-food reward was the encouragement of theories of reward learning in which reward-expectancy played a central role. Of course, Tolman's (1932) theory had that theoretical emphasis from the outset (see Chapter 6) but S-R theory did not. To accommodate the new data, S-R theory was changed in important ways.

Confronted with mounting evidence from the latent-learning and contrast-effect literatures that S^V functions as a performance variable and that expectancy-like processes seem involved in reward training, Spence (1956) proposed an S-R theory of reward learning in which these emphases were central. He completely rejected the S-R reinforcement principle and asserted that S-R contiguity is sufficient to strengthen associations between stimuli and the

instrumental response. Instead of acting as a "reinforcer," S^V
was assigned the dual roles of directing and motivating instrumental
performance. A very influential assertion was that S^V accomplished
its twin functions by influencing r_G-s_G, S-R theory's analogue of
expectancy. Spence assumed that r_G-s_G directs instrumental responding when its feedback stimulus becomes associated with the
instrumental response (i.e., r_G-s_G--$\rightarrow$$R_{Inst}$). The motivational
function of S^V was captured in the new assumption that the conditioned strength of r_G-s_G determines the strength of incentive
motivation, the theoretical construct introduced by Hull (1952)
to allow S^V to act as a performance variable. The background of
these theoretical developments was outlined in Chapter 6. In
summary, Spence's (1956) theory invoked a learned-expectancy construct as the mediator of the effects of S^V on performance and
asserted that S-R contiguity is sufficient for the formation of
associations between the stimulus conditions and the instrumental
response. In these respects, Spence's version of S-R theory
blurred some of the historically important differences between S-R
theory and Tolman's cognitive theory.

Recent attempts to formulate an account of successive contrast effects have been cast in terms of sensory-perceptual theory
(Bevan & Adamson, 1963; Helson, 1964), the Wagner-Rescorla model
(McHose & Moore, 1976), and various modifications of the traditional learned-expectancy hypothesis (e.g., Black, 1968, 1976).
Unfortunately, none of these approaches accounts for the range
of species-response-reward interactions in negative contrast,
and the nature of positive contrast is too poorly understood to
submit to effective theoretical analysis. Thus, while successive
contrast effects have encouraged much theoretical activity, a
satisfactory account of these effects is not yet available.
Present evidence indicates that a successful theory will incorporate learned-expectancies, sensory-perceptual interactions, and
the S-R reinforcement principle in some configuration.

Simultaneous Contrast

Contrast effects observed when the value of S^V is repeatedly
shifted during a single training session have been termed simultaneous contrast. The basic experiment involves an experimental group
that receives a high-valued S^V for responding in the presence of
one stimulus and a low-valued S^V for responding when a different
stimulus is present. Two control groups are also trained with
the two stimulus conditions but receive either a high-valued S^V or
a low-valued S^V for responding in the presence of both stimuli.
The stimulus conditions change frequently during each training ses-

sion so that the experimental group experiences repeated shifts in
the value of S^v. A sample of contrast effects obtained with different responses, species and rewards in this type of experiment
is shown in Figure 7. These data indicate the general rule that
simultaneous negative contrast is found under a wide variety of
conditions, but that positive contrast occurs only in some conditions.

Figure 7 illustrates the fact that simultaneous negative contrast in the rat is not restricted to the limited response-reward
combinations that appear necessary for successive negative contrast
to occur in this species. This fact implies that negative contrast
arises from different processes in the simultaneous- and successive-
contrast experiments. In panels A and B there is a sizable negative contrast effect in the locomotion response whether solid-food
(Bower, 1961) or sucrose-concentration (Flaherty, Riley & Spear, 1973)
is used as reward. In these experiments, all rats were trained in
two runways, one painted black and one white. On any given trial
the rats ran only in one runway and the order of presentation of
black and white runways was irregular across trials. Negative contrast in both experiments is evident in a comparison of running
speed to the lower-valued reward by the experimental and control
groups (i.e., group 1 (8) vs. group 1 in panel A; group 6% (64%)
vs. group 6% in panel B). Striking support for the argument that
simultaneous and successive negative contrast effects are controlled
by different variables was found in another part of the Flaherty,
et al. (1973), experiment where successive negative contrast failed
to occur in the locomotion response following a downshift in the 64%
control group's sucrose concentration. While the latter result has
been reported previously (e.g., Homzie & Ross, 1962), the Flaherty
et al., finding is unique in showing that in a single experiment
using a single response-reward combination, simultaneous but not
successive negative contrast can be demonstrated.

Further evidence that simultaneous negative contrast occurs
across a broad range of response-reward combinations in rats is
shown by comparing the two lower curves in panel D (Flaherty &
Largen, 1975, Exp. 3A). In this experiment, sucrose solution was
available for licking different tubes and consummatory behavior
was measured. Throughout each session one drinking tube was present at a time and its location alternated between left and right
positions on one wall of the test chamber after the rat had 60-sec
continuous access to it in one location. In the experimental group
the two locations were correlated with different sucrose concentrations; the control groups received either the high or the low
concentration in both locations. The data show a strong negative

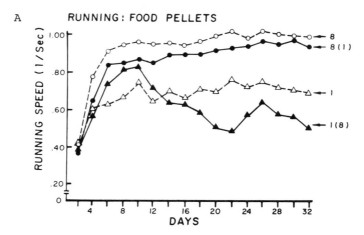

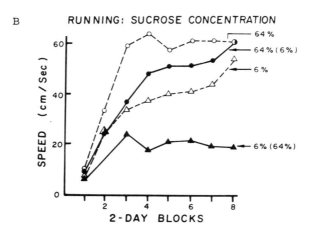

Figure 7.7 A sample of simultaneous contrast effects. Each panel
shows data from three groups of animals: the two curves
labelled with a single number show performance by the two
groups trained with a single value of S^V (e.g., in panel
A, 8 pellets or 1 pellet); the two curves labelled with a
double number are from the third group trained with both
values of S^V (the first number indicates the value of S^V
for which performance is plotted, the number in parentheses indicates the other reward-value experienced in the
experiment: e.g., in panel A the curves labelled 8(1)
and 1(8), respectively, show speed in different runways
which provided 8-pellet and 1-pellet rewards). Panels A
and B show contrast in rats' locomotion response when food
pellets (after Bower, 1961) or sucrose concentrations (after
Flaherty, Riley & Spear, 1973) are employed as reward.

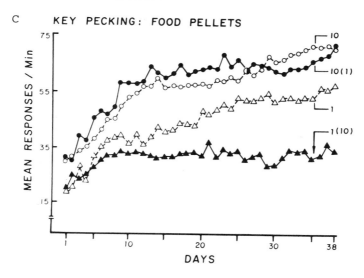

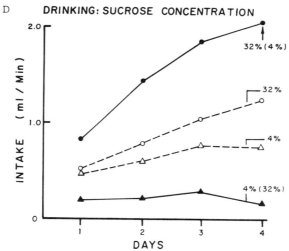

Figure 7.7 (continued) Panel C shows contrast in the pigeons' keypecking response when responding on a VI 60-sec schedule
was rewarded with 1 or 10 food pellets (after Gonzalez &
Champlin, 1974).   Panel D shows contrast in rats' consummatory response when licking a tube produced 32% or 4%
sucrose concentrations (after Flaherty & Largen, 1975).
The outstanding feature of the data is that negative contrast is seen in every panel (compare the lower two
curves in each panel) but positive contrast is seen only
in panel D (compare upper two curves in each panel).

contrast effect similar to that seen with rats in the locomotion measure when either food pellets or sucrose solution served as reward (Panels A and B).

It is also clear that simultaneous negative contrast has much broader phyletic generality than does successive negative contrast. It has been reported in goldfish (Burns, Woodard, Henderson & Bitterman, 1974; Cochrane, Scobie & Fallon, 1973; Gonzalez & Powers, 1973), in turtles (Pert & Gonzalez, 1974), in pigeons (Brownlee & Bitterman, 1968; Gonzalez & Champlin, 1974) and in monkeys (Schrier, 1958). Panel C of Figure 7 illustrates the effect in pigeons (Gonzalez & Champlin, 1974, Exp. 2). In this experiment pigeons were trained to keypeck on an operant schedule (VI 60-sec) and the two stimulus conditions were signalled by different colored keylights projected on the response key for 60-sec intervals in an irregular sequence. The key was dark for 2 sec between stimulus presentations. The high-valued S^V was 10 food pellets, the low-valued S^V was 1 pellet. Negative contrast is evident in keypecking rate for 1-pellet reward which was lower in the group experiencing both values of S^V than in the control group receiving only the 1-pellet reward.

Figure 7 also shows that simultaneous negative contrast appears to be permanent, persisting over an extended training period in the various experimental arrangements. The persistence of simultaneous negative contrast is not matched by successive negative contrast which is typically a transient effect.

The differences between simultaneous and successive negative contrast identified above suggest that these two negative contrast effects do not arise from the same processes. For one thing, if the effects did arise from a single process it would be expected that both effects or neither effect would be demonstrated reliably in the same species (Bitterman, 1975). This is not the case for goldfish which show the simultaneous effect but not the successive effect (e.g., Gonzalez, et al., 1974; Gonzalez & Powers, 1973). Nor is it the case for rats which show the simultaneous effect but not the successive effect in running speed when the reward is sucrose concentration (e.g., Flaherty, et al., 1973; Homzie & Ross, 1962). Another argument against a single-process account of both negative contrast effects is that the effect is permanent in the simultaneous case and transient in the successive case. The transience of successive negative contrast has been attributed to the elimination of emotional/frustrative reactions and (particularly in massed-trials procedures) recovery from generalization decrement that was occasioned by the downshift in reward. But neither factor provides a satisfactory account for the permanence of the simultaneous effect. As it is currently understood, frustration would not be expected to persist much beyond the clear development of differential responding to the two stimuli in the experimental group (e.g., Amsel & Ward, 1965), and generalization decrement would be overridden by the development of discriminative control over responding by the stimuli signalling large and small reward.

Unlike negative contrast, positive contrast is not found widely in the simultaneous contrast experiment. Figure 7 indicates that it does not occur when rats' running speed is measured (panels A and B); in fact, in these experiments the group receiving both values of reward ran slightly _slower_ to the large reward than did the control group that received large reward only. The figure also shows no positive contrast in the pigeon-keypecking experiment (panel C). However, sizable positive contrast reliably occurs when the rats' consummatory behavior is measured and the concentration of sucrose solution is varied within a session. Panel D of Figure 7 shows one example of this result (Flaherty & Largen, 1975; see also Flaherty & Avdzej, 1974; Flaherty & Lombardi, 1977). The reasons for the occurrence of simultaneous positive contrast in this procedure but not the others are not well understood at present. Some obvious possibilities may be noted. For example, there is some evidence that measures of consummatory activity provide more reliable and more robust evidence of contrast than do measures of latency to initiate the consummatory response (Flaherty & Largen, 1975). This finding implies differential sensitivity of response measures for simultaneous positive contrast when sucrose-solution reward is employed. Another possibility is that the procedure is successful in showing positive contrast because it arranges relatively long periods of uninterrupted access to the solutions, and very short time intervals between termination of access to one solution and the availability of the other. This arrangement may allow relatively substantial opportunities for sensory contrast between the two solutions. A pertinent finding in this regard is that the duration of access to the solutions is not an important variable over the range 15 to 60 sec since strong simultaneous positive (and negative) contrast effects are shown within this range (Flaherty & Avdzej, 1974; Flaherty & Lombardi, 1977).

There is no satisfactory theoretical account of simultaneous contrast effects. Some attempts have been made to account for successive and simultaneous contrast with the same set of ideas (e.g., Black, 1968), but the available literature does not encourage this strategy. Much of the present evidence suggests that simultaneous contrast requires no more than a sensory-perceptual account. According to such a view, the repeated exposures to different reward events in close succession during a training session allows sensory-perceptual contrast to alter the perceived value of these reward events. Such an account would be consistent with the wide generality of simultaneous negative contrast in various species-response-reward combinations, and with the apparent permanence of the effect across many training sessions. Of course, without additional considerations (e.g., ceiling effects, Bower, 1961), it would not account for the absence of simultaneous positive contrast in many situations. Nevertheless, as Flaherty and Largen (1975) have noted, the sensory-perceptual account suggests the importance of manipulating the temporal interval between exposures to the two reward events in the simultaneous-contrast experiment. Such manipulations could shed considerable light on the viability of this interpretation of simultaneous contrast effects.

Behavioral Contrast

A much-studied operant conditioning procedure that yields contrast effects is known as a multiple schedule. In it, an animal is repeatedly exposed to two (or more) stimuli in succession, each correlated with an operant schedule of reinforcement. Under certain conditions, performance on a given operant schedule is strongly influenced by the other schedule(s) that occur during the training session. It has been traditional to term these contrast effects behavioral contrast (Reynolds, 1961a, b; Skinner, 1938), and positive and negative behavioral contrasts are illustrated in Figure 8. Each panel shows keypecking by an individual pigeon when a color projected on the response key signalled that pecks would be reinforced with grain on a VI 3-min schedule. Positive contrast is shown in the increased strength of keypecking when reinforcement frequency associated with the alternating key color decreased (Reynolds, 1961a). Negative contrast (Reynolds, 1961b, 1963) is characterized by decreased keypecking rate when the reinforcement frequency provided by the alternating scheduled increased (Schwartz, 1975).

It may be noted that these contrast effects are based on within-subject comparisons between performance on a baseline condition (when both key colors signal the same reinforcement schedule) and performance in later sessions (when the schedules are different). Although the within-subject experimental design has been favored in studies of behavioral contrast, it does not accurately estimate the magnitude of behavioral contrast in many cases. The problem is that this design does not allow the influence of the changed schedule to be separated from that of the additional training sessions. For example, if responding on the baseline is pre-asymptotic when the contrast condition is introduced, the size of positive contrast will be exaggerated and the size of negative contrast reduced. One approach to this problem is to attempt to recover baseline performance after the contrast condition, as shown in the positive contrast data in Figure 8. As often as not, however, it proves difficult to recover baseline, possibly as a consequence of the contrast condition itself. A more satisfactory approach is to

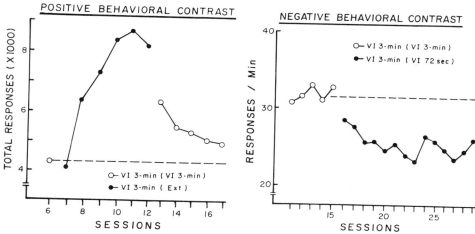

Figure 7.8   Positive and negative behavioral contrast in pigeons.
Every point in each panel shows keypecking when the
stimulus projected on the response-key signalled that
pecks would be reinforced on a VI 3-min schedule.
The open-circle points show baseline pecking when the
stimulus alternated with another that also signalled
VI 3-min reinforcement. The closed-circle points
show pecking when a different schedule was signalled
by the other stimulus: for the pigeon in the left
panel, the other stimulus signalled that no reinforcement was available (i.e., Extinction); for the pigeon
in the right panel the other stimulus signalled reinforcement for keypecks every 72 sec on the average
(i.e., VI 72-sec). Keypecking during this other stimulus
is not shown. Relative to the VI 3-min baseline, however,
keypecking to the other stimulus declined during Extinction and increased during VI 72 sec. The dotted line in
each panel is an estimate of the baseline level of responding when the schedule was the same for both stimuli.
Each data point in the left panel shows the total number
of responses in the 30 180-sec presentations of the VI 3-
min stimulus in each session (after Reynolds, 1961a).
Each data point in the right panel is the rate of responding over the 18 100-sec presentations of the VI 3-
min stimulus in each session (after Schwartz, 1975). In
both cases, the alternating stimulus was presented for
the same duration and as many times as the VI 3-min
stimulus in each session.

employ a between-subject design in which a control group is trained
only on the baseline condition, as in the simultaneous contrast experiment discussed earlier. Experiments using this design have indicated that the within-subject design is prone to overestimate the
size of positive behavioral contrast (Boakes, Halliday & Mole, 1976;
Gonzalez & Champlin, 1974; Mackintosh, Little & Lord, 1972).

The contrast effects in Figure 8 occurred in performance measures taken over the entire time the stimulus signalling VI 3-min
was present on the response key. Contrast is also evident in more
localized periods of the stimulus presentations, however, particularly in the period immediately after the transition from one
stimulus to another (Boneau & Axelrod, 1962; Catania & Gill, 1964).
Figure 9 illustrates these "local" contrast effects in the performance of an individual pigeon pecking on a key-color that signalled
a VI 8-min reinforcement schedule (Nevin & Shettleworth, 1966).
Depending upon whether the immediately preceding schedule signalled
a relatively low (extinction) or high (VI 2-min) frequency of reinforcement, responding in the initial portions of the VI 8-min
stimulus was higher or lower, respectively, than in the later portions of that stimulus. In Nevin and Shettleworth's experiment
these local contrast effects developed after a few sessions of
training and disappeared once responding to the three stimuli became well differentiated. Because it was short-lived, they termed
this effect "transient" contrast and distinguished it from the "sustained" contrast, shown in Figure 8, which persists over many sessions and is evident in performance measures taken over the entire
stimulus period (e.g., Nevin & Shettleworth, 1966; Terrace, 1966).
In many training conditions, these contrast effects are not transient (e.g., Arnett, 1973; Catania & Gill, 1964; Staddon, 1969),
so that the term "local" contrast seems more descriptive than
"transient" contrast (Malone & Staddon, 1973). The present discussion will use the terms overall and local contrast to refer to
phenomena similar to those in Figures 8 and 9, respectively.

Temporal variables have an important influence on behavioral
contrast and the nature of this influence provides some clues to
the origins of these contrast phenomena. One finding is that the
magnitude of positive behavioral contrast (both overall and local)
is reduced when more than a few seconds intervene between stimulus
conditions associated with the low- and high-reinforcement frequencies. In one experiment, contrast was evident at 10-sec but not at
60-sec interstimulus intervals (Mackintosh, et al., 1972); an experiment with 23-hour interstimulus intervals showed no evidence of contrast (Wilton & Clements, 1972), another showed that it was attenuated (Bloomfield, 1967b). Another finding is that when contrast
effects do occur, the magnitude of local contrast seen during one

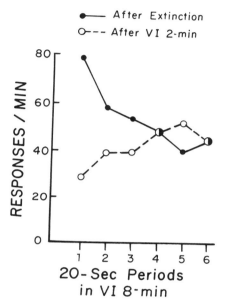

Figure 7.9   "Local" positive and negative contrast occasioned by
the transition from a stimulus signalling Extinction
or VI 2-min reinforcement to a stimulus signalling VI
8-min reinforcement.   The data are from one pigeon and
show rate of keypecking in successive 20-sec periods
of the time the stimulus signalling VI 8-min was projected on the response key.   Each data point is an
average over the 24 stimulus presentations in session
10 of discrimination training (after Nevin &
Shettleworth, 1966)

stimulus is positively related to the duration of the preceding
stimulus in both positive (Wilton & Clements, 1971) and negative
(Bernheim & Williams, 1967) local contrast.

The importance of temporal variables implies that behavioral
contrast may arise, at least in part, from processes analogous to
those underlying contrast effects in sensory-perceptual systems.
In fact, several writers have noted that neural models of sensory
contrast (e.g., Békésy, 1967; Ratliff, 1965) seem applicable to
these effects (Catania & Gill, 1964; Malone & Staddon, 1973;
Williams, 1965). Such models conceptualize sensory contrast in
terms of interacting processes of neural excitation and inhibition,
interactions that are limited by both the temporal and spatial parameters of these processes. Attempts to explore the implications of
casting behavioral contrast in this framework have produced some
interesting results (Malone, 1975, 1976; Malone & Staddon, 1973).
One important finding is that a given stimulus may produce either
positive or negative local contrast during a subsequent stimulus
presentation; the contrast effect that occurs is determined by the
relative conditioned strengths of the two stimuli, not by some absolute classification of whether the stimuli are excitatory or
inhibitory, as is implied in some treatments (e.g., Nevin & Shettleworth, 1966). When the conditioned strength of the second stimulus
is relatively strong, the prior stimulus will cause positive local
contrast; when it is relatively weak, the same prior stimulus will
cause negative local contrast (Malone & Staddon, 1973). The sensory-perceptual interpretation of behavioral contrast may turn out
to have considerable theoretical utility in conceptualizing behavioral contrast phenomena.

Over the years, a number of different hypotheses have been proposed to account for behavioral contrast. Although none of them
alone provides a satisfactory account, they have stimulated research that has led to an improved understanding of the determinants
of behavioral contrast. The three most important hypotheses and the
major findings to which they have led will be discussed below.

Reynolds (1961a,c) proposed an account of overall behavioral
contrast in terms of the relative frequency of reinforcement associated with the stimulus in which contrast is observed. The idea is
that in baseline training half the reinforcers occur in the presence
of that stimulus, and so its relative rate of reinforcement is 0.5.
When the reinforcement frequency associated with the other stimulus
is changed in the contrast condition, the relative frequency of reinforcement will change and, according to the hypothesis, the strength
of responding to the stimulus is directly related to the value of
this variable. For example, if reinforcement frequency were reduced
to zero in the other stimulus, positive contrast should occur because the relative frequency of reinforcement associated with the

reinforced stimulus would increase from 0.5 to 1.0. Negative contrast is accounted for in similar terms. Although there are many demonstrations that relative reinforcement frequency is a good predictor of overall behavioral contrast in a multiple schedule (e.g., Bloomfield, 1967a; Catania, 1961; Nevin, 1968; Reynolds, 1961c), it does not adequately account for the attenuation of contrast under many conditions. For example, it does not predict that contrast will occur more readily when the absolute rate of reinforcement in the stimulus showing contrast is low or moderate (Reynolds, 1963), or that response rate and relative reinforcement frequency come closest to matching each other when the absolute durations of the stimuli are short (i.e., 5 sec, Shimp & Wheatley, 1971). Other problems for this hypothesis will be described below.

A different hypothesis was prompted by the finding that certain variations in the training procedure prevent positive behavioral contrast in multiple schedules. Terrace (e.g., 1963, 1966) found positive contrast in pigeons trained in the usual way with VI and extinction components in a multiple schedule. Early in training these pigeons typically made hundreds of nonreinforced responses in the presence of the stimulus signalling extinction (i.e., S-); after several sessions, however, responding in the presence of S- was more or less completely suppressed. Terrace failed to find positive contrast in pigeons trained with procedures designed to prevent the pigeons from ever responding during S-. The failure of pigeons to show positive behavioral contrast in what Terrace called "errorless discrimination learning" procedures is not predicted by the relative frequency of reinforcement hypothesis; whether discrimination learning occurs with errors or not, the relative frequency of reinforcement in the stimulus associated with the VI schedule is 1.0.

In exploring the differences between error and errorless learning, Terrace found little evidence that S- became inhibitory or aversive in the errorless procedure, although it acquired both properties in normal discrimination training (e.g., Terrace, 1966, 1971, 1972). These findings suggested the hypothesis that positive behavioral contrast is a "by-product" of those discrimination procedures that allow one stimulus (i.e., S-) to acquire inhibitory/aversive properties. The idea is that positive contrast occurs because each presentation of an inhibitory/aversive S- leaves a relatively short-duration aftereffect that invigorates responding when the other stimulus is presented (cf. the "frustration effect," Amsel & Roussel, 1952; and, "positive induction," Pavlov, 1927). The hypothesis goes on to specify that an S- acquires inhibitory/aversive properties when the reinforcement schedule associated with it causes the animal to withhold responses. The prototype of such a schedule is Extinction in which generalized responding to S- is suppressed by repeated nonreinforcement of responses. In errorless discrimination learning, however, the animal is never allowed to respond to S- by means of special training procedures and, so, it cannot <u>learn</u> to suppress responding to S-. Accordingly, in this case S- displays neither aversive nor inhibitory properties and positive behavioral contrast will not be found. Terrace's hypothesis is often referred to as the <u>response suppression hypothesis</u> of positive behavioral contrast.

The response-suppression hypothesis accounted for many aspects of positive behavioral contrast that seemed difficult to reconcile with the relative-reinforcement-frequency hypothesis. Besides accounting for the absence of contrast in errorless learning, its emphasis on S- aftereffects as the source of contrast provided a rationale for the occurrence of local positive contrast and for the finding that contrast is abolished or attenuated by long temporal delays between presentation of the stimuli on a multiple schedule. It also accounted for demonstrations of positive contrast when reinforcement frequency in S- was not reduced but responding in the presence of S- was suppressed by adding electric shocks (Brethower & Reynolds, 1962; Terrace, 1968) or by scheduling reinforcment for spaced responding, or for no responding at all (Terrace, 1968; Weisman, 1969, 1970).

Unfortunately, the hypothesis has not withstood more recent empirical scrutiny. For example, contrary to the assertion of the hypothesis, errorless learning does not prevent positive behavioral contrast in all cases (Kodera & Rilling, 1976). In fact, there seems to be growing evidence that under many conditions an S- in errorless learning has inhibitory/aversive properties similar to those associated with S- in normal discrimination learning (e.g., Karpicke & Hearst, 1975; Rilling, 1977; Rilling, Caplan, Howard & Brown, 1975). Why errorless learning has different results in different experiments remains to be worked out (see Kodera & Rilling, 1976; Rilling, 1977, for a discussion of this issue). Another problem is posed by a well designed experiment that maintained closely matched reinforcement frequencies during the two stimuli of a multiple schedule while insuring suppression of responding to S- by a schedule that reinforced <u>not</u> responding (Boakes, <u>et al</u>., 1976). Although the hypothesis predicts that positive contrast will occur because the S- schedule forces response suppression, positive contrast failed to occur in **this** experiment. This finding implies that earlier demonstrations of positive contrast with similar procedures (Terrace, 1968; Weisman, 1970) resulted from imprecise matching of reinforcement frequencies in the two schedules. Because of these (and other) developments the usefulness of the response-suppression hypothesis as an account of positive behavioral contrast has been markedly reduced.

A popular recent account of behavioral contrast is known as the additivity hypothesis (Gamzu & Schwartz, 1973; Hemmes, 1973; Rachlin, 1973; Schwartz & Gamzu, 1977). It asserts that positive behavioral contrast occurs when the stimulus-reinforcer relation in one component of a multiple schedule causes the stimulus to evoke classically conditioned responses of the same topography as those required to obtain reinforcement on the operant schedule.

The basic idea was expressed in a paper by Gamzu and Schwartz (1973). They trained pigeons in sessions where the response key was alternately illuminated red or green for 27-sec periods at a time. Brief access to a grain hopper was provided intermittently and always independently of the pigeons' behavior. Gamzu and Schwartz found that the pigeons pecked one of the key colors at a high rate when it was differentially associated with a relatively high frequency of grain presentations (e.g., grain every 33 sec on the average in red periods vs. no grain in green periods). When grain was presented at the same frequency in the presence of both colors, little or no keypecking occurred. Because there was no response-reinforcer relation in the experiment, this result implies that the keypecking was controlled by the stimulus-reinforcer relation which became strong for the stimulus that was differentially associated with reinforcement (Gamzu & Schwartz, 1973; Hearst & Jenkins, 1974; cf. Rescorla, 1967). The application of the idea to a multiple schedule simply assumes that in baseline training keypecking is controlled primarily by the operant response-reinforcer relation because the two stimuli are associated with identical reinforcement schedules. When one stimulus subsequently signals extinction, however, the other (S+) is now differentially associated with the reinforcer. Thus, even though S+ will continue to control keypecking as before because of the operant schedule, its now strong stimulus-reinforcer relation should cause additional keypecks to be evoked, and positive contrast should occur. Obviously, the hypothesis predicts positive contrast only when the response evoked by the stimulus-reinforcer relation is topographically compatible with the response required by the operant schedule and when the evoked response is directed towards the place where the operant response is made.

Because the additivity hypothesis explains positive behavioral contrast in terms of the stimulus-reinforcer relation and its effect on directed motor actions, the hypothesis makes some unique predictions about the conditions under which positive contrast should occur. For example, in autoshaping with food reinforcers it is known that pigeons' pecking responses are directed almost solely towards a localized signal for the food (e.g., a stimulus projected on a response key), but that more-varied and relatively undirected behaviors occur when the food-signal is not localized (e.g., when an auditory

stimulus or a diffuse "houselight" signals food, Schwartz, 1973;
Wasserman, 1973). Accordingly, the additivity hypothesis predicts
that positive behavioral contrast will occur in pigeons reinforced
with grain provided that the stimuli in the multiple schedule are
projected on the key and that the operant schedule requires that
keypecks be made to obtain reinforcement. It also predicts that
contrast will not occur in the same training situation if the stimuli are presented off the key.

Data in support of this prediction are shown in Figure 10.
Schwartz (1975) trained pigeons to keypeck for grain reinforcement
on a VI 1-min schedule. For several sessions a baseline was established during which two stimuli were alternately presented during
successive 100-sec periods of the session. Then, the reinforcement
schedule in the presence of one of the stimuli was changed to extinction in an attempt to induce positive behavioral contrast. The manipulation of interest was the effect of different types of stimuli on
contrast. Figure 10 shows the general features of the result obtained. No matter what stimuli were used in baseline training, responding to them was strong and not differential, as would be expected if the operant schedule alone controlled responding. In the
contrast condition, responding to the extinction stimulus declined to
a low level in every case, but response rate to the stimulus that
continued to signal reinforcement on VI 1-min increased only when the
stimuli were projected on the response key (middle panel), not when
the stimuli were a change in houselight illumination or in auditory
stimulation. This result is in complete agreement with the prediction of the additivity hypothesis (see also Redford & Perkins, 1974).

The additivity hypothesis finds support in other experiments
as well. For example, there is evidence that pigeons do not show
positive behavioral contrast if signals for the VI and Extinction
schedules are projected on a different response key than the one on
which the pigeon is required to peck to obtain reinforcement (Keller,
1974; Schwartz, 1975). In fact, in these experiments the pigeons
often pecked at the signal key in periods when the differential signal for food was on it, and when these pecks were added to the number of pecks made on the operant key during the same periods the
total indicated a positive contrast effect. Another supportive
finding is that positive behavioral contrast is difficult to achieve
when the operant schedule requires pigeons to perform responses such
as treadle-hopping which are topographically dissimilar from the
(pecking) response evoked in pigeons by a strong stimulus-(food) reinforcer relation (Hemmes, 1973; Westbrook, 1973). Also, pigeons
keypecking on identical operant reinforcement schedules in the presence of two colors projected alternately on the response key show
increased rate of keypecking in the presence of one color (i.e.,
positive contrast) when extra reinforcers are presented independently
of responding in the presence of that key color (Boakes, Halliday &

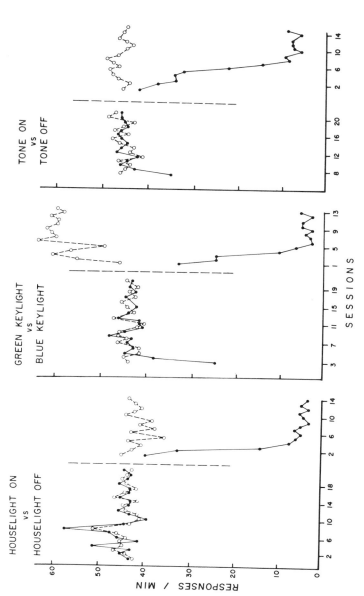

Figure 7.10   Positive behavioral contrast in a pigeon as a function of the type of stimulus that signals the two reinforcement conditions on a multiple schedule. In each panel the data to the left of the dashed vertical line show keypecking rate when both stimuli signalled VI 1-min; the curves to the right show performance when the schedule associated with one stimulus was changed to extinction (closed circles). The signals for the two schedules are identified at the top of each panel. They were located on the response key only in the center panel. The two stimuli alternated throughout each session and each data point shows response rate over the 18 100-sec presentations of one stimulus in each session. (after Schwartz, 1975)

Poli, 1975; Rachlin, 1973). According to the additivity hypothesis,
this latter result occurs because the response-independent food
presentations strengthen the stimulus-reinforcer relation sufficiently that added pecks are evoked by the key color.

It is clear that the additivity hypothesis correctly predicts
positive behavioral contrast in some situations where the other
hypotheses reviewed above fail. For example, both the relative-
reinforcement-frequency hypothesis and the response-suppression
hypothesis would predict positive contrast in all the cases shown
in Figure 10. Nevertheless, the additivity hypothesis appears to
have some important limitations. For example, there are some
demonstrations of positive behavioral contrast under conditions
where the additivity hypothesis predicts failure. In particular,
pigeons trained to keypeck when the VI and Extinction schedules are
signalled by stimuli located away from the response key (e.g., different colored houselights) show contrast in some experiments (e.g.,
Hemmes, 1973), and strong behavioral contrast has been reported in
rats under conditions unfavorable to additivity (e.g., Gutman, 1977;
Gutman, Sutterer & Brush, 1975; cf. Schwartz & Gamzu, 1977). Also,
it may be that the additivity account should be limited to positive
local contrast effects because it appears that most of the pecks
resulting from the strong stimulus-reinforcer relation occur shortly
after a transition to the stimulus signalling the higher frequency
of reinforcement (Schwartz & Gamzu, 1977; Schwartz, Hamilton &
Silberberg, 1975). If this proves to be the case, it might be
argued that at least some apparent failures of the additivity hypothesis may result from application of the hypothesis to situations
in which overall contrast effects are measured, situations for which
the hypothesis may not be suited (Schwartz & Gamzu, 1977). Finally
the additivity hypothesis does not account for negative behavioral
contrast (Schwartz, 1975; Schwartz & Gamzu, 1977). In fact, Schwartz
(1975) has shown that negative contrast occurs strongly in pigeons
under conditions where the opportunity for additivity is poor.
Thus, at best, "additivity" accounts for only positive contrast
effects and, perhaps, only for local positive contrast, in some
versions of the behavioral contrast experiment.

This survey of three hypotheses provides a glimpse of the diverse variables and processes that might be involved in behavioral
contrast. At the moment it seems appropriate to acknowledge that the
term "behavioral contrast" may not refer to a unitary phenomenon
with a single set of controlling variables (e.g., Arnett, 1973).
It is not yet clear along what lines the phenomena are best subdivided, however, and the construction and testing of hypotheses such
as those discussed here provide the only sure way to arrive at an
appropriate categorization.

Finally, throughout this chapter we have been concerned with species generality of contrast phenomena. In particular, it will be recalled that successive negative contrast has not been reported in goldfish or painted turtles but that simultaneous negative contrast is readily found in these species. Positive behavioral contrast seems to have considerable phyletic generality. In addition to pigeons and rats (discussed above), it has been reported in goldfish (Ames & Yarczower, 1965; Bottjer, Scobie & Fallon, 1977) and in painted turtles (Pert & Gonzalez, 1974). In fact, Bottjer, et al., (1977) have shown that the effect does not occur when the stimuli are located away from the disc fish strike to obtain food on the operant schedule. This result is similar to some of the pigeon data taken as evidence for the additivity hypothesis (e.g., Keller, 1974; Schwartz, 1975). Given the available evidence, it seems to be only the successive negative contrast effect which is governed, at least in part, by different processes than the other contrast effects. The possibility raised above was that it involves the operation of representational processes which may be relatively primitive in gold-fish and turtles (Bitterman, 1975). Simultaneous negative contrast and positive behavioral contrast may be relatively general phenomena because they depend more heavily on sensory-perceptual comparisons.

Conclusion

This survey of successive, simultaneous and behavioral contrast effects illustrates the complexity of the factors that determine performance in reward training. At the present there is no easy reduction of these factors to a few well-defined principles. However, it seems that at least some of the contrast effects surveyed demand a concept with the properties of a learned-expectancy, while others seem to be adequately accounted for by sensory-perceptual principles. The problem for future research and theory is to construct an appro-priate classification of these various contrast phenomena and a satisfactory theoretical account of each.

The interested reader may find it useful to consult the various analytical reviews of the contrast literature that have appeared over the years (Black, 1968, 1976; Cox, 1975; Dunham, 1968; Freeman, 1971; Mackintosh, 1974; McHose & Moore, 1976; Schwartz & Gamzu, 1977).

REFERENCES

Ames, L. L., & Yarczower, M. Some effects of wavelength discrimi-
    nation on stimulus generalization in the goldfish. PSYCHO-
    NOMIC SCIENCE, 1965, 3, 311-312.

Amsel, A, & Roussel, J. Motivational properties of frustration: I.
    Effect on a running response of the addition of frustration
    to the motivational complex. JOURNAL OF EXPERIMENTAL
    PSYCHOLOGY, 1952, 43, 363-368.

Amsel, A., & Ward, J. S. Frustration and persistence: Resistance
    to discrimination following prior experience with the dis-
    criminanda. PSYCHOLOGICAL MONOGRAPHS, 1965, 79 (4, Whole
    No. 597).

Arnett, F. B. A local-rate-of-response and interresponse-time
    analysis of behavioral contrast. JOURNAL OF THE EXPERIMENTAL
    ANALYSIS OF BEHAVIOR, 1973, 20, 489-498.

Békésy, G. V. SENSORY INHIBITION. Princeton, N.J.: Princeton
    University Press, 1967.

Benefield, R., Oscós, A., & Ehrenfreund, D. Role of frustration in
    successive positive contrast. JOURNAL OF COMPARATIVE AND
    PHYSIOLOGICAL PSYCHOLOGY, 1974, 86, 648-651.

Bernheim, J. M., & Williams, D. R. Time-dependent contrast effects
    in a multiple schedule of food reinforcement. JOURNAL OF THE
    EXPERIMENTAL ANALYSIS OF BEHAVIOR, 1967, 10, 243-249.

Bevan, W., & Adamson, R. E. Internal referents and the concept of
    reinforcement. In N. F. Washburn (Ed.), DECISIONS, VALUES,
    AND GROUPS, Vol. 2. New York: Pergamon, 1963.

Bitterman, M. E. The comparative analysis of learning. SCIENCE,
    1975, 188, 699-709.

Bitterman, M. E. Incentive contrast in honey bees. SCIENCE, 1976,
    192, 380-382.

Black, R. W. Shifts in magnitude of reward and contrast effects in
    instrumental and selective learning: A reinterpretation.
    PSYCHOLOGICAL REVIEW, 1968, 75, 114-126.

Black, R. W. Reward variables in instrumental conditioning: A
    theory. In G. H. Bower (Ed.), THE PSYCHOLOGY OF LEARNING AND
    MOTIVATION, Vol. 10. New York: Academic Press, 1976.

Bloomfield, T. M. Behavioral contrast and relative reinforcement
    frequency in two multiple schedules. JOURNAL OF THE EXPERI-
    MENTAL ANALYSIS OF BEHAVIOR, 1967, 10, 151-158. (a)

Bloomfield, T. M. Some temporal properties of behavioral contrast. JOURNAL OF THE EXPERIMENTAL ANALYSIS OF BEHAVIOR, 1967, 10, 159-164. (b)

Boakes, R. A., Halliday, M. S., & Mole, J. S. Successive discrimination training with equated reinforcement frequencies: Failure to obtain behavioral contrast. JOURNAL OF THE EXPERIMENTAL ANALYSIS OF BEHAVIOR, 1976, 26, 65-78.

Boakes, R. A., Halliday, M. S., & Poli, M. Response additivity: Effects of superimposed free reinforcement on a variable-interval baseline. JOURNAL OF THE EXPERIMENTAL ANALYSIS OF BEHAVIOR, 1975, 23, 177-191.

Boneau, C. A., & Axelrod, S. Work decrement and reminiscence in pigeon operant responding. JOURNAL OF EXPERIMENTAL PSYCHOLOGY, 1962, 64, 352-354.

Bottjer, S. W., Scobie, S. R., & Wallace, J. Positive behavioral contrast, autoshaping, and omission responding in the goldfish (*Carassius auratus*). ANIMAL LEARNING & BEHAVIOR, 1977, 5, 336-342.

Bower, G. H. A contrast effect in differential conditioning. JOURNAL OF EXPERIMENTAL PSYCHOLOGY, 1961, 62, 196-199.

Bower, G. H. The influence of graded reductions in reward and prior frustrating events upon the magnitude of the frustration effect. JOURNAL OF COMPARATIVE AND PHYSIOLOGICAL PSYCHOLOGY, 1962, 55, 582-587.

Brethower, D. M., & Reynolds, G. S. A facilitative effect of punishment on unpunished behavior. JOURNAL OF THE EXPERIMENTAL ANALYSIS OF BEHAVIOR, 1962, 5, 191-199.

Brownlee, A., & Bitterman, M. E. Differential reward conditioning in the pigeon. PSYCHONOMIC SCIENCE, 1968, 12, 345-346.

Burns, R. A., Woodard, W. T., Henderson, T. B., & Bitterman, M. E. Simultaneous contrast in the goldfish. ANIMAL LEARNING & BEHAVIOR, 1974, 2, 97-100.

Capaldi, E. D., & Singh, R. Percentage body weight and the successive negative contrast effect in rats. LEARNING AND MOTIVATION, 1973, 4, 405-416.

Capaldi, E. J. The successive negative contrast effect: Intertrial interval, type of shift and four sources of generalization decrement. JOURNAL OF EXPERIMENTAL PSYCHOLOGY, 1972, 96, 433-438.

Capaldi, E. J., & Lynch, D. Repeated shifts in reward magnitude: Evidence in favor of an associational and absolute (noncontextual) interpretation. JOURNAL OF EXPERIMENTAL PSYCHOLOGY, 1967, 75, 226-235.

Capaldi, E. J., & Ziff, D. R.   Schedule of partial reward and the
    negative contrast effect. JOURNAL OF COMPARATIVE AND PHYSIO-
    LOGICAL PSYCHOLOGY, 1969, 68, 593-596.

Catania, A. C.   Behavioral contrast in a multiple and concurrent
    schedule of reinforcement. JOURNAL OF THE EXPERIMENTAL ANALY-
    SIS OF BEHAVIOR, 1961, 4, 335-342.

Catania, A. C., & Gill, C. A.   Inhibition and behavioral contrast.
    PSYCHONOMIC SCIENCE, 1964, 1, 257-258.

Cochrane, T. L., Scobie, S. R., & Fallon, D.   Negative contrast in
    goldfish (Carassius auratus). BULLETIN OF THE PSYCHONOMIC
    SOCIETY, 1973, 1, 411-413.

Collier, G., Knarr, F. A., & Marx, M. H.   Some relations between the
    intensive properties of the consummatory response and rein-
    forcement. JOURNAL OF EXPERIMENTAL PSYCHOLOGY, 1961, 62, 484-
    495.

Cox, W. M.   A review of recent incentive contrast studies involving
    discrete-trial procedures. THE PSYCHOLOGICAL RECORD, 1975,
    25, 373-393.

Crespi, L. P.   Quantitative variation of incentive and performance
    in the white rat. AMERICAN JOURNAL OF PSYCHOLOGY, 1942, 55,
    467-517.

Crespi, L. P.   Amount of reinforcement and level of performance.
    PSYCHOLOGICAL REVIEW, 1944, 51, 341-357.

Cross, H. A., & Boyer, W. N.   Evidence of a primary frustration
    effect following quality reduction in the double runway.
    JOURNAL OF EXPERIMENTAL PSYCHOLOGY, 1974, 102, 1069-1075.

Daly, H. B.   Learning of hurdle-jump response to escape cues paired
    with reduced reward or frustrative nonreward. JOURNAL OF
    EXPERIMENTAL PSYCHOLOGY, 1969, 79, 146-157.

Daly, H. B.   Reinforcing properties of escape from frustration
    aroused in various learning situations. In G. H. Bower (Ed.),
    THE PSYCHOLOGY OF LEARNING AND MOTIVATION, Vol. 8. New York:
    Academic Press, 1974.

DiLollo, V., & Beez, V.   Negative contrast effects as a function of
    magnitude of reward decrements. PSYCHONOMIC SCIENCE, 1966, 5,
    99-100.

Dunham, P. J.   Contrasted conditions of reinforcement: A selective
    critique. PSYCHOLOGICAL BULLETIN, 1968, 69, 295-315.

Ehrenfreund, D.   Effect of drive on successive magnitude shift in
    the rats. JOURNAL OF COMPARATIVE AND PHYSIOLOGICAL PSYCHOLOGY,
    1971, 76, 418-423.

Elliott, M. H. The effect of change of reward on the maze performance of rats. UNIVERSITY OF CALIFORNIA PUBLICATIONS IN
PSYCHOLOGY, 1928, 4, 19-30.

Flaherty, C. F., & Avdzej, A. Bidirectional contrast as a function
of rate of alternation of two sucrose solutions. BULLETIN OF
THE PSYCHONOMIC SOCIETY, 1974, 4, 505-507.

Flaherty, C. F., & Caprio, M. The dissociation of instrumental and
consummatory measures of contrast. AMERICAN JOURNAL OF
PSYCHOLOGY, 1976, 89, 485-498.

Flaherty, C. F., & Kelly, J. Effect of deprivation state on successive negative contrast. BULLETIN OF THE PSYCHONOMIC SOCIETY,
1973, 1, 365-367.

Flaherty, C. F., & Largen, J. Within-subjects positive and negative
contrast effects in rats. JOURNAL OF COMPARATIVE AND PHYSIO-
LOGICAL PSYCHOLOGY, 1975, 88, 653-664.

Flaherty, C. F., & Lombardi, B. R. Effect of prior differential
taste experience on retention of taste quality. BULLETIN OF
THE PSYCHONOMIC SOCIETY, 1977, 9, 391-394.

Flaherty, C. F., Capobianco, S., & Hamilton, L. W. Effect of septal
lesions on retention of negative contrast. PHYSIOLOGY AND
BEHAVIOR, 1973, 11, 625-631.

Flaherty, C. F., Riley, E. P., & Spear, N. E. Effects of sucrose
concentration and goal units on runway behavior in the rat.
LEARNING AND MOTIVATION, 1973, 4, 163-175.

Freeman, B. J. Behavioral contrast: Reinforcement frequency or
response suppression? PSYCHOLOGICAL BULLETIN, 1971, 75, 347-
356.

Gamzu, E., & Schwartz, B. The maintenance of key pecking by stimulus-
contingent and response-independent food presentation. JOURNAL
OF THE EXPERIMENTAL ANALYSIS OF BEHAVIOR, 1973, 19, 65-72.

Gleitman, H., & Steinman, F. Retention of runway performance as a
function of proactive inhibition. JOURNAL OF COMPARATIVE AND
PHYSIOLOGICAL PSYCHOLOGY, 1963, 56, 834-838.

Gleitman, H., & Steinman, F. Depression effect as a function of retention interval before and after shift in reward magnitude.
JOURNAL OF COMPARATIVE AND PHYSIOLOGICAL PSYCHOLOGY, 1964, 57,
158-160.

Gonzalez, R. C. Patterning in goldfish as a function of magnitude
of reinforcement. PSYCHONOMIC SCIENCE, 1972, 28, 53-55.

Gonzalez, R. C., & Champlin, G. Positive behavioral contrast, negative simultaneous contrast and their relation to frustration
in pigeons. JOURNAL OF COMPARATIVE AND PHYSIOLOGICAL PSYCHOL-
OGY, 1974, 87, 173-187.

Gonzalez, R. C., & Powers, A. S.   Simultaneous contrast in goldfish.
     ANIMAL LEARNING & BEHAVIOR, 1973, 1, 96-98.

Gonzalez, R. C., Fernhoff, D., & David, F. G.   Contrast, resistance
     to extinction, and forgetting in rats.   JOURNAL OF COMPARATIVE
     AND PHYSIOLOGICAL PSYCHOLOGY, 1973, 84, 562-571.

Gonzalez, R. C., Ferry, M., & Powers, A. S.   The adjustment of gold-
     fish to reduction in magnitude of reward in massed trials.
     ANIMAL LEARNING & BEHAVIOR, 1974, 2, 23-26.

Gonzalez, R. C., Gleitman, H., & Bitterman, M. E.   Some observations
     on the depression effect.   JOURNAL OF COMPARATIVE AND PHYSI-
     OLOGICAL PSYCHOLOGY, 1962, 55, 578-581.

Gonzalez, R. C., Potts, A., Pitcoff, K., & Bitterman, M. E.   Runway
     performance of goldfish as a function of complete and incom-
     plete reduction in amount of reward.   PSYCHONOMIC SCIENCE,
     1972, 27, 305-307.

Gutman, A.   Positive contrast, negative induction, and inhibitory
     stimulus control in the rat.   JOURNAL OF THE EXPERIMENTAL
     ANALYSIS OF BEHAVIOR, 1977, 27, 219-233.

Gutman, A., Sutterer, J. R., & Brush, F. R.   Positive and negative
     behavioral contrast in the rat.   JOURNAL OF THE EXPERIMENTAL
     ANALYSIS OF BEHAVIOR, 1975, 23, 377-383.

Hearst, E., & Jenkins, H. M.   SIGN-TRACKING: THE STIMULUS-REINFORCER
     RELATION AND DIRECTED ACTION.   Austin, Texas: The Psychonomic
     Society Press, 1974.

Helson, H.   ADAPTATION-LEVEL THEORY: AN EXPERIMENTAL AND SYSTEMATIC
     APPROACH TO BEHAVIOR.   New York: Harper & Row, 1964.

Hemmes, N. S.   Behavioral contrast in pigeons depends upon the
     operant.   JOURNAL OF COMPARATIVE AND PHYSIOLOGICAL PSYCHOLOGY,
     1973, 85, 171-178.

Homzie, M. J., & Ross, L. E.   Runway performance following a reduc-
     tion in the concentration of a liquid reward.   JOURNAL OF
     COMPARATIVE AND PHYSIOLOGICAL PSYCHOLOGY, 1962, 55, 1029-1033.

Hull, C. L.   Goal attraction and directing ideas conceived as habit
     phenomena.   PSYCHOLOGICAL REVIEW, 1931, 38, 487-506.

Hull, C. L.   PRINCIPLES OF BEHAVIOR.   New York: Appleton-Century-
     Crofts, 1943.

Hull, C. L.   A BEHAVIOR SYSTEM.   New Haven: Yale University Press,
     1952.

Hulse, S. H.   Partial reinforcement, continuous reinforcement, and
     reinforcement shift effects.   JOURNAL OF EXPERIMENTAL PSYCHOL-
     OGY, 1962, 64, 451-459.

Karpicke, J., & Hearst, E. Inhibitory control and errorless dis-
    crimination learning. JOURNAL OF THE EXPERIMENTAL ANALYSIS
    OF BEHAVIOR, 1975, 23, 159-166.

Keller, K. The role of elicited responding in behavioral contrast.
    JOURNAL OF THE EXPERIMENTAL ANALYSIS OF BEHAVIOR, 1974, 21,
    249-257.

Kodera, T. L., & Rilling, M. Procedural antecedents of behavioral
    contrast: A re-examination of errorless learning. JOURNAL
    OF THE EXPERIMENTAL ANALYSIS OF BEHAVIOR, 1976, 25, 27-42.

Lehr, R. Partial reward and positive contrast effects. ANIMAL
    LEARNING & BEHAVIOR, 1974, 2, 221-224.

Lowes, G., & Bitterman, M. E. Reward and learning in the goldfish.
    SCIENCE, 1967, 157, 455-457.

Mackintosh, N. J. Reward and aftereffects of reward in the learning
    of goldfish. JOURNAL OF COMPARATIVE AND PHYSIOLOGICAL PSY-
    CHOLOGY, 1971, 76, 225-232.

Mackintosh, N. J. THE PSYCHOLOGY OF ANIMAL LEARNING. London:
    Academic Press, 1974.

Mackintosh, N. J., Little, L., & Lord, J. Some determinants of
    behavioral contrast in pigeons and rats. LEARNING AND MOTI-
    VATION, 1972, 3, 148-161.

Malone, J. C., Jr. Stimulus-specific contrast effects during operant
    discrimination learning. JOURNAL OF THE EXPERIMENTAL ANALYSIS
    OF BEHAVIOR, 1975, 24, 281-289.

Malone, J. C., Jr. Local contrast and Pavlovian induction. JOURNAL
    OF THE EXPERIMENTAL ANALYSIS OF BEHAVIOR, 1976, 26, 425-440.

Malone, J. C., Jr., & Staddon, J. E. R. Contrast effects in main-
    tained generalization gradients. JOURNAL OF THE EXPERIMENTAL
    ANALYSIS OF BEHAVIOR, 1973, 19, 167-179.

Maxwell, F. R., Calef, R. S., Murray, D. W., Shepard, J. C., &
    Norville, R. A. Positive and negative successive contrast
    effects following multiple shifts in reward magnitude under
    high drive and immediate reinforcement. ANIMAL LEARNING &
    BEHAVIOR, 1976, 4, 480-484.

McHose, J. H., & Moore, J. N. Expectancy, salience, and habit: A
    noncontextual interpretation of the effects of changes in the
    conditions of reinforcement on simple instrumental responses.
    PSYCHOLOGICAL REVIEW, 1976, 83, 292-307.

McHose, J. H., & Peters, D. P. Partial reward, the negative contrast
    effect, and incentive averaging. ANIMAL LEARNING & BEHAVIOR,
    1975, 3, 239-244.

Mellgren, R. L. Positive contrast in the rat as a function of num-
    ber of preshift trials in the runway. JOURNAL OF COMPARATIVE
    AND PHYSIOLOGICAL PSYCHOLOGY, 1971, 77, 329-333.

Mellgren, R. L. Positive and negative contrast effects using delayed
    reinforcement. LEARNING AND MOTIVATION, 1972, 3, 185-193.

Mellgren, R. L., Seybert, J. A., Wrather, D. M., & Dyck, D. G. Pre-
    shift reward magnitude and positive contrast in the rat.
    AMERICAN JOURNAL OF PSYCHOLOGY, 1973, 86, 383-387.

Mikulka, P. J., Lehr, R., & Pavlik, W. B. Effect of reinforcement
    schedules on reward shifts. JOURNAL OF EXPERIMENTAL PSYCHOL-
    OGY, 1967, 74, 57-61.

Nation, J. R., Roop, S. S., & Dickinson, R. W. Positive contrast
    following gradual and abrupt increases in reward magnitude
    using delay of reinforcement. LEARNING AND MOTIVATION, 1976,
    7, 571-579.

Nevin, J. A. Differential reinforcement and stimulus control of not
    responding. JOURNAL OF THE EXPERIMENTAL ANALYSIS OF BEHAVIOR,
    1968, 11, 715-726.

Nevin, J. A., & Shettleworth, S. J. An analysis of contrast effects
    in multiple schedules. JOURNAL OF THE EXPERIMENTAL ANALYSIS
    OF BEHAVIOR, 1966, 9, 305-315.

Pavlov, I. P. CONDITIONED REFLEXES. London: Oxford University
    Press, 1927.

Pert, A, & Bitterman, M. E. Reward learning in the turtle. LEARNING
    AND MOTIVATION, 1970, 1, 121-128.

Pert, A., & Gonzalez, R. C. Behavior of the turtle (*Chrysemys picta
    picta*) in simultaneous, successive, and behavioral contrast
    situations. JOURNAL OF COMPARATIVE AND PHYSIOLGOICAL PSYCHOL-
    OGY, 1974, 87, 526-538.

Peters, D. P., & McHose, J. H. Effects of varied preshift reward
    magnitude on successive negative contrast effects in rats.
    JOURNAL OF COMPARATIVE AND PHYSIOLOGICAL PSYCHOLOGY, 1974, 86,
    85-95.

Rachlin, H. Contrast and matching. PSYCHOLOGICAL REVIEW, 1973, 80,
    217-234.

Ratliff, F. MACH BANDS: QUANTITATIVE STUDIES ON NEURAL NETWORKS IN
    THE RETINA. San Francisco: Holden-Day, 1965.

Redford, M. E., & Perkins, C. C., Jr. The role of autopecking in
    behavioral contrast. JOURNAL OF THE EXPERIMENTAL ANALYSIS OF
    BEHAVIOR, 1974, 21, 145-150.

Rescorla, R. A. Pavlovian conditioning and its proper control procedures. PSYCHOLOGICAL REVIEW, 1967, 74, 71-80.

Reynolds, G. S. Behavioral contrast. JOURNAL OF THE EXPERIMENTAL
ANALYSIS OF BEHAVIOR, 1961, 4, 57-71. (a)

Reynolds, G. S. An analysis of interactions in a multiple schedule.
JOURNAL OF THE EXPERIMENTAL ANALYSIS OF BEHAVIOR, 1961, 4,
107-117. (b)

Reynolds, G. S. Relativity of response rate and reinforcement frequency in a multiple schedule. JOURNAL OF THE EXPERIMENTAL
ANALYSIS OF BEHAVIOR, 1961, 4, 179-184. (c)

Reynolds, G. S. Some limitations on behavioral contrast and induction during successive discrimination. JOURNAL OF THE EXPERI-
MENTAL ANALYSIS OF BEHAVIOR, 1963, 6, 131-139.

Rilling, M. Stimulus control and inhibitory processes. In W. K.
Honig & J. E. R. Staddon (Eds.), HANDBOOK OF OPERANT BEHAVIOR.
Englewood Cliffs, N. J.: Prentice-Hall, 1977.

Rilling, M. Caplan, H. J., Howard, R. C., & Brown, C. H. Inhibitory
stimulus control following errorless discrimination learning.
JOURNAL OF THE EXPERIMENTAL ANALYSIS OF BEHAVIOR, 1975, 24,
121-133.

Rosen, A. J. Incentive-shift performance as a function of magnitude
and number of sucrose rewards. JOURNAL OF COMPARATIVE AND
PHYSIOLOGICAL PSYCHOLOGY, 1966, 62, 487-490.

Rosen, A. J., & Ison, J. R. Runway performance following changes in
sucrose rewards. PSYCHONOMIC SCIENCE, 1965, 2, 335-336.

Rosen, A. J., Glass, D. H., & Ison, J. R. Amobarbitol sodium and
instrumental performance changes following reward reduction.
PSYCHONOMIC SCIENCE, 1967, 9, 129-130.

Schrier, A. M. Comparison of two methods of investigating the effect of amount of reward on performance. JOURNAL OF COMPARA-
TIVE AND PHYSIOLOGICAL PSYCHOLOGY, 1958, 49, 117-125.

Schwartz, B. Maintenance of key pecking by response-independent food
presentation: The role of the modality of the signal for food.
JOURNAL OF THE EXPERIMENTAL ANALYSIS OF BEHAVIOR, 1973, 20,
17-22.

Schwartz, B. Discriminative stimulus location as a determinant of
positive and negative behavioral contrast in the pigeon.
JOURNAL OF THE EXPERIMENTAL ANALYSIS OF BEHAVIOR, 1975, 23,
167-176.

Schwartz, B., & Gamzu, E. Pavlovian control of operant behavior.
In W. K. Honig & J. E. R. Staddon (Eds.), HANDBOOK OF OPERANT
BEHAVIOR. Englewood Cliffs, N.J.: Prentice-Hall, 1977.

Schwartz, B., Hamilton, B., & Silberberg, A. Behavioral contrast in the pigeon: A study of the duration of key pecking maintained on multiple schedules of reinforcement. JOURNAL OF THE EXPERIMENTAL ANALYSIS OF BEHAVIOR, 1975, 24, 199-206.

Shanab, M. E., Sanders, R., & Premack, D. Positive contrast in the runway obtained with delay of reward. SCIENCE, 1969, 164, 724-725.

Shimp, C. P., & Wheatley, K. L. Matching to relative reinforcement frequency in multiple schedules with a short component duration. JOURNAL OF THE EXPERIMENTAL ANALYSIS OF BEHAVIOR, 1971, 15, 205-210.

Skinner, B. F. THE BEHAVIOR OF ORGANISMS. New York: Appleton-Century-Crofts, 1938.

Spear, N. E., & Spitzner, J. H. Simultaneous and successive contrast effects of reward magnitude in selective learning. PSYCHOLOGICAL MONOGRAPHS, 1966, 80 (10, Whole No. 618).

Spence, K. W. BEHAVIOR THEORY AND CONDITIONING. New Haven: Yale University Press, 1956.

Staddon, J. E. R. Multiple fixed-interval schedules: Transient contrast and temporal inhibition. JOURNAL OF THE EXPERIMENTAL ANALYSIS OF BEHAVIOR, 1969, 12, 583-590.

Terrace, H. S. Discrimination learning with and without "errors." JOURNAL OF THE EXPERIMENTAL ANALYSIS OF BEHAVIOR, 1963, 6, 1-27.

Terrace, H. S. Stimulus control. In W. K. Honig (Ed.), OPERANT BEHAVIOR: AREAS OF RESEARCH AND APPLICATION. New York: Appleton-Century-Crofts, 1966.

Terrace, H. S. Discrimination learning, the peak shift, and behavioral contrast. JOURNAL OF THE EXPERIMENTAL ANALYSIS OF BEHAVIOR, 1968, 11, 727-741.

Terrace, H. S. Escape from S-. LEARNING AND MOTIVATION, 1971, 2, 148-163.

Terrace, H. S. By-products of discrimination learning. In G. H. Bower (Ed.), THE PSYCHOLOGY OF LEARNING AND MOTIVATION, Vol. 5. New York: Academic Press, 1972.

Tinklepaugh, O. L. An experimental study of representative factors in monkeys. JOURNAL OF COMPARATIVE PSYCHOLOGY, 1928, 8, 197-236.

Tolman, E. C. PURPOSIVE BEHAVIOR IN ANIMALS AND MEN. New York: The Century Co., 1932.

Vogel, J. R., Mikulka, P. J., & Spear, N. E. Effect of interpolated extinction and level of training on the "depression" effect. JOURNAL OF EXPERIMENTAL PSYCHOLOGY, 1966, 72, 51-60.

Vogel, J. R., Mikulka, P. J., & Spear, N. E. Effects of shifts in sucrose and saccharine concentrations on licking behavior in the rat. JOURNAL OF COMPARATIVE AND PHYSIOLOGICAL PSYCHOLOGY, 1968, 66, 661-666.

Wasserman, E. A. The effect of redundant contextual stimuli on auto-shaping the pigeon's keypeck. ANIMAL LEARNING & BEHAVIOR, 1973, 1, 198-206.

Weinstein, L. Negative incentive contrast with sucrose. PSYCHONOMIC SCIENCE, 1970, 19, 13-14.

Weinstein, L. Negative incentive contrast effects with saccharin vs. sucrose and partial reinforcement. PSYCHONOMIC SCIENCE, 1970, 21, 276-278.

Weisman, R. G. Some determinants of inhibitory stimulus control. JOURNAL OF THE EXPERIMENTAL ANALYSIS OF BEHAVIOR, 1969, 12, 443-450.

Weisman, R. G. Factors influencing inhibitory stimulus control: Differential reinforcement of other behavior during discrimination training. JOURNAL OF THE EXPERIMENTAL ANALYSIS OF BEHAVIOR, 1970, 14, 87-91.

Westbrook, R. F. Failure to obtain positive contrast when pigeons press a bar. JOURNAL OF THE EXPERIMENTAL ANALYSIS OF BEHAVIOR, 1973, 20, 499-510.

Williams, D. R. Negative induction in instrumental behavior reinforced by central stimulation. PSYCHONOMIC SCIENCE, 1965, 2, 341-342.

Wilton, R. N., & Clements, R. O. Behavioral contrast as a function of the duration of an immediately preceding period of extinction. JOURNAL OF THE EXPERIMENTAL ANALYSIS OF BEHAVIOR, 1971, 16, 425-428.

Wilton, R. N., & Clements, R. O. A failure to demonstrate behavioral contrast when the S+ and S- components of a discrimination schedule are separated by about 23 hours. PSYCHONOMIC SCIENCE, 1972, 28, 137-139.

8. REWARD TRAINING: EXTINCTION

M. E. Rashotte

Florida State University
Tallahassee, Florida, U.S.A.

A response established in reward training weakens following a downshift in the value of S^v (Chapter 7). It is not surprising, therefore, that complete elimination of S^v from the training situation also results in weakened responding. The procedure of eliminating S^v after the response has been established through training and the behavioral result of that procedure are both commonly termed extinction. For example, it might be said that "extinction began after four sessions of continuous reinforcement" (the procedure) or that "the response extinguished (i.e., weakened) very rapidly" (the result). In the latter case, it would also be common for the response to be described as having little "resistance to extinction."

Why does responding weaken in extinction? A straightforward possibility is that the associations established in reinforced training are erased when responding is non-rewarded. Unfortunately, the evidence indicates that this simple explanation is not viable. Among other things, it implies that an animal will be functionally untrained after extinction. Yet, after responding has been weakened to its pre-training level in extinction, it will recover a substantial amount of its former strength after an interval without training (e.g., Ellson, 1938; Miller & Stevenson, 1936). This result represents an instance of the "spontaneous recovery" of an extinguished response (Pavlov, 1927), and it implies that weakened responding in extinction is not solely attributable to the erasure of associations established in training. This is not to say that no associative strength is lost in extinction, but only that other

241

factors seem to be influential determinants of the observed response decrement. The present chapter reviews the most important empirical evidence and theoretical developments that have resulted from the search for these other factors.

BASIC EMPIRICAL RELATIONSHIPS

It will be helpful to identify basic empirical relationships around which a discussion of extinction in reward training must be organized. These relationships involve three variables which have a particularly important influence on performance in extinction: the amount of training prior to extinction, the magnitude of reward, and the schedule on which the reward is presented during training.

Consider, first, a central finding about reward magnitude and the amount of continuously reinforced training. In an experiment that was influential in its time, Williams (1938) trained rats to press a lever for food reward on a continuous reinforcement schedule and varied the number of reinforced presses made before extinction began. He reasoned that the more often the response was reinforced, the more often it would be repeated without reinforcement in extinction. Figure 1 (panel A) shows the number of presses made by the different groups in extinction before a criterion of 5 min without a press was reached. The data show that, as Williams expected, resistance to extinction was positively related to the number of reinforced occurrences of the response. For many years this result was viewed as representative of the relationship between resistance to extinction and amount of continuously reinforced training (Hull, 1943, Perin, 1942). However, later experiments obtained precisely the opposite result. The first such report was by North and Stimmel (1960) who found that rats reached an extinction criterion in fewer trials if they had received 90 or 135 continuously rewarded trials in a runway than if they had received 45 trials. Figure 1 (panel B) shows the result of Ison's (1962) extension of the North and Stimmel experiment. It is clear from these data that responding becomes less resistant to extinction with larger numbers of reinforced training trials.

While there are a large number of procedural differences between the Williams and Ison experiments, it turns out that the variable most responsible for the conflicting results is the magnitude of reward employed. A large body of evidence indicates that in rats resistance to extinction is inversely related to food-reward magnitude after training on a continuous reinforcement schedule

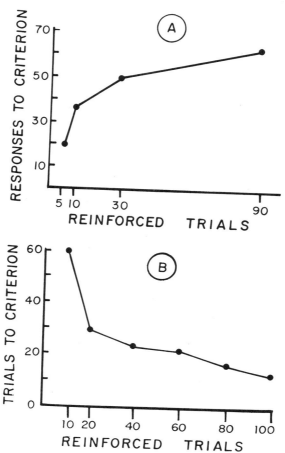

Figure 8.1 Conflicting relationships between number of continuously reinforced training trials and resistance to extinction in rats. Panel A shows the number of lever presses made by four groups of rats before reaching a criterion of 5 min without a press. (After Williams, 1938) Panel B shows the number of non-rewarded runs in a runway by six groups of rats before reaching a criterion of 1 trial without completing the run in 2 min or more. (After Ison, 1962) In each panel, the values on the abscissa indicate the number of continuously reinforced responses made by the group before extinction began.

(e.g., Armus, 1959; Hulse, 1958; Wagner, 1961). In Williams'
experiment, reward was small (one 0.045 g pellet); in Ison's experiment, reward was relatively large (0.4 g wet mash). It does not
seem to matter that the responses were different in the two experiments since results similar to Williams' are obtained in the
runway when reward is small (e.g., Hill & Spear, 1962), and a result
similar to Ison's is obtained with lever pressing when reward is
large (e.g., Traupmann & Porter, 1971). Thus, the functions shown
in Figure 1 illustrate the important fact that resistance to extinction after continuously reinforced training is determined by
an interaction between amount of training and magnitude of reward
in rats trained with food reward. A major theoretical problem is
to account for this interaction.

Consider a second important relationship in extinction. It is
well demonstrated that the rate of performance decrement is
greatly influenced by the schedule of reward used in training. This
finding was reported early (e.g., Humphreys, 1939a,b; Skinner, 1938)
and it has stimulated much research and theorizing in the intervening years. The so-called partial reinforcement extinction effect
(PRE) is the most studied case. It is the finding that training in
which S^V is omitted on a random subset of trials results in more
resistance to extinction than does training on a continuous reinforcement schedule. It turns out that the PRE, too, is strongly
dependent on magnitude of reward.

To illustrate, consider an experiment by Wagner (1961) in which
rats' running speeds were measured as they ran to food reward in a
runway. Only one trial was given each day. The top panel of Figure
2 shows performance in extinction by four groups of rats trained with
a small food reward. S^V was presented either on all trials or on a
random 50% of trials, and within each reinforcement condition, two
groups received different amounts of training prior to extinction
(16 or 60 trials). The results indicated no significant effect of
reinforcement schedule but an overall higher level of performance
by the groups given the larger number of training trials. In contrast, the bottom panel of Figure 2 shows that the same training
conditions with large food reward result in a sizeable PRE but no
effect of the different numbers of acquisition trials.

Several points may be noted about Wagner's results. First,
they are representative of a large number of findings which indicate that the size of the PRE is positively related to magnitude
of S^V (e.g., Gonzalez & Bitterman, 1969; Hulse, 1958; Ratliff &
Ratliff, 1971). Second, that resistance to extinction was positively related to amount of training when S^V was small is consistent with many other findings (e.g., D'Amato, Schiff, & Jagoda,
1962; Hill & Spear, 1962; Williams, 1938). Third, Wagner's data

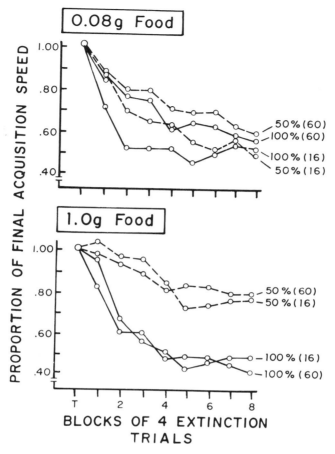

Figure 8.2   Resistance to extinction of the rat's running response
as a function of reward magnitude, schedule of reinforcement and amount of training. Each panel shows
four groups, two trained on a 50% reinforcement schedule,
two on a 100% schedule. Within each reinforcement
schedule, one group received 16 training trials prior
to extinction, the other received 60 trials. Data in
the top panel are from groups trained with small food
reward (0.08 g pellet); the bottom panel shows groups
trained with large reward (1.0 g pellet). The data are
from a measure taken in the middle section of a runway
and are expressed as a proportion of terminal speed at
the end of reinforced training. The points above "T"
on the abscissa are for terminal acquisition performance. (After Wagner, 1961)

indicate that when reward is large the magnitude of the PRE is not greatly increased by extended training. The PRE has been demonstrated with large food reward after as few as five or six trials (e.g., McCain, 1966), especially when the food is presented in multiple pellets or in some other form that insures repeated approaches to the food cup on rewarded trials (Amsel, Hug, & Surridge, 1968). Fourth, to some extent the effects of reward magnitude and amount of training shown in Figure 2 were dependent upon where performance was measured in the runway. In particular, in a measure close to the goalbox the PRE was found with both reward sizes, but it was larger in the large-reward condition. Also in the goal measure, amount of training did not influence resistance to extinction in the small-reward condition. These findings show that the effects of certain variables may depend upon the place in the response chain where performance is measured. Related findings were noted in the discussion of the effects of partial reinforcement schedules on performance during training (Chapter 5). Finally, a particularly important feature of Wagner's procedure is that the trials were separated by 24-hr intervals. This result is representative of many which show the PRE in spaced-trials training (e.g., Weinstock, 1954), even when trials are spaced 72 hr apart (Rashotte & Surridge, 1969).

Taken together, these points indicate that a successful theoretical account of the PRE must explain how resistance to extinction is influenced by reward schedule, magnitude of reward, amount of training, and distance of the performance measure from the site of reinforcement. Furthermore, it must do so by appealing to processes that can operate over very long intertrial intervals. Needless to say, this is not a trivial theoretical problem. To complicate matters, it will become clear later in this chapter that other variables must be incorporated by a successful theory as well. For example, the variables cited above do not include the sequence of rewarded and non-rewarded trials during partial reinforcement, which turns out to be very important. Also, the empirical relationships described above are based on experiments with rats; the evidence shows differences in some of these relationships as a function of species.

THEORETICAL APPROACHES TO EXTINCTION

Theoretical proposals about the source of weakened responding in extinction have been greatly influenced by the empirical relationships described above. The theories, in turn, have stimulated additional discoveries about these relationships and others which indicate the complexity of the factors influencing performance in extinction. In the present section, the sources of weakened responding identified in the major theories of extinction are considered and new findings to which these theories have led are illustrated.

Inhibition

Pavlov's (1927) influential analysis of extinction suggested to many theorists that an inhibitory process might account for extinction in reward training. Pavlov proposed that non-reinforced presentations of CS result in the accumulation of internal inhibition which counteracts conditioned excitatory tendencies to respond to that CS (See Chapter 2, this volume). In his view, the growth of inhibition in extinction accounted for the observed loss in response strength; the labile nature of inhibition allowed rapid recovery of response strength following extinction, as found in spontaneous recovery.

Hull's (1943) account is the best known of the inhibition analyses that applied to extinction in instrumental training. He proposed that every occurrence of a response produces a tendency not to make that same response again. He termed this tendency "reactive inhibition," and viewed it as being akin to effector fatigue. Reactive inhibition was assigned two notable properties. First, its effect on performance was minimized whenever the response produced a reinforcer that increased the tendency to respond. In the absence of reinforcement in extinction, reactive inhibition would occur unchecked and would suppress responding. Second, reactive inhibition dissipated across a short time interval following each response so that it could be a source of response decrement only when extinction trials are massed. Because performance weakens when trials are widely spaced, too, Hull included a second, more permanent, source of inhibition. He proposed that reactive inhibition becomes conditioned to the apparatus cues. In its conditioned form, inhibition would be relatively independent of trial spacing.

Hull's theory encountered many difficulties over the years, not the least of which was the problem of clearly specifying the source of reinforcement for the conditioning of inhibition (Hull, 1943; Koch, 1954). Although this and other logical difficulties might have been circumvented by theoretical revisions, a more serious problem was posed by the nature of performance in extinction. For example, his theory cannot account for the empirical relationships described in Figures 1 and 2. The theory assumed that the stronger the associative tendency established in training, the greater the inhibition (and, therefore, the greater the number of non-reinforced responses) necessary to overcome that tendency. Since Hull's was an S-R reinforcement theory, he must expect that a stronger association would be established when reward is large rather than small and when reward occurs on every trial rather than on 50% of the trials. Obviously, the theory makes the wrong predictions in these cases. Furthermore, extinction is often characterized by the appearance of distinctive new responses which cause the trained response to weaken (e.g., Miller & Stevenson, 1936). This result, too, is not expected from inhibition theory which implies a gradual weakening of the response across trials as the strength of the inhibitory tendency grows.

As the nature of extinction in reward training was better understood, it became obvious that an inhibitory process of the sort described by Pavlov and by Hull could not account for the facts. Attention turned to other sources of response decrement around which a more suitable theory could be developed.

Change in Stimulus Conditions

An important possibility is that the difference in stimulus conditions between reinforced training and extinction contributes to performance decrement in the same way that a change in stimulus conditions results in weakened responding in a generalization test (Chapter 12). This approach to extinction has been developed by S-R theorists, and is often characterized as the "generalization-decrement hypothesis" of extinction (e.g., Kimble, 1961).

The idea was first proposed to account for the PRE and for the effects of various operant schedules of reinforcement on extinction in massed-trials and free-responding training procedures (Hull, 1952; V. Sheffield, 1949; Skinner, 1950). To illustrate, consider its application in a PRE experiment. The continuously reinforced group receives food on every trial, and it is reasonable to suppose that food particles in the mouth and traces of proprioceptive stimulation from the consummatory response linger for a short time after a trial. When trials are highly massed, the stimulus complex present on each reinforced trial will include these aftereffects of reward and, therefore, they should become associated with the instrumental response. In extinction, the stimulus complex on each trial will be changed in at least two significant ways from that present during training; the aftereffects of reward will no longer be present, and new aftereffects characteristic of non-reward will be introduced which may include stimulation arising from emotional reactions occasioned by extinction (e.g., Miller & Stevenson, 1936; Skinner, 1950). The argument is that performance weakens quickly in extinction after continuous reinforcement because the stimulus to which the animal learned to respond during reinforced training is rather suddenly changed in an important way.

In the partial reinforcement group, the instrumental response will be rewarded in an unpredictable sequence, sometimes on a trial following a reward, sometimes following a non-reward. In this case, the response should become associated with the after-effects of rewarded and non-rewarded trials. Thus, it should continue to be performed relatively strongly when only the after-effects of non-reward occur in extinction.

There are several aspects of this theoretical approach which deserve comment. First, it provides a reasonable account of the PRE when trials are highly massed, but it cannot deal with the spaced-trials PRE without further assumptions about the longevity of trial aftereffects. Second, the theory makes the assumption of S-R reinforcement theory that an association is strengthened only on rewarded trials. Thus, it focuses attention on the stimulus conditions present on those trials. Third, it is possible to use the generalization-decrement logic to account for the effects of reward magnitude on extinction (Capaldi, 1967, 1970). The details of this treatment will not be elaborated here, but a central notion is that the shift from large reward to no reward represents a greater change in the stimulus conditions than does a shift from small to no reward. Fourth, it is intuitively appealing that trial aftereffects should be recognized as determinants of performance in massed-trials training, and there is empirical evidence for this view as well (e.g., see discussion of patterned schedules of reward, Chapter 5). Finally, in Chapter 7, stimulus change arising from short-lived trial aftereffects was presented as an important determinant of successive negative contrast when trials are highly massed. Capaldi (1967) has provided a formal theoretical account of this contrast effect in terms of generalization decrement. This is but one of several links between extinction and the successive negative contrast effect to be noted in this chapter.

The important question remaining is whether the type of theoretical account outlined here applies solely to massed-trials procedures. In the discussion of contrast effects in Chapter 7, it was argued that when trials are widely spaced the occurrence of successive negative contrast implies a learned-expectancy process, but that a stimulus-change account seems sufficient for massed-trials contrast phenomena. Essentially, the same case has been

made in the extinction literature (e.g., Amsel, 1967; Gonzalez &
Bitterman, 1969). However, Capaldi (e.g., 1966, 1967) has argued
that stimulus-change theory also applies to spaced-trials training
and, most important, that it makes correct predictions which are
not forthcoming from other approaches. The key assumption required by Capaldi's analysis is that the aftereffects of a trial
include relatively permanent memories of the events that occurred
on that trial as well as short-lived sensory aftereffects of those
events.

Capaldi reasons in the following way. The events of a given
trial (including the reward received, the characteristics of the
response performed, etc.) will be remembered on a trial occurring
much later if the trial stimuli provide effective cues for
"storing" and, later, "retrieving" these memories. Suppose, for
example, that in the context of a series of rewarded trials, a
rat experiences non-reward in the white goal box of a runway. When
the rat is next placed in a white runway, even several hours later,
the white stimulus should act as a "retrieval cue" for the memory
of non-reward. These remembered events of the previous trial will
then be available as stimuli that function in exactly the same way
as do transient sensory aftereffects on trials in massed-trials
training.

Capaldi's theory has been very influential in recent years,
especially as an account of the PRE. Its workings can be illustrated by considering its main assumptions and showing how they
have stimulated new empirical evidence about the determinants of
performance in extinction.

A key assumption is that extinction should be viewed as an
instance of changing stimulus conditions in whose presence the
instrumental response is performed. The idea is that the stimulus
conditions change from one trial to the next because the aftereffects of successive non-rewards cumulate and form discriminably
different stimuli. For example, on the second extinction trial,
the response will be performed in the presence of the aftereffect
arising from the first non-reward (S^N1, in the theory's terminology);
on the next trial, the stimulus will be characteristic of two successive non-rewards (S^N2); and so on. The theory asserts that the
strength of responding on any given extinction trial will greatly
depend on the strength of the associative tendency between S^Nk and
the instrumental response (where k assumes values of 1 through
the maximum number of trials given). There are two main determinants
of the strength of that tendency. One is the amount of associative
strength established between S^Nk and the instrumental response on
reinforced trials during training. The other is the generalization
of associative strength to S^Nk from similar stimuli (e.g., S^Nk-1).

Capaldi's position on the establishment of associative strength between $S^N k$ and the instrumental response is that of Hull's (1943) S-R reinforcement theory. It is assumed that an $S^N k$-Response association can form only on reinforced trials in training, and that the size of the increment in associative strength will be positively related to the magnitude of reinforcement on that trial. The theory's position on the generalization of associative strength between different values of $S^N k$ is, simply, that it occurs in the same way that generalization occurs between similar exteroceptive stimuli (see Chapter 12). Unfortunately, little is known about how $S^N k$ changes across extinction trials or about the discriminability of its different values. Capaldi (e.g., 1964) has made the assumption that the modification of $S^N k$ across successive extinction trials is described by a simple positive growth function.

In explaining the PRE, the theory focuses attention on two so-called "sequential" variables, the length of the sequences of non-rewarded trials arranged by a partial reinforcement schedule (i.e., "N-length"), and the number of times each of the N-lengths has been followed by a reinforced trial (i.e., "N-R transitions"). Theoretically, N-length determines the values of $S^N k$; N-R transitions produce increments in the strength of association between $S^N k$ and the instrumental response.

Consider an early experiment by Capaldi (1964) which established the importance of these variables in massed-trials procedures. Rats were trained to run to food reward in a runway on either of two schedules, both of which provided reward on 50% of the trials. The schedules differed in the number of consecutive non-rewarded trials that preceded a rewarded trial and, therefore, in the values of $S^N k$ conditioned to running. One schedule ($S^N 3$) arranged for only $S^N 3$ to be conditioned, the other ($S^N 1, 2, 3$) arranged for $S^N 1$, $S^N 2$, and $S^N 3$ to be conditioned. Table 1 illustrates how this was accomplished. It shows the sequences of rewards (R) and non-rewards (N) as they occurred across trials given on each of the first 4 days of the experiment. The values of $S^N k$ assumed to be conditioned in the N-R transitions on each day are also shown.

The experiment also varied the number of N-R transitions before extinction began. This was accomplished by arranging 4, 10, or 20 days of training on the $S^N 1$ and $S^N 1, 2, 3$ schedules. The present discussion is confined to the 4- and 20-day conditions. It can be seen from the table that after 4 days' training a group on the $S^N 3$ schedule will have had four N-R transitions on which $S^N 3$ will be conditioned to the response; a group trained on the $S^N 1, 2, 3$ schedule will have had only two opportunities for each of the values of $S^N k$ to be conditioned.

TABLE 1

Daily Sequences of Reward and Nonrewards
for Two Schedules in
Capaldi's (1964) Experiment

S^N3 SCHEDULE:

Trial:	1	2	3	4	5	6	S^Nk Conditioned
Day: 1	R	N	N	N	R	R	S^N3
2	R	R	N	N	N	R	S^N3
3	R	R	N	N	N	R	S^N3
4	R	N	N	N	R	R	S^N3

S^N1, 2, 3 SCHEDULE:

Trial:	1	2	3	4	5	6	S^Nk Conditioned
Day: 1	R	N	N	N	R	R	S^N3
2	N	R	N	N	R	R	S^N1, S^N2
3	R	R	N	N	N	R	S^N3
4	N	N	R	N	R	R	S^N2, S^N1

Capaldi's reasoning is shown schematically in Figure 3. After
4 days of training, the relative strength of the S^Nk-Response
connections for the S^N3 and S^N1, 2, 3 schedules should be approximately as shown in the upper left panel of the figure. Capaldi
arrived at values for the S^Nk-Response associations by employing
quantitative assumptions from Hull's (1943) theory about the
increments in associaitve strength derived from each N-R transition.
These values are solely intended to illustrate the _relative_
strengths of the associations established by the two schedules.
After 20 training days, associative strength should be asymptotic
in all cases, as shown in the lower left panel of the figure.
For simplicity, these panels omit reference to the assumption
that there will be a generalization gradient of associative
strength extending from each training value of S^Nk to nearby values
of S^Nk.

To arrive at predictions about extinction, Capaldi assumes
that all sources of associative strength at a given value of S^Nk
summate. Using Hull's (1943) quantitative assumptions about
generalization and summation of associative strength, Capaldi

derived estimates of generalized associative strength to values of S^N_k greater than S^N3. These estimates are shown as generalization gradients for each schedule and for the different amounts of training in the right panels of Figure 3. The figure shows that after 4 days of training the groups should respond about the same early in extinction, but the S^N3 group should eventually show greater resistance to extinction. After 20 days, however, the groups should separate relatively early in extinction and the S^N1, 2, 3 group should be more resistant to extinction. Figure 4 shows that this prediction was confirmed in Capaldi's experiment.

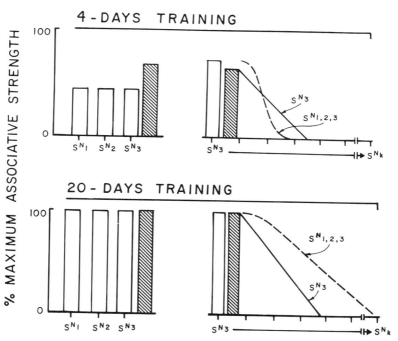

Figure 8.3 Application of Capaldi's theory to extinction after 4 days and 20 days of training on 50% reinforcement schedules in which N-length varies. The open bars show estimated associative strength at different values of S^N_k for groups trained on the S^N1, 2, 3 schedule. The closed bars are for groups trained on the S^N3 schedule. The sloping lines in the right panels show estimated values of generalized associative strength to values of S^N_k greater than S^N3 which occur for the first time in extinction. See text for additional explanation. (After Capaldi, 1964)

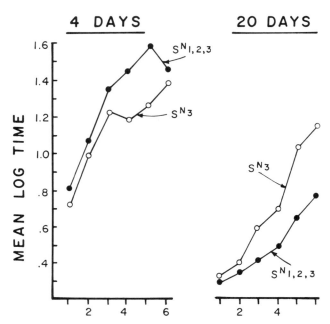

Figure 8.4   Performance on successive days of extinction by S^N1, 2,
3, and S^N3 groups given 4 or 20 days of reinforced
training.   Each data point is the average of the logged
running times on six extinction trials.   In comparing
these data to the theoretical curves in Figure 3
keep in mind that the strength of the response is
inversely related to the time to perform the response.
(After Capaldi, 1964)

        This result (and many others, e.g., Tyler, Wortz, & Bitterman,
1953) shows that a complete theory of extinction must recognize
the role of sequential variables.   The latter statement is demon-
strably true for massed-trials training but is more problematic
when intertrial intervals are very long.   There is conflicting
evidence about the replicability of Capaldi's (1964) result when
trials are more widely spaced (Gonzalez & Bitterman, 1969; Jobe
& Mellgren, 1974).   More generally, sequential effects have been
reported in some experiments with 24-hr intertrial intervals
(Mellgren & Seybert, 1973; Seybert, Mellgren, & Jobe, 1973), but
others have found no sequential effects at comparable intertrial
intervals (e.g., Amsel, Hug, & Surridge, 1969; Surridge & Amsel,
1966) or shorter ones (e.g., Haggbloom & Williams, 1971; Kotesky,
1969; Mackintosh & Little, 1970).   There is recent evidence that

sequential effects fail to occur when trials are widely spaced if the experiment provides inadequate opportunity for the animal to "store" the trial events in memory and to "retrieve" them on subsequent trials (Jobe, Mellgren, Feinberg, Littlejohn, & Rigby, 1977). This is an important possibility which is consistent with Capaldi's emphasis on memorial processes. In any case, Capaldi's (1964) massed-trials experiment discussed here yields a result that is understandable only in terms of sequential variables, and Capaldi's account of it makes an impressive showing. It is re-assuring that the assumptions made in this account about generali-zation, etc., have remained essentially unchanged in subsequent applications of the theory to other extinction phenomena (e.g., Capaldi, 1966, 1967).

Capaldi's theory has stimulated a number of other fruitful lines of inquiry. Suppose, for example, that an animal has just experienced a non-rewarded trial. Capaldi's theory implies that placing the animal directly back in the goalbox and giving reward will replace the non-reward aftereffect with one characteristic of reward. According to this reasoning, it should be possible to manipulate N-length (and thereby resistance to extinction) by interpolating rewarded (or non-rewarded) goalbox placements between trials.

Considerable evidence supports this reasoning. For example, rewarded placements following non-rewarded trials reduce resist-ance to extinction (e.g., Capaldi, 1964; Capaldi, Hart, & Stanley, 1963). However, this finding is not obtained if many placement trials are given during extended training (Homzie, Rudy, & Carter, 1970; Spence, Platt, & Matsumoto, 1965). A complemen-tary finding is that the occurrence of non-rewarded placements among continuously rewarded trials increases resistance to extinc-tion if limited training is given (Homzie, et al., 1970). The probable reason for the reduction in the effectiveness of place-ments with extended training is that placements are eventually discriminated from trials and, therefore, the events are stored differentially in memory. Placements are known to be most ef-fective when the placement- and trial-stimuli are highly similar, even when only a few trials are given (e.g., Capaldi & Spivey, 1963). Some data show that partially rewarded placements (i.e., some non-rewarded and some rewarded placements) interspersed among continuously reinforced trials increases resistance to extinction more so than if non-rewarded placements alone are given (e.g., E. D. Capaldi, 1971; Theios & Polson, 1962). Surprisingly, this result seems to occur only if there are N-R transitions on placement trials (E. D. Capaldi, 1971).

An experiment by Capaldi and Morris (1974) illustrates a recent direction in the theory's application of memorial concepts to extinction phenomena. In this experiment, the first trial of each day always occurred after about a 24-hr intertrial interval the other trials of the day were run highly massed. The idea is that stimuli associated with the longer intertrial interval (termed "STl") will be discriminably different from those associated with the highly massed trials ("STM"). According to the theory, the events on the first trial of the day should be stored in memory in the context of STl; events on subsequent trials should be stored in the context of STM. If this reasoning is correct, it should be possible to show independent effects of different reinforcement schedules programmed on the first and the later trials of the day by arranging for extinction to occur in the presence of the different retrieval cues, STl and STM.

Four groups were trained in all. Two groups always received reward on the first trial of the day, but the subsequent daily trials were rewarded on a partial reinforcement schedule (Group C-P). The other groups received the opposite training; reward and non-reward occurred unpredictably on the first daily trial, all of the remaining trials were rewarded (Group P-C). Figure 5 shows that when extinction trials were given at 24-hr intervals, a PRE occurred that was appropriate for the reinforcement schedule on the first trial of the day (Group P-C showed greater resistance to extinction than group C-P). When extinction trials were highly massed on a single day, however, the relationship between the groups was reversed and the PRE was appropriate for the schedules on the massed trials of each day (Group C-P showed greater resistance to extinction than Group P-C). The results are consistent with the logic of the theory.

In concluding this discussion, it may be noted that Capaldi's work has been a major source of new proposals about the stimuli that regulate instrumental behavior. His theory has adapted and expanded the original Hull-Sheffield version of generalization-decrement theory to include not only sensory aftereffects of trials but long-lasting memories of these trials. The present discussion has emphasized aftereffects generated by the reward and non-reward outcomes of trials. However, Capaldi has also included in his thinking other events that occur on the trial such as distinctive exteroceptive stimuli and the actual characteristics of the response (e.g., Capaldi & Morris, 1974, 1976). The theory focuses attention on conditions conducive to the accurate storage and retrieval of all these events and, in this way, constitutes a unique blend of S-R reinforcement theory with more recent interests in the nature of animal memory (e.g., Medin, Roberts, & Davis, 1976).

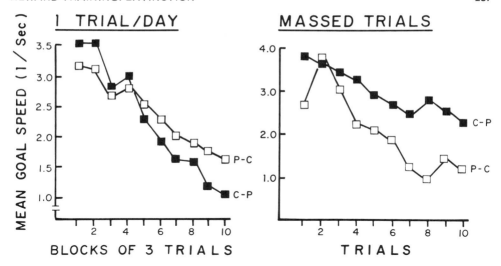

Figure 8.5   Resistance to extinction determined by the nature of
              "retrieval" cues provided by the intertrial interval.
              The two groups in the left panel received one extinction
              trial every day; the groups in the right panel received
              highly massed extinction trials on a single day. Other
              details are given in the text. (After Capaldi & Morris,
              1974)

        Despite its obvious role, stimulus change is not the only
source of response decrement recognized in contemporary theories
of extinction. Another possibility which has been very influential
in terms of S-R theory is discussed in the next section.

Competing Responses

        Several accounts of extinction have emphasized that perform-
ance decrement results from new responses which compete with the
instrumental response (e.g., Guthrie, 1935; Wendt, 1936). These
competing responses often appear to be emotional in nature, and
it is commonly suggested that they are symptomatic of a frustrative reaction similar to that experienced by humans when a response
abruptly ceases to produce its accustomed reward. In a runway
experiment with rats, for example, extinction is often marked by
the appearance of agitated behavior, sudden stopping in the runway followed by abrupt turning and running back towards the startbox (termed "retracing"), urination, vocalizations, biting at the
runway, and so on (e.g., Miller & Stevenson, 1936).

The involvement of competing responses in the PRE is illustrated in an experiment by Amsel, Rashotte, and Mackinnon (1966, Exp. 4). Retraces were measured in rats during extinction of a food-reinforced running response. A large food pellet was used as the reward and the groups received 160 acquisition trials. Figure 6 shows that retracing began earlier in extinction and occurred at an overall higher rate in the continuously reinforced group. Since retracing competes with running, these groups showed a sizeable PRE in the traditional running speed measures of performance. Furthermore, there was evidence that in the continuously reinforced group retracing occurred first near the goalbox and then spread into more remote parts of the runway as extinction progressed. These data imply that the PRE observed in running-response measures results, at least in part, from the differential occurrence of competing responses following partial and continuous reinforcement training. They also suggest that a successful competing response theory of the PRE will make provision for competing responses to occur initially near the former site of reward, and then to migrate into other parts of the response chain.

There have been a number of attempts to construct an explanation of extinction in reward training around the idea of competing emotional responses (Adelman & Maatsch, 1955; Amsel, 1958;

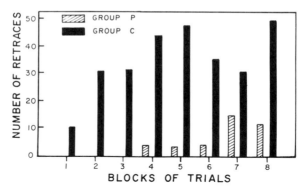

Figure 8.6   Competing responses ("retraces") during extinction of a food rewarded running response in rats trained on continuous reinforcement (Group C) or 50% partial reinforcement (Group P). Each bar shows the number of retraces made by the rats in each group in successive blocks of eight extinction trials. Only the first retrace on any trial was scored for an animal. The maximum score on a block (all animals retracing on all trials) was 72 for each group. (After Amsel, Rashotte, & Mackinnon, 1966)

Rohrer, 1949; Skinner, 1950). The most influential account has been Amsel's (1958, 1967), in which these responses are viewed as a consequence of the violation of a learned expectancy of reward. In his theory, an expectancy is conceptualized as a classically conditioned goal response (r_G-s_G) in the tradition of S-R theory. Thus, Amsel assumes that runway stimuli are analogous to CS in a classical conditioning experiment, and that S^V obtained in the goalbox is analogous to US. As training progresses, the runway stimuli should come to evoke a conditioned response (r_G-s_G) whose properties are specific to the reward obtained. In these respects, the theory requires no assumptions beyond those already articulated in S-R theory (see Chapter 6).

Amsel's fundamental new assumption is that when the learned expectancy of reward (r_G-s_G) is strong, non-reward will evoke an aversive emotional reaction ("frustration"). If r_G-s_G is weak or non-existent, there can be little or no frustration. Because r_G-s_G is said to be classically conditioned, its strength will depend on variables such as the magnitude of US, number of trials, CS-US interval, and so on. For example, r_G-s_G (and therefore frustration) should be greater after many than after few trials with large reward; it may be very weak or absent when reward is small, even after many trials.

A second assumption is that the frustrative reactions evoked in the goalbox when the animal encounters the empty foodcup will themselves become classically conditioned to the runway stimuli if the situation is favorable for conditioning. By this process, a conditioned form of frustration (r_F-s_F) can occur in portions of the apparatus where the animal normally performs the instrumental response. Frustration may also occur there solely through generalization from the goalbox to the runway stimuli.

Finally, Amsel assumes that frustration (in its conditioned or its generalized form) occurring in the runway is the principal source of the emotional responses which compete with the instrumental response.

Consider Amsel's application of these assumptions in explaining the competing response data shown in Figure 6. The continuously reinforced group experiences non-reward in the goalbox for the first time in extinction. If the training conditions have favored the development of strong r_G-s_G, extinction should occasion strong frustrative reactions in the goalbox and conditioned and/or generalized frustration should eventually occur in the runway. Competing responses evoked by frustration should then interfere with performance, and running will weaken rapidly. Note that this logic implies that the competing responses should occur near the goalbox

initially but should spread into the runway later as frustration
begins to occur there. Note also that if r_G-s_G were weak at the
end of continuously reinforced acquisition (e.g., when a small
reward is employed), frustration should be mild or non-existent
in extinction and the instrumental response should weaken relatively
slowly, for reasons other than the presence of competing responses.
Obviously, Amsel's reasoning would account for the differential
effects of reward magnitude in the Ison (1962) and Williams (1938)
experiments discussed earlier.

The situation is quite different for the group trained on
partial reinforcement. It experiences non-rewarded trials randomly intermixed among rewarded trials throughout training. Very
early in training, non-reward cannot be frustrative since r_G-s_G
will still be weak. If the training conditions favor the strengthening of r_G-s_G, a frustrative reaction will eventually be evoked
on non-rewarded trials. Through generalization and conditioning,
frustration should occur in the runway where, weak at first, it
should produce mild disruption of the instrumental response. Given
this situation, Amsel maintains that continued partial reinforcement training will result in fewer competing responses in extinction because stimuli associated with frustration will, themselves,
come to be associated with the instrumental response. The idea is
that r_F-s_F and generalized frustration are present on trials when
the instrumental response is performed, and this allows the formation of r_F-s_F⟶Run associations, for example. Because of this
conditioning process, when frustration occurs on every trial in
extinction it will evoke the instrumental response rather than
competing responses.

Amsel's competing-response account of the PRE has stimulated
a large body of experimentation and has provided the model for a
general theory of persistence which also emphasizes competing
responses (Amsel, 1972). Some of the evidence which led to that
theory will be described here, then the general theory itself, and
finally, some recent evidence about the determinants of competing
responses in extinction which comes from experiments related to
the general theory.

New evidence about the role of competing responses in the PRE
has come from a series of experiments inspired by Amsel's theory
(Amsel & Rashotte, 1969; Rashotte, 1968; Rashotte & Amsel, 1968;
Ross, 1964). The idea was to test the theory's assertion that the
PRE depends on the conditioning to r_F-s_F of a response whose form
does <u>not</u> compete with the instrumental response later subjected
to extinction. In the usual experiment, the partially reinforced
response and the extinguished response have identical topographies.

However, it is possible to dissociate these two response forms.

For example, in a two-phase experiment, three groups of rats
were trained to run to large food reward in a runway (Rashotte
& Amsel, 1968).   Group C received food on every trial and served
as a continuously reinforced reference group. Figure 7 shows that
this group learned to run quickly to the goalbox in the first
phase of the experiment. The other groups received equivalent
experiences with partial reinforcement but in an atypical way.

The rats in Group SLOW obtained reward only when they took
5 sec or more to run from the startbox to the goalbox, a reward
schedule first employed by Logan (1960).   Rats normally traverse
the runway in a second or so on each trial, and to meet the slow-
running criterion they learned idiosyncratic responses which
consumed time. For example, some learned to wait in the startbox
and then to run quickly to the goalbox; others ran quickly to a
place in the runway, stopped and performed distinctive head
movements towards some feature of the runway for several seconds,
and finally ran on. Figure 7 shows that during Phase 1 Group SLOW
took a relatively long time to reach the goalbox. On the average,
these rats met the slow-running criterion on about 45% of the

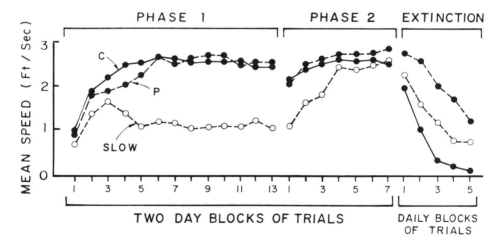

Figure 8.7   Illustration of the fact that the form of the response
              learned during partial reinforcement training influ-
              ences the resistance to extinction of other response
              forms learned later under continuous reinforcement.
              Each group received four trials/day. Other details
              are in the text. (After Rashotte & Amsel, 1968)

trials; on most non-rewarded trials, they ran only fractions of
a second too fast to obtain reward. Thus in Phase 1, the animals
in Group SLOW experienced rewarded and non-rewarded trials in an
irregular order, not unlike that arranged by a partial reinforcement schedule. In terms of Amsel's theory, these animals should
have had their distinctive slow-response rituals associated with
r_F-s_F during Phase 1.

In the other partial reinforcement group (P), each animal
was yoked to one in Group SLOW to receive an identical order of
rewarded and non-rewarded trials irrespective of their running
speed. Figure 7 shows that Group P learned to run quickly to the
goalbox in Phase 1 of the experiment. Amsel's theory states that
r_F-s_F will be conditioned to uninterruped fast running for this
group, as in conventional partial reinforcement training.

In the second phase of the experiment, the groups were
trained in a different colored runway on a continuous reinforcement schedule. Figure 7 shows that all groups had learned to run
quickly in the new runway by the end of Phase 2. Then, extinction
was given in the new runway.

Comparison of Groups C and P in extinction shows the familiar
PRE. However, running extinguished at an intermediate rate for
Group SLOW and, most important, other performance measures showed
that this occurred because the idiosyncratic slow-response rituals
learned by that group in the black runway in Phase 1 reappeared
after a few extinction trials in the white runway. The result
is clearly in line with Amsel's theory. The idea is that when
r_F-s_F occurred during extinction in the white runway it evoked
the response previously associated with it during training in the
black runway. When the evoked response was fast, uninterrupted
running (as it was in Group P), running continued to be relatively
strong for many trials. When the evoked response was idiosyncratic
slow-running (as it was in Group SLOW), extinction was facilitated.
In Group C, r_F-s_F was evoked for the first time in extinction and
the competing responses produced by it resulted in the fastest
performance decrement.

This result and others (particularly Ross, 1964) imply that
the emotional consequences of non-reward can be potent mediators
of responses learned at other times and in other places in the
presence of similar emotional reactions. Amsel's theory provides
a reasonable account of these results and, in fact, has led to
other discoveries about the emotional nature of extinction. For
example, a number of experiments have demonstrated that scheduling
electric shocks intermittently in the goalbox during continuously
reinforced training mimics the effects of partial reinforcement

on resistance to extinction of running (e.g., Brown & Wagner, 1964);
Wagner, 1966). This result is important in showing that experience
with electric shock can alter the disruptive effect of a different
aversive stimulus, frustrative non-reward. Another finding is
that rats will learn to perform a new response to escape from
goalbox cues associated with frustrative non-reward (e.g., Daly,
1974). The implication is that frustration is sufficiently aversive
that its offset will reinforce instrumental responding.

     In view of the evidence, there is no doubt that Amsel's theory
has identified an important constellation of factors which influence
extinction in reward training. It has provided an effective logic
for understanding their varied effects in terms of the operation
of competing responses. Amsel (1972) has now suggested that his
theory of the PRE offers a model for a general theory of persistence, defined as continued responding in the presence of stimuli
which evoke competing responses. The idea is simply expressed by
reference to Figure 8 which describes the steps in developing
response persistence.

     The first stage shows that a response (R_0) is performed in
the presence of an ongoing stimulus situation (S_0). R_0 can be
made persistent by arranging a second stage in which a new stimulus
(S_x) is introduced into the ongoing stimulus situation. S_x is a
stimulus which evokes a competitive-disruptive response (R_x) that

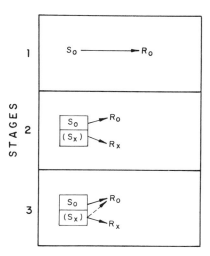

Figure 8.8 Schematic diagram of the stages in making a response
            more persistent, as conceptualized in Amsel's general
            theory of persistence.

interferes with the normally smooth performance of R_O. If R_O continues to be performed, an association will form between S_x and R_O, as represented by the broken line between them in Stage 3. Persistence results from this association because when the original distracting stimulus or another from the class S_x is introduced into the training situation, it, as well as S_O, will evoke the ongoing response, making it predominate over the competing response evoked by S_x. The process of establishing persistence may be described as behavioral habituation to the disruptive stimulus. In fact, Amsel asserts that counterconditioning of this sort underlies behavioral habituation.

Amsel's general theory of persistence is obviously relaxed about specifics such as the members of the class S_x, the variables which influence formation of the association in Stage 3 (e.g., is reinforcement necessary or is contiguity sufficient?), and so on. Nevertheless, it has demonstrated heuristic value. For example, Amsel, Glazer, Lakey, McCuller, and Wong (1973) have studied the effects of presenting a moderately disruptive auditory stimulus during responding on a fixed ratio schedule that produced food reward. They found that this treatment increased resistance to extinction of the lever pressing response, and thereby demonstrated transfer between the effects of two stimuli whose effects are not normally thought comparable (i.e., tone and frustrative non-reward).

More recently, Chen and Amsel (1977) have presented data showing that persistence is increased by extended exposure to stimuli from the class S_x presented independently of R_O. In one of their experiments, rats were initially trained to run to food reward in a runway. One group was trained with continuous reinforcement (C), the other with partial reinforcement (P). When the running response was at full strength, half the animals in each group received extended exposure to intermittent, inescapable electric shocks in a different apparatus. During shock treatment, there was noticeable evidence of behavioral habituation to the disruptive effects of the shocks. However, since the instrumental response was not performed, there was no opportunity for the formation of a shock-->running association which Stage 3 of the theory prescribes as the basis of persistence.

Figure 9 shows how the groups differed in susceptibility to competing responses induced by frustrative non-reward when the running response was subsequently extinguished. It is obvious that the non-shocked groups (C-NS, P-NS) showed the familiar pattern of retracing after training on partial and continuous reinforcement schedules. The new finding is that the groups shocked before extinction began retraced significantly less often than the non-shocked groups. (Of course, resistance to extinction of the running response was inversely related to the strength of re-tracing.) This result indicates that conditioning of S_x to R_O is not the sole basis of increased persistence. This finding portends further developments in the general theory of persistence.

In concluding this discussion, there are several comments which should be made about Amsel's theory, particularly the original theory designed specifically to account for the PRE (Amsel, 1958, 1967).

An aspect of that theory which proved to be a source of some difficulty, as well as a decided virtue, is its reliance on classical conditioning of goalbox reactions. One difficulty is that the theory becomes afflicted with problems common to all such mediational theories (Chapter 11). Another is that the effects of sequential variables are not readily conceptualized in terms of classically conditioned mediating processes (Capaldi, 1967), and this restricts the theory's application considerably. It may be noted that, at this time, sequential theory seems potentially applicable to a wider range of phenomena than does frustration theory, especially given Capaldi's assumption about the memorial character of trial aftereffects.

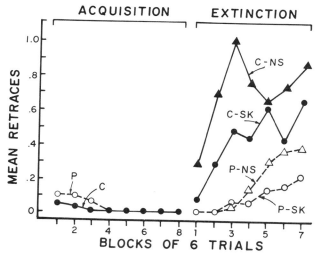

Figure 8.9 Increased persistence by response-independent exposure to electric shock between acquisition and extinction of a running response. The data shown are the mean number of retraces by the various groups in acquisition and extinction. Details of the experiment are discussed in the text. (After Chen & Amsel, 1977, Exp. 2)

On the positive side, however, Amsel's emphasis on classical conditioning allowed the theory to account readily for spaced-trials extinction phenomena, such as the PRE. The classical conditioning literature indicates that conditioned responses develop more quickly and may have a more robust effect on "instrumental" behaviors when the intertrial interval is long relative to the duration of the CS (e.g., Terrace, Gibbon, Farrell, & Baldock, 1975; Woodruff, Conner, Gamzu, & Williams, 1977). On this basis alone, we might expect classically conditioned goalbox reactions to be very influential when reward training trials are widely spaced and, possibly, not to be a factor at all when trials are massed. In fact, for many years, Amsel's theory seemed to provide the most reasonable account of the spaced-trials PRE while sequential theory seemed better suited for massed-trials extinction phenomena (e.g., Amsel, 1967; Gonzalez & Bitterman, 1969). In view of recent evidence about the role of sequential variables in spaced-trials training (see previous section), it is not clear that this distinction can be maintained.

Also on the positive side, it should be noted that the theory's appeal to classical conditioning of goalbox reactions provides a reasonable account of the fact that competing responses spread gradually into the early sections of the runway, and it offers a potential solution to the problem of accounting for different relationships found in performance measures taken at various distances from the goalbox [see discussion of Wagner's (1961) data above]. Other theories do not lend themselves well to these aspects of extinction performance.

A second distinctive aspect of the theory is that it emphasizes that non-reward has important emotional consequences when it occurs in a situation where reward is expected. This fact has been known for many years (e.g., Amsel & Roussel, 1952; Tinklepaugh, 1928). However, Amsel's theoretical formulation has allowed many implications of this aspect of extinction to be readily seen. A case in point is the implication that extinction and successive negative contrast seem to arise from the same processes, which include an emotional basis for competing responses (see ahead). In fact, the learned-expectancy account of negative contrast presented in Chapter 7 is closely fashioned after the logic of Amsel's theory.

A feature of the theory which is sometimes overlooked is that it does not assert that competing responses are the only source of weakened responding in extinction. For example, the running speed of a partially reinforced group eventually weakens even though competing responses are relatively rare. Also, responding deteriorates in extinction after training with small reward even

though frustration-mediated competing responses should be weak or
absent. The clear implication is that a variety of factors contribute to extinction. Competing responses seem foremost when
reward is large and training has been on continuous reinforcement.

Finally, it may be noted that the theory takes no specific
stand on certain basic issues such as whether reinforcement is
necessary for the formation of associations. Thus, there is some
uncertainty about the factors which regulate growth of associative
strength between the runway stimuli and r_G-s_G or r_F-s_F and between
the feedback stimuli of these classically conditioned responses
and the instrumental response. The theory is cast in the general
context of Spence's (1956) version of S-R theory which, itself,
was tentative on these points. In this respect, the theory is
open-ended and stands in contrast to the explicit acknowledgement
of the principles of Hull's S-R reinforcement theory in Capaldi's
theory of extinction.

Successive Negative Contrast

The preceding discussion points to the conclusion that massed-
trials extinction phenomena are interpretable by a single theory,
but that spaced-trials extinction seems to be regulated by a
variety of processes. The adequacy of generalization-decrement
theory (e.g., Capaldi, 1966) in the massed-trials case is not dis-
puted; some evidence favors that theory as well as competing-
response theory (e.g., Amsel, 1958) when trials are widely spaced.
The situation is particularly uncertain when theoretical accounts
of the spaced-trials PRE are considered.

Gonzalez and Bitterman (1969) have proposed that the spaced-
trials PRE is solely determined by reward magnitude effects in the
continuously reinforced group. The idea is that as reward magnitude
increases, a continuously reinforced group (C) becomes more likely
to show a successive negative contrast effect when S^V is downshifted to a zero value in extinction. The partial reinforcement
group (P) will be protected from contrast by partial reinforcement
training (see Chapter 7). When extinction follows training with
large reward, the PRE will occur because contrast causes responding
to weaken rapidly in Group C relative to Group P which extinguishes
at the normal rate without benefit of contrast. When reward is
small, the downshift to zero-value S^V in extinction will not generate contrast in either group, and the PRE will not occur.

Before discussing implications of the contrast theory of the
spaced-trials PRE, it is important to show that successive nega-

tive contrast occurs when S^V is completely eliminated in extinction. In the usual contrast experiment, S^V is downshifted to some small but above-zero value of S^V. It is conceivable, for example, that contrast only occurs when the animal encounters a smaller reward in the post-shift period. Were this the case, the value of treating extinction as a contrast phenomenon would be problematic.

Gonzalez and Bitterman (1969) provided the evidence that contrast occurs in spaced-trials extinction. They trained two groups of rats to run for food reward on a continuous reinforcement schedule. One group received small reward, the other large reward. Since successive negative contrast is more sizeable, the larger the discrepancy between the pre- and post-shift values of S^V (Chapter 7), it was expected that contrast would be more evident in the large-reward group. Of course, a demonstration of successive negative contrast requires a control group trained only with the post-shift value of S^V, so Gonzalez and Bitterman included a third group trained throughout with no food reward in the runway.

Figure 10 shows that in acquisition response latency was lower with 18-pellet than with 1-pellet reward, and that the control group ran most slowly. In extinction, responding weakened gradually across several days in the 1-pellet group, and ultimately reached the level maintained by the control group. However, in the 18-pellet group, responding weakened much more rapidly (its curve actually crossed over the 1-pellet curve) and eventually overshot the performance of the reference group as is required in a successful demonstration of successive negative contrast. Thus, we may take it as demonstrated that, in rats trained with food reward, successive negative contrast is a factor regulating performance in spaced-trials extinction following continuous reinforcement with large but not small reward.

The contrast hypothesis of the spaced-trials PRE has stimulated a number of new discoveries about the determinants of performance in extinction and has provided a framework in which some apparently unrelated extinction findings can be organized. The heuristic value of the theory comes from its assertion that the variables which influence the spaced-trials successive negative contrast effect will also influence resistance to extinction in spaced-trials training. Consider two cases.

First, in Chapter 7 it was shown that successive negative contrast in rats is weakened by a long delay between training with large food reward and the shift to small food reward. It follows that extinction phenomena which depend on contrast will be eliminated by imposing a long delay between acquisition and extinction. Gonzalez, Fernhoff, and David (1973) tested this implication in two ways. In one experiment, they trained different groups of rats to run to large or small food reward on a continuous reinforcement schedule and began extinction 1 or 68 days after the last acquisition trial. Figure 11 shows that large reward produced lower response latencies than small reward at the end of acquisition. When extinction began 1 day later, running extinguished more rapidly in the large-reward group, whose curve actually crossed the small-reward group's curve. This result replicates the finding of Gonzalez and Bitterman shown in Figure 10. After a 68-day delay, however, running extinguished at equivalent rates in the small- and large-reward groups, mainly because the longer delay <u>increased</u> resistance to extinction of the large-reward group. It is entirely consistent with the contrast hypothesis that the long delay should primarily influence resistance to extinction following training with large reward.

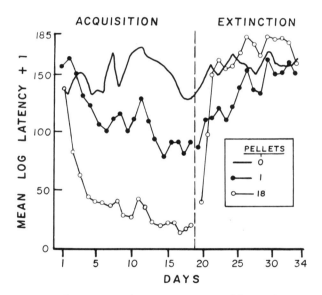

Figure 8.10 Successive negative contrast effect in rats during extinction as a function of reward magnitude. For the first 19 days of training, one group received 18 (0.045 g) food pellets following each run, another received 1 pellet, and a control group received no pellets. Beginning on Day 20, all groups received no pellets. There were four training trials on each day with a 1-hr intertrial interval. (After Gonzalez & Bitterman, 1969)

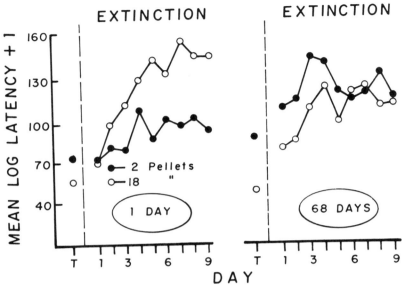

Figure 8.11   Resistance to extinction in rats following training
with different amounts of reward when 1 or 68 days
intervene between acquisition and extinction.   The
data points above "T" in each panel show terminal
acquisition performance of groups trained with 2-
or 18-pellet (0.045 g) reward.   The remaining data
points show performance on each day of extinction.
There were four trials each day with 1-hr intertrial
intervals.   (After Gonzalez, Fernhoff, & David, 1973)

        In another experiment, Gonzalez, et al. (1973), showed that a
68-day delay between acquisition and extinction abolishes the
spaced-trials PRE but not the massed-trials PRE.   In the spaced-
trials case, the principal effect of delay was on resistance to
extinction in the continuously reinforced group.   The implication is
that the spaced-trials and massed-trials PRE are regulated by different factors.   When trials are spaced, negative contrast seems
necessary for the PRE.

The second case to be considered concerns the fact that species differences in resistance to extinction are predicted by contrast theory. Specifically, successive negative contrast has not been demonstrated in goldfish and turtles (see Chapter 7) and, in the absence of contrast, these species should not show contrast-related extinction phenomena. As expected, goldfish and painted turtles (Chrysemys picta picta) do not show the inverse relationship between resistance to extinction and reward magnitude found so reliably in rats following continuous reinforcement. Goldfish show a positive relation between these variables (Gonzalez & Bitterman, 1967; Gonzalez, Holmes, & Bitterman, 1967; Gonzalez, Potts, Pitcoff, & Bitterman, 1972; Mackintosh, 1971); extinction in turtles seems unaffected by reward magnitude (Pert & Bitterman, 1970; Pert & Gonzalez, 1974). Furthermore, the PRE is not found in spaced-trials training in either species (e.g., goldfish: Schutz & Bitterman, 1969; turtles: Pert & Bitterman, 1970), although a massed-trials PRE is readily found (e.g., goldfish: Gonzalez & Bitterman, 1967; turtles: Murillo, Diercks, & Capaldi, 1961). The implication is that the PRE is regulated by different factors in massed- and spaced-trials extinction in these species, as in others. Generalization decrement is the factor in the massed-trials case.

The above discussion illustrates some ways in which contrast theory has provided new perspectives on spaced-trials extinction effects. It would seem very useful to make further comparisons between the variables controlling spaced-trials contrast and extinction. Of course, an appeal to contrast in the account of these extinction effects leaves the origins of spaced-trials contrast itself to be explained. Gonzalez and Bitterman (1969) favor the learned-expectancy account of contrast outlined in Chapter 7. That account has many conceptual ties to Amsel's (1958) competing-response theory, suggesting in yet another way that a competing response account of extinction is useful.

CONCLUSION

At the beginning of this chapter the question posed was, "Why does responding weaken in extinction?" The possibility that associations are simply erased by extinction was rejected in favor of the view that a number of factors are important determinants of performance in extinction. The intervening pages have documented the most significant developments in the search for these factors.

The evidence indicates that the source of response weakening in extinction is highly dependent on the parameters of reward training and the conditions of extinction. Clearly, the magnitude and scheduling of the reward during training are critical, and the temporal spacing between trials seems very important, too. The stimulus-change, competing-response, and contrast theories have provided a conceptual framework for understanding at least some of these effects. However, a full answer to the question posed above is far from being available. It is not at all certain that the final answer will exclude inhibition and the erasure of associations as factors that operate under some conditions. Given the complexity of extinction in reward training, it will probably include other factors as well.

REFERENCES

Adelman, H. M., & Maatsch, J. L. Resistance to extinction as a function of the type of response elicited by frustration. JOURNAL OF EXPERIMENTAL PSYCHOLOGY, 1955, 50, 61-65.

Amsel, A. The role of frustrative nonreward in noncontinuous reward situations. PSYCHOLOGICAL BULLETIN, 1958, 55, 102-119.

Amsel, A. Partial reinforcement effects on vigor and persistence. In K. W. Spence & J. T. Spence (Eds.), THE PSYCHOLOGY OF LEARNING AND MOTIVATION. Vol. 1. New York: Academic Press, 1967.

Amsel, A. Behavioral habituation, counterconditioning, and a general theory of persistence. In A. H. Black & W. F. Prokasy (Eds.), CLASSICAL CONDITIONING II: CURRENT RESEARCH AND THEORY. New York: Appleton-Century-Crofts, 1972.

Amsel, A., Glazer, H., Lakey, J. R., McCuller, T., & Wong, P. T. P. Introduction of acoustic stimulation during acquisition and resistance to extinction in the normal and hippocampally damaged rat. JOURNAL OF COMPARATIVE AND PHYSIOLOGICAL PSYCHOLOGY, 1973, 84, 176-186.

Amsel, A., Hug, J. J., & Surridge, C. T. Number of food pellets, goal approaches, and the partial reinforcement effect after minimal acquisition. JOURNAL OF EXPERIMENTAL PSYCHOLOGY, 1968, 77, 530-534.

Amsel, A., Hug, J. J., & Surridge, C. T. Subject-to-subject trial sequence, odor trials, and patterning at 24-h ITI. PSYCHONOMIC SCIENCE, 1969, 15, 119-120.

Amsel, A., & Rashotte, M. E. Transfer of experimenter-imposed slow-response patterns to extinction of a continuously rewarded response. JOURNAL OF COMPARATIVE AND PHYSIOLOGICAL PSYCHOLOGY, 1969, 69, 185-189.

Amsel, A., Rashotte, M. E., & MacKinnon, J. R. Partial reinforcement effects within subject and between subjects. PSYCHOLOGICAL MONOGRAPHS, 1966, 80 (20, Whole No. 628).

Amsel, A., & Roussel, J. Motivational properties of frustration: I. Effect on a running response of the addition of frustration to the motivational complex. JOURNAL OF EXPERIMENTAL PSYCHOLOGY, 1952, 43, 363-368.

Armus, H. L. Effect of magnitude of reinforcement on acquisition and extinction of a running response. JOURNAL OF EXPERIMENTAL PSYCHOLOGY, 1959, 58, 61-63.

Brown, R. T., & Wagner, A. R. Resistance to punishment and extinction following training with shock or nonreinforcement. JOURNAL OF EXPERIMENTAL PSYCHOLOGY, 1964, 68, 503-507.

Capaldi, E. D. Effect of nonrewarded and partially rewarded placements on resistance to extinction in the rat. JOURNAL OF COMPARATIVE AND PHYSIOLOGICAL PSYCHOLOGY, 1971, 76, 483-490.

Capaldi, E. J. Effect of N-length, number of different N-lengths, and number of reinforcements on resistance to extinction. JOURNAL OF EXPERIMENTAL PSYCHOLOGY, 1964, 68, 230-239.

Capaldi, E. J. Partial reinforcement: A hypothesis of sequential effects. PSYCHOLOGICAL REVIEW, 1966, 73, 459-477.

Capaldi, E. J. A sequential hypothesis of instrumental learning. In K. W. Spence & J. T. Spence (Eds.), THE PSYCHOLOGY OF LEARNING AND MOTIVATION, Vol. 1. New York: Academic Press, 1967.

Capaldi, E. J. An analysis of the role of reward and reward magnitude in instrumental learning. In J. H. Reynierse (Ed.), CURRENT ISSUES IN ANIMAL LEARNING: A COLLOQUIUM. Lincoln, Nebraska: University of Nebraska Press, 1970.

Capaldi, E. J., Hart, D., & Stanley, L. R. Effect of intertrial reinforcement on the aftereffect of nonreinforcement and resistance to extinction. JOURNAL OF EXPERIMENTAL PSYCHOLOGY, 1963, 65, 70-74.

Capaldi, E. J., & Morris, M. D. Reward schedule effects in extinction: Intertrial interval, memory and memory retrieval. LEARNING AND MOTIVATION, 1974, 5, 473-483.

Capaldi, E. J., & Morris, M. D. A role of stimulus compounds in eliciting responses: Relatively spaced extinction following massed acquisition. ANIMAL LEARNING & BEHAVIOR, 1976, 4, 113-117.

Capaldi, E. J., & Spivey, J. E. Effect of goal-box similarity on the aftereffect of nonreinforcement and resistance to extinction. JOURNAL OF EXPERIMENTAL PSYCHOLOGY, 1963, 66, 461-465.

Chen, J., & Amsel, A. Prolonged, unsignaled, inescapable shocks increase persistence in subsequent appetitive instrumental learning. ANIMAL LEARNING & BEHAVIOR, 1977, 5, 377-385.

D'Amato, M. R., Schiff, D., & Jagoda, H. Resistance to extinction after varying amounts of discriminative or nondiscriminative instrumental training. JOURNAL OF EXPERIMENTAL PSYCHOLOGY, 1962, 64, 526-532.

Daly, H. B. Reinforcing properties of escape from frustration aroused in various learning situations. In G. H. Bower (Ed.), THE PSYCHOLOGY OF LEARNING AND MOTIVATION, Vol. 8. New York: Academic Press, 1974.

Ellson, D. G. Quantitative studies of the interaction of simple habits. I. Recovery from specific and generalized effects of extinction. JOURNAL OF EXPERIMENTAL PSYCHOLOGY, 1938, 23, 339-358.

Gonzalez, R. C., & Bitterman, M. E. Partial reinforcement effect in the goldfish as a function of amount of reward. JOURNAL OF COMPARATIVE AND PHYSIOLOGICAL PSYCHOLOGY, 1967, 64, 163-167.

Gonzalez, R. C., & Bitterman, M. E. Spaced-trials partial reinforcement effect as a function of contrast. JOURNAL OF COMPARATIVE AND PHYSIOLOGICAL PSYCHOLOGY, 1969, 67, 94-103.

Gonzalez, R. C., Fernhoff, D., & David, F. G. Contrast, resistance to extinction, and forgetting in rats. JOURNAL OF COMPARATIVE AND PHYSIOLOGICAL PSYCHOLOGY, 1973, 81, 562-571.

Gonzalez, R. C., Holmes, N. K., & Bitterman, M. E. Asymptotic resistance to extinction in fish and rat as a function of interpolated retraining. JOURNAL OF COMPARATIVE AND PHYSIOLOGICAL PSYCHOLOGY, 1967, 63, 342-344.

Gonzalez, R. C., Potts, A., Pitcoff, K., & Bitterman, M. E. Runway performance of goldfish as a function of complete and incomplete reduction in amount of reward. PSYCHONOMIC SCIENCE, 1972, 27, 305-307.

Guthrie, E. R. THE PSYCHOLOGY OF LEARNING. New York: Harper, 1935.

Haggbloom, S. J., & Williams, D. T. Increased resistance to extinction following partial reinforcement: A function of N-length or percentage of reinforcement. PSYCHONOMIC SCIENCE, 1971, 24, 16-18.

Hill, W. F., & Spear, N. E. Resistance to extinction as a joint function of reward magnitude and the spacing of extinction trials. JOURNAL OF EXPERIMENTAL PSYCHOLOGY, 1962, 64, 636-639.

Homzie, M. J., Rudy, J. W., & Carter, E. N. Runway performance in rats as a function of goal-box placements and goal-event sequence. JOURNAL OF COMPARATIVE AND PHYSIOLOGICAL PSYCHOLOGY, 1970, 71, 283-291.

Hull, C. L. PRINCIPLES OF BEHAVIOR: AN INTRODUCTION TO BEHAVIOR THEORY. New York: Appleton-Century-Crofts, 1943.

Hull, C. L. A BEHAVIOR SYSTEM: AN INTRODUCTION TO BEHAVIOR THEORY CONCERNING THE INDIVIDUAL ORGANISM. New Haven: Yale University Press, 1952.

Hulse, S. H. Amount and percentage of reinforcement and duration of goal confinement in conditioning and extinction. JOURNAL OF EXPERIMENTAL PSYCHOLOGY, 1958, 56, 48-57.

Humphreys, L. G. The effect of random alternation of reinforcement on the acquisition and extinction of conditioned eyelid reactions. JOURNAL OF EXPERIMENTAL PSYCHOLOGY, 1939, 25, 141-158. (a)

Humphreys, L. G. Acquisition and extinction of verbal expectations in a situation analogous to conditioning. JOURNAL OF EXPERIMENTAL PSYCHOLOGY, 1939, 25, 294-301. (b)

Ison, J. R. Experimental extinction as a function of number of reinforcements. JOURNAL OF EXPERIMENTAL PSYCHOLOGY, 1962, 64, 314-317.

Jobe, J. B., & Mellgren, R. L. Successive nonreinforcements (N-length) and resistance to extinction at spaced trials. JOURNAL OF EXPERIMENTAL PSYCHOLOGY, 1974, 103, 652-657.

Jobe, J. B., Mellgren, R. L., Feinberg, R. A., Littlejohn, R. L., & Rigby, R. L. Patterning, partial reinforcement, and N-length effects at spaced trials as a function of reinstatement of retrieval cues. LEARNING AND MOTIVATION, 1977, 8, 77-97.

Kimble, G. A. HILGARD & MARQUIS' CONDITIONING AND LEARNING. New York: Appleton-Century-Crofts, 1961.

Koch, S. Clark L. Hull. In W. K. Estes, S. Koch, K. MacCorquodale,
    P. Meehl, C. G. Mueller, W. N. Schoenfeld, & W. S. Verplanck.
    MODERN LEARNING THEORY. New York: Appleton-Century-Crofts,
    1954.

Kotesky, R. L. The effect of unreinforced-reinforced sequences on
    resistance to extinction following partial reinforcement.
    PSYCHONOMIC SCIENCE, 1969, 14, 34-36.

Logan, F. A. INCENTIVE. New Haven: Yale University Press, 1960.

Mackintosh, N. J. Reward and aftereffects of reward in the learning
    of goldfish. JOURNAL OF COMPARATIVE AND PHYSIOLOGICAL PSY-
    CHOLOGY, 1971, 76, 225-232.

Mackintosh, N. J., & Little, L. Effects of different patterns of
    reinforcement on performance under massed or spaced extinc-
    tion. PSYCHONOMIC SCIENCE, 1970, 20, 1-2.

McCain, G. Partial reinforcement effects following a small number
    of acquisition trials. PSYCHONOMIC MONOGRAPH SUPPLEMENTS,
    1966, 1, 251-270.

Medin, D. L., Roberts, W. A., & Davis, R. T. PROCESSES OF ANIMAL
    MEMORY. Hillsdale, N. J.: Lawrence Erlbaum Associates, 1976.

Mellgren, R. L., & Seybert, J. A. Resistance to extinction at
    spaced trials using the within-subject procedure. JOURNAL
    OF EXPERIMENTAL PSYCHOLOGY, 1973, 100, 151-157.

Miller, N. E., & Stevenson, S. S. Agitated behavior of rats during
    experimental extinction and a curve of spontaneous recovery.
    JOURNAL OF COMPARATIVE PSYCHOLOGY, 1936, 21, 205-231.

North, A. J., & Stimmel, D. T. Extinction of an instrumental re-
    sponse following a large number of reinforcements. PSYCHO-
    LOGICAL REPORTS, 1960, 6, 227-234.

Pavlov, I. P. CONDITIONED REFLEXES. London: Oxford University
    Press, 1927.

Perin, C. T. Behavior potentiality as a joint function of the
    amount of training and degree of hunger at the time of ex-
    tinction. JOURNAL OF EXPERIMENTAL PSYCHOLOGY, 1942, 30, 93-
    113.

Pert, A., & Bitterman, M. E. Reward and learning in the turtle.
    LEARNING AND MOTIVATION, 1970, 1, 121-128.

Pert, A., & Gonzalez, R. C. Behavior of the turtle (*Chrysemys picta
    picta*) in simultaneous, successive, and behavioral contrast
    situations. JOURNAL OF COMPARATIVE AND PHYSIOLOGICAL PSY-
    CHOLOGY, 1974, 87, 526-538.

Rashotte, M. E. Resistance to extinction of the continuously re-warded response in within-subject partial reinforcement ex-periments. JOURNAL OF EXPERIMENTAL PSYCHOLOGY, 1968, 76, 206-214.

Rashotte, M. E., & Amsel, A. Transfer of slow-response rituals to extinction of a continuously rewarded response. JOURNAL OF COMPARATIVE AND PHYSIOLOGICAL PSYCHOLOGY, 1968, 66, 432-443.

Rashotte, M. E., & Surridge, C. T. Partial reinforcement and par-tial delay of reinforcement effects with 72-hour intertrial intervals and interpolated continuous reinforcement. QUAR-TERLY JOURNAL OF EXPERIMENTAL PSYCHOLOGY, 1969, 21, 156-161.

Ratliff, R. G., & Ratliff, A. R. Runway acquisition and extinction as a joint function of magnitude of reward and percentage of rewarded acquisition trials. LEARNING AND MOTIVATION, 1971, 2, 289-295.

Rohrer, J. H. A motivational state resulting from non-reward. JOURNAL OF COMPARATIVE AND PHYSIOLOGICAL PSYCHOLOGY, 1949, 42, 476-485.

Ross, R. R. Positive and negative partial reinforcement effects carried through continuous reinforcement, changed motivation, and changed response. JOURNAL OF EXPERIMENTAL PSYCHOLOGY, 1964, 68, 492-502.

Schutz, S. L., & Bitterman, M. E. Spaced-trials partial reinforce-ment and resistance to extinction in the goldfish. JOURNAL OF COMPARATIVE AND PHYSIOLOGICAL PSYCHOLOGY, 1969, 68, 126-128.

Seybert, J. A., Mellgren, R. L., & Jobe, J. B. Sequential effects on resistance to extinction at widely-spaced trials. JOURNAL OF EXPERIMENTAL PSYCHOLOGY, 1973, 101, 151-154.

Sheffield, V. F. Extinction as a function of partial reinforcement and distribution of practice. JOURNAL OF EXPERIMENTAL PSY-CHOLOGY, 1949, 39, 511-526.

Skinner, B. F. THE BEHAVIOR OF ORGANISMS: AN EXPERIMENTAL ANALYSIS. New York: Appleton-Century-Crofts, 1938.

Skinner, B. F. Are theories of learning necessary? PSYCHOLOGICAL REVIEW, 1950, 57, 193-216.

Spence, K. W. BEHAVIOR THEORY AND CONDITIONING. New Haven: Yale University Press, 1956.

Spence, K. W., Platt, J. R., & Matsumoto, R. Intertrial reinforce-ment and the partial reinforcement effect as a function of the number of training trials. PSYCHONOMIC SCIENCE, 1965, 3, 205-206.

Surridge, C. T., & Amsel, A. Acquisition and extinction under
    single alternation and random partial-reinforcement conditions
    with a 24-hour intertrial interval. JOURNAL OF EXPERIMENTAL
    PSYCHOLOGY, 1966, 72, 361-368.

Terrace, H. S., Gibbon, J., Farrell, L., & Baldock, M. D. Temporal
    factors influencing the acquisition and maintenance of an
    auto-shaped keypeck. ANIMAL LEARNING & BEHAVIOR, 1975, 3,
    53-62.

Theios, J., & Polson, P. Instrumental and goal responses in non-
    response partial reinforcement. JOURNAL OF COMPARATIVE AND
    PHYSIOLOGICAL PSYCHOLOGY, 1962, 55, 987-991.

Tinklepaugh, O. L. An experimental study of representative factors
    in monkeys. JOURNAL OF COMPARATIVE PSYCHOLOGY, 1928, 8,
    197-236.

Traupmann, K., & Porter, J. J. The overlearning-extinction effect
    in free-operant bar pressing. LEARNING AND MOTIVATION, 1971,
    2, 296-304.

Tyler, D. W., Wortz, E. C., & Bitterman, M. E. The effect of random
    and alternativing partial reinforcement on resistance to
    extinction in the rat. AMERICAN JOURNAL OF PSYCHOLOGY, 1953,
    66, 57-65.

Wagner, A. R. Effects of amount and percentage of reinforcement and
    number of acquisition trials on conditioning and extinction.
    JOURNAL OF EXPERIMENTAL PSYCHOLOGY, 1961, 62, 234-242.

Wagner, A. R. Frustration and punishment. In R. N. Haber (Ed.),
    CURRENT RESEARCH IN MOTIVATION. New York: Holt, Rinehart
    & Winston, 1966.

Weinstock, S. Resistance to extinction of a running response fol-
    lowing partial reinforcement under widely spaced trials.
    JOURNAL OF COMPARATIVE AND PHYSIOLOGICAL PSYCHOLOGY, 1954,
    47, 318-322.

Wendt, G. R. An interpretation of inhibition of conditioned re-
    flexes as competition between reactions systems. PSYCHOLOGICAL
    REVIEW, 1936, 43, 258-281.

Williams, S. B . Resistance to extinction as a function of the num-
    ber of reinforcements. JOURNAL OF EXPERIMENTAL PSYCHOLOGY,
    1938, 23, 506-521.

Woodruff, G., Conner, N., Gamzu, E., & Williams, D. R. Associative
    interaction: Joint control of key pecking by stimulus-
    reinforcer and response-reinforcer relationships. JOURNAL OF
    THE EXPERIMENTAL ANALYSIS OF BEHAVIOR, 1977, 28, 133-144.

9. PUNISHMENT

J. B. Overmier

University of Minnesota
Minneapolis, Minnesota, USA

In the immediately preceding sections we have been concerned with the instrumental reward learning paradigm, such learning being characterized by the delivery of hedonically positive events contingent upon the occurrence of a specified response. We turn now to the punishment paradigm wherein the occurrence of a specified response results in the delivery of a hedonically negative, noxious event which typically decreases the probability of future occurrences of that response. Thus, the behavioral outcome of the punishment operation is usually the inverse of that of the reward operation. Thorndike (1913) captured this relationship in his statement of the symmetrical Law of Effect. With respect to punishment, he argued that whenever a response resulted in an annoying state of affairs this weakened the stimulus response association which led to that response

That punishment can permanently alter the behavior of an organism has not always been uniformly accepted. Indeed, Thorndike, himself even changed his view on the effectivensss of punishment. This change was based upon some experiments on verbal learning during which students who were told "No" for wrong answers failed to show significant reductions in their errors (Thorndike, 1932). Later experimenters revealed that Thorndike's verbal learning data were artifactual, but it was too late because the myth that punishment was ineffective in weakening responses had been born. This myth was bolstered by some data on punishment reported by Skinner (1938). Skinner studied the effects of mild punishment (a slap on a rat's paw) delivered after just a few bar pressing responses upon the course of long term, complete extinction of that response. He reported that the punishments temporarily suppressed the rate of responding. After the punishment contingency was removed, however, there was compensatory recovery such that after two days the total number of responses generated in extinction by the punished and un-punished groups were equal. This led Skinner to argue that the punishment could not have affected the strength of the stimulus-response association, and whatever suppression of responding that was observed during the punishment must be attributed to some other mechanism, such as a temporary increase in general emotional arousal.

As we shall see in this chapter, the myth that punishment is ineffective in producing permanent alterations in behavioral tend-encies is wrong. Punishment can have very powerful, long-lasting effects (see Boe & Church, 1967). For example, Masserman (1943) reported that if a gust of air were directed into a cat's face while eating, the cat would refuse to eat in that apparatus for months on end -- even though it was severely deprived of food. On the other hand, there exist a number of instances of <u>counter-intuitive</u> effects of punishment. For example, punishment of avoid-ance response often facilitates the behavior, which in turn in-creases the frequency of punishment (Carlsmith, 1961). We will consider several of these counter-intuitive phenomena late in the chapter. But first, we must focus upon the traditional phenomena and the behavioral mechanisms which underlie them.

We shall consider several factors which modulate the outcomes of punishment experiments. In gross summary, three factors deter-mine the effectiveness of punishment:

(1)   the punisher itself, including its physical parameters
       and features of the contingency;
(2)   response strength based upon the conditions of acquisi-
       tion and maintenance of the to-be-punished response; and
(3)   prior experience with punishment, including the method
       of introduction of the punisher.

Space will not allow a thorough treatment of each of these. Our focus will be upon the first of these factors, the punisher, with a general restriction to the effects of punishment upon positively reinforced behavior. This is because aversively maintained re-sponses constitute a special case. This special case will be con-sidered when we finally take up the last factor.

THE PUNISHER

What is a punisher? Earlier we suggested it was a hedonically

negative event, a noxious event. We all have certain intuitions
about what constitute such noxious events and occasionally even
go so far as to test our intuitions by showing that response contingent termination of such events will support escape learning.
But we need to recognize that there is likely some species specificity with respect to the noxiousness of some events. The sight
of a toy snake may prove a very effective punisher for a monkey
but not a mongoose. For reasons of ease of specification, control, and manipulation, plus its broad species applicability, psychologists have chosen <u>electric</u> <u>shock</u> as their prototypic punisher.
This choice has been criticized for "unnaturalness"; after all,
lightening strikes few animals. But the choice is defended by
those experiments which show that parallel effects are obtained
with other, seemingly more natural punishing agents. Even the
use of highly controllable electric shock has problems. For example, the electrical resistance of the animal may change during
experiments thus altering the physical properties of the punisher.

Parameters of Punishment

    With any punisher, including electric shock which is our
prototype, there are a number of parameters to consider. These
include intensity, duration, schedule, and delay. All of these
except delay are related to the overall severity of the punishment operation. Let us consider the important effects of variation in each of these.

    <u>Intensity and duration</u>. A variety of different effects are
observable as a function of the intensity of punishment (Church,
1963). This variety may well be the source of much of the controversy about the effects of punishment, because those engaged in the
controversy are not taking into account how the results of punishment operations differ as a function of intensity. Also, it is
important to recognize that an electric shock intensity that is
just barely detectable by one species may be severely intense for
another. The just detectable intensities of punishers result only
in alerting and "orienting" reactions, sometimes with no disruption
in the baseline rate of responding. Weak punishers lead to <u>tempor-
ary suppression</u>, a suppression of responding followed by nearly
total recovery. Moderately intense punishers result in permanent
<u>partial suppression</u>, while intense punishers result in permanent
and <u>total suppression</u> of responding.

    An experiment by Azrin (1961) obtained results which closely
follow this pattern. Following training to peck a key for food,
pigeons were punished for pecking with various intensities of shocks
delivered to the first response after each 5 minutes (FI 5-min).

The obtained data are presented in Figure 1; especially salient in the figure are the graded suppression effects as a function of intensity, the partial recoveries of responding following introduction of moderate level punishers, and the lack of any recovery during more intense punishment. The figure also illustrates two other points. The first is that prior experiences with punishments influence the effectiveness of a given punishment. In Figure 1, we see that the effects of the first and second applications of a 60-volt punisher are quite different as a function of the intervening experience with the severe 90-volt shock. We shall return to this sequential effect later. The second point is that removal of a punishment contingency results in a rebound or contrast effect, with the rate of responding going to a higher rate than before any punishments. Such contrast effects can even be observed upon removal of a punishment contingency during which the animal has so fully recovered that it appears as if the punisher is no longer having any effect. Such rebounds belie the idea that the punishments no longer are having any effect because the response rate is fully recovered.

It is important to note, however, that removal of a punishment contingency does not always result in a rebound, especially when the punisher is quite severe. As one example, Boe and Church (1967) have shown that after a few minutes of punishment with an intensity of 220 volts, rats not only show no rebound -- they never recover responding. Their results when compared to those of Skinner (1938) re-emphasize the importance of relative intensity of the punisher in determining the pattern of effect that punishments produce.

The effects of various durations of punishers have also been systematically explored. Church, Raymond, and Beauchamp (1967) compared the effects of variation in duration of punishers of up to 3 seconds. The punishers were delivered according to a VI 2-min schedule to rats which had been trained to respond for occasional food rewards. Their results are presented in Figure 2 and are plotted in terms of rate of responding during punishment relative to the rate before punishment plus the rate during punishment (smaller numbers reflect more suppression). A graded effect of duration was observed, with longer duration punishers yielding greater suppression of responding. Also observable with variation in duration are phenomena similar to that obtained with various intensities. Examples from Figure 2 are those of temporary suppression with full recovery (0.3-sec duration case) and that of partial suppression with partial recovery (0.5-sec duration case).

Given that variations in intensity and in duration of the punisher yield similar phenomena, it is not suprising that one can

establish trade-off relationships between these two parameters
(Church, Raymond & Beauchamp, 1967). These we might call iso-
severity functions in which decreases in intensity can be compensated for in terms of behavioral effect by increases in duration.
Similar iso-severity functions may be obtained between intensity
and scheduled frequency of punishment as well (Azrin et al, 1963).
Such iso-severity functions are, of course, unique to each species
and each apparatus.

   <u>Delay</u>. Delay is a parameter of more than empirical interest.
If the effects of punishment are attributable to the contingency
between the response and the punisher, then one would expect that
introducing a delay would reduce the effect of the punisher. And

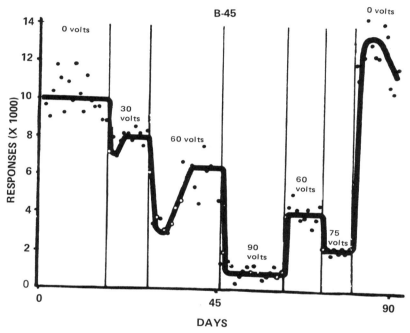

Figure 9.1   Effects of sequential shifts in intensity of shock
             punishments upon responding for food by pigeons. Note
             in particular (a) the large temporary suppression fol-
             lowed by partial recovery upon initial exposure to
             moderate intensities of punishment (first 30 and 60
             volts), (b) the greater degree of suppression to a
             moderate degree of punishment following exposure to
             more severe punishment (second exposure to 60 volts),
             and (c) the "overshoot" of recovery following re-
             moval of punishment (second 0 volts). (After Azrin &
             Holz, 1961)

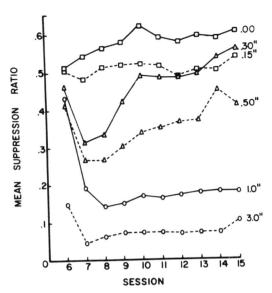

Figure 9.2   Effects of different durations of punishment upon
lever-press responding for food by rats as indexed
by suppression ratio (values of 0.5 imply no suppression, less than 0.5 suppression with value of
0 being complete suppression). Note that the effects
of very brief punishers, like mild punishers, produce temporary suppression followed by recovery,
while longer durations produce increasing levels of
permanent suppression.   (After Church, Raymond &
Beauchamp, 1967)

this seems to be the case; just as delay of presentation of reward
has deleterious effects upon acquisition, delay of application of
a punisher reduces the effectiveness of the punisher.

      The effect of delayed punishment is demonstrated in a study
by Baron (1965). In his experiment, thirsty rats ran down a runway for water reward. After reaching their asymptotic levels of
performance, brief shocks were administered in the goal box after
a delay (of 0-30 sec) which differed across groups; group C received no punishment. The results for the several groups are
shown in Figure 3 which presents the running speeds as a function
of runway section for the several groups. Two facts are clear.
First and foremost, the longer the delay, the less the suppression
of running. Second, it was the running response in the section
nearest to the punishment that was most suppressed.

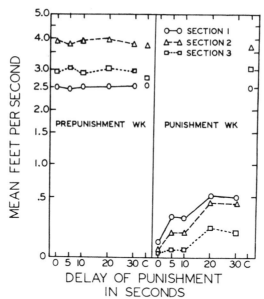

Figure 9.3   Degree of suppression of running to food by rats as
a function of distance from the reward/punishment in
the goalbox and temporal delay of punishment upon
completion of the response. Note that punishment
(a) is less effective with longer delays and (b)
suppresses the more spatially remote responding
(running in Section 1) the least. (After Baron, 1965)

Caution is required in the interpretation of Baron's or
any other experiment which studies the effect of delayed punishment when the punished behavior also continues to be rewarded as
before. This is because confoundings may arise as a by-product
of the delay interval. One is that with different delays, the
punisher can stand in different temporal relations to the consumption of the reward. With 0-sec delay, the animal runs down the
runway, enters the goalbox, gets punished, and consumes the reward; with 15-sec delay, consumption of reward will antedate the
punisher. This shift in temporal sequence leads to yet a second
posible confounding. With 0-sec delay, the running response is
directly followed by the punisher; but, with longer delays, the
running response is separated from the action of the punisher by
other responses, i.e., by "not-running" and by the consumptive response. Thus, with a delayed punisher one may be punishing only
the consumptive response which results in reduction of the instrumental running response only indirectly. Finally, we ought to note

that if the different delays of the punisher result in different
total goalbox exposure times, the result is a third confounding
that will lead to underestimation of the punishment effect. This
is the fact that longer goalbox exposures even with a fixed reward,
result in greater reward effects (Czeh, in Spence, 1956).

Since we have now raised the issue of the contingency between
response and punisher and the role of the contingency in producing
the response suppression commonly observed, let us turn to theoretical conceptions of punishment. Such "theories" are attempts to
specify the necessary and sufficient conditions for the punishment
effect. Which among the response-punisher contiguity, stimulus-
punisher contiguity, and the response-punisher contingency or
dependency is (are) responsible for the behavioral effects observed?

Theories of Punishment

The recent rapid growth of the empirical literature (see Boe,
1969) has not been accompanied by the development of powerful theoretical models or insights. All current theories are little more
than restatements of theories offered 30 or more years ago. Even
then, they are almost afterthoughts tagged onto major statements
about learning in other paradigms. Indeed, punishment is sometimes completely overlooked (Hull, 1943; Tolman, 1932)! In any
case, few of the extant theories purport to explain the variety
of punishment phemomena known. With this grim caveat, let us consider the available conceptualizations of punishment.

There are three major classes of theory to be found in the
literature. They are the (1) Negative Law of Effect, (2) conditioned emotional response, and (3) competing motor response views.
The "purist" might subdivide some of these further. We will comment on this as we proceed with our discussions.

Negative Law of Effect. As noted earlier, this theory comes
to us from Thorndike (1913) who viewed the punishment effect as a
direct weakening of S-R associations. Thus, the action of punishers
was seen as the opposite of that of reward. The view is intuitively
appealing. Logan (1969) has offered an updated version of the
theory which speaks not in terms of associative bonds being altered
but in terms of the conditioned "incentive value" which leads to
response selection. Rewards contribute to incentive while punishers
result in subtractions from incentive. The words are different,
the principle of symmetry essentially the same.

Rachlin and Herrnstein (1969) have provided several demonstrations that "responding is controlled by contingent punishment [in]

... a way that is analogous to the control of responding by positive reward, except that the direction of control is reversed" (p.104). However, one can accept the _empirical_ generalization without agreeing that the mechanisms underlying the two directions of behavioral change are the same. (As a case in point, extinction is not usually assumed to proceed by breaking associative bonds.)

The data that have proved most difficult for the Negative Law of Effect are those in which recovery of responding (either partial or complete) occurs after the introduction of punishment. Especially embarrassing are those cases when recovery occurs during extinction after _both_ rewards and punishers have been withdrawn. Such compensatory recovery is not uncommon with moderate to weak punishers (Skinner, 1938; Boe & Church, 1967, with 75 volt punisher)!

Conditioned Emotional Response. The conditioned emotional response (CER) theory has its roots in Estes and Skinner's (1941) observation that when tone presentations are repeatedly paired with shocks while an animal is responding to get food, the animal soon slows or stops responding during the tone. This slowing of responding during the tone occurs even though the tone and shock presentations are scheduled to occur independently of the animals responding. This is of course, the well known conditioned suppression phenomenon, an effect Estes and Skinner attributed to the conditioning of emotional responses which disrupt the appetitive behavior. Estes was led to propose the CER theory of punishment based upon his experimental observations. In Experiment I (Estes, 1944), he observed that the delivery of shocks independent of responding but in the presence of the "discriminative" stimuli for responding produced about the same degree of suppression as did response-contingent shock punishments. Hence, Estes concluded that the effects of the punisher were not attributable to the contingency between responses and shocks but to the contiguity between the response-controlling stimuli and shocks.

Substantial data exist that confirm the reliability of the CER effect (see Davis, 1968). We need not review it here except to note that it shares with punishment many parametric functional relationships. For example, the amount of conditioned suppression produced by noncontingent shocks is a positive function of intensity and duration of the shocks (see Church, 1969). One contrary observation worth noting is a study by Annau and Kamin (1961); they reported that a response contingent 0.28 mA shock was sufficient to produce response decrements in bar-pressing for food by rats (a punishment effect) but non-contingent presentations would not yield conditioned suppression (no CER effect). This result is a serious challenge to the CER theory of punishment.

Most of the research aimed at testing the CER theory of punishment has revolved about the relative importance of two contingencies: (1) the response-shock contingency or dependency versus (2) a stimulus-shock contiguity. In contrasting these two, one has a problem in the general operant paradigm because it is difficult to identify and manipulate the stimuli which may become associated with the shocks. Therefore, the best tests of this theory use a discriminative paradigm so as to be certain there are discriminative stimuli present and to control the relationship between these stimuli and shocks. Two kinds of experiments have sought to assess the importance of the response-shock contingency; these are delay of punishment and discriminative punishment experiments. Let us consider each.

We have already noted that imposing a temporal delay between the response and the delivery of the punisher reduces the magnitude of the punishment effect observed. Using free-operant baselines, some have reported that long-delayed shocks yield effects not different from non-contingent shocks. Camp, Raymond, and Church (1967) found only a transient effect of a punisher delayed 30 seconds. But there is a problem with these free-operant, delay of punishment tests, because they really are only assessing the contribution of response-shock contiguity not the contingency (dependency).

The second kind of test compares the magnitudes of response decrements under discriminative punishment and CER procedures. Let us consider the logic of this comparison. In the CER procedure, the delivery of the shock is contingent only upon the presence of the signal (giving rise to stimulus-shock contiguity). In the discriminative punishment procedure, shock is contingent upon both the presence of the signal and the prior occurrence of a response. Contrasting these two treatments, then, ought to determine whether or not the response contingency contributes to the observed response decrementing.

Church, Wooten, & Mathews (1970) trained two groups of rats to press a bar to earn food (VI 1-min). Meanwhile, a 3-min stimulus was occassionally presented. For one group, brief shocks occurred randomly once a minute during the stimulus; for the other, shocks were "primed" on exactly the same schedule, but their delivery was contingent upon the next response. The former is CER; the latter is a discriminative punishment treatment. The results of their experiment are shown in Figure 4. They found that the punishment contingency in which shocks were response dependent led to greater decrements in responding than did the CER treatment.

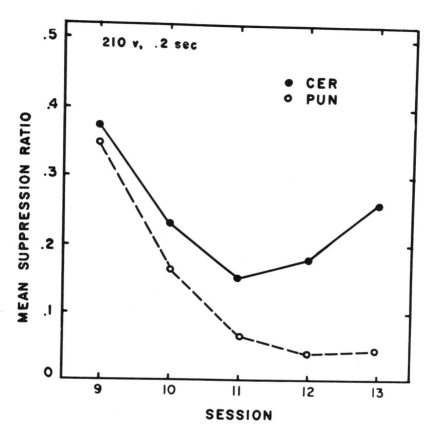

Figure 9.4 Comparison of the amounts of suppression of responding
            for food by rats as a function of two procedures: the
            presentation of a CS for response independent shocks
            (i.e., the Pavlovian CER procedure) and the signalled
            punishment procedure. Shocks were scheduled on a 1-min
            random interval schedule for both groups. Note that
            the addition of a response-shock contingency (i.e.,
            punishment, PUN) during the stimulus resulted in
            greater behavioral suppression than did the stimulus-
            shock contingency (CER) alone. (After Church, Wooten,
            & Matthews, 1970)

While the results of their experiment are clear, the experimental contingencies continue to confound response-shock contiguity and response-shock dependency. In the CER procedure there is neither (necessarily) contiguity of response and shock nor dependency. In Church, et al's discriminative punishment treatment, both were present. Frankel (1975) sought to unconfound contiguity and dependency with a novel discriminative punishment procedure. The design of his experiment was very similar to Church, et al (1970), except for the delivery of shocks. In Frankel's experiment, all shocks were delivered at the end of the stimulus period independent of when during the stimulus they were earned. For the CER group, all stimulus trials ended in shocks. For the punishment group, shocks were only delivered if the rat responded sometime during the stimulus; thus if a rat did not respond during the discriminative stimulus, no punishers were given. Thus, the two groups only necessarily differed in the response-shock dependency. Figure 5 presents Frankel's findings showing that the punishment group was more suppressed than the CER group. Note that this result was obtained even though the punishment group received fewer total shocks. This is a powerful demonstration of the importance of the response-shock contingency and stands as evidence against a CER theory of punishment.

The body of data represented by the experiments just presented would seem to make untenable the CER theory of punishment as a full account of response decrements observed in the punishment paradigm. On the other hand, to the extent that there is any CER effect, failure to control for it may lead to an overestimation of the effectiveness of the punisher qua punisher.

If the simple CER explanation is not adequate to encompass the importance of the contingency, how then do we account for the punishment effect? One need not be terribly imaginative, it turns out, to resurrect the CER account. Consider a sophisticated CER theory which includes in the range of stimuli that can become signals for the shocks the response produced internal and external stimuli necessarily accompanying occurrences of the punished response. With such a theory, the operational response contingency, of necessity, will always be important although the theoretical mechanism invoked is one of stimulus-shock contiguity and not response-shock dependency. However, such a theory is so sophisticated that it is virtually immune to falsification because of the unavoidable confounding between the response itself and the response-produced stimuli!

Let us then turn to the last class of theories.

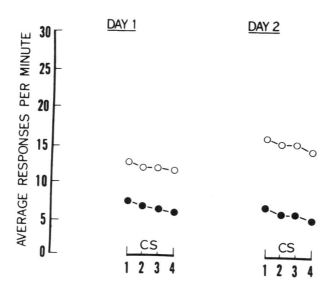

Figure 9.5   Comparison of rates of responding for food by rats
under two conditions of shock presentations with response shock contiguity equated: response dependent
delivery of shock but with the shock at the end of
the signal (CONTINGENT, punishment) and response independent delivery of shocks at the end of the signal
(NONCONTINGENT, CER). (Adapted from Frankel, 1975.)

Competing Response Theories. The competing response, or interference, theories are of several types. They differ primarily
in terms of the source of the competing response.

One version of the competing response theory (Guthrie, 1934)
argues that the punisher is really just a powerful US which when
presented necessarily elicits certain URs. These URs may become
classically conditioned to the stimuli which antedate the delivery
of the punisher. Finally, when these CRs are incompatible with
the to-be-punished response, that response is suppressed through
response competition. This classical conditioning concept of the
origin of competing response seems relatively straightforward. A
first approximation test of this version of the theory is to ask
what responses are the URs to the shock presentations.

Consider a rat pressing a bar to earn either food or water.
What happens if we periodically deliver a "free" shock? The behavior that is reliably elicited by the shock is certainly the
most likely condidate for the response to become classically conditioned. Under such circumstances, we commonly observe that the
shock "elicits" a burst of rapid lever-presses. One instance of
this relationship is shown in Figure 6; presented there are the
data of Church, Wooten, and Mathews (1970). If in a food reinforced lever-pressing task the UR to shock is an increase in
lever-pressing, it becomes most awkward to account for the effects
of punishment through classical conditioning of the shock elicited
response. Of course, it is well-known that a UR and CR need not
be identical, but if one invokes this as an "escape clause" then
the elegance of the classical conditioning theory becomes ephemeral.

A second version of competing response theory suggests that
the competing response is learned through instrumental reinforcement. There are at least two subdivisions of this version of the
theory. These subdivisions differ in their identification of the
reinforcer for the competing instrumental response. The first
suggests that the source of reinforcement which strengthens the
competing response lies in the termination of the punisher. Thus,
the response that escapes (or is contiguous with the termination
of) the punisher will be learned. The second suggests that the
stimulus elements closely attendant upon the delivery of the punisher through conditioning become aversive themselves or come to elicit
fear. Responses which serve to remove or reduce these conditioned
aversive or fear stimuli will be secondarily reinforced through,
say, fear reduction. Since those stimulus elements closely attendant upon delivery of the punisher are necessarily going to be
those correlated with the occurrence of the punisher, the only way
of preventing those stimuli is to not make the to-be-punished

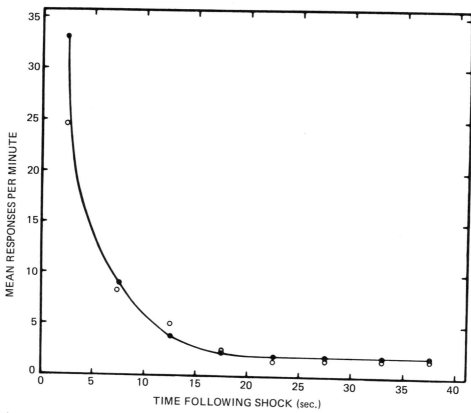

Figure 9.6 Lever-press response rate as a function of time following a shock; shocks were presented on a random
schedule. The function indicates that responses are
<u>most</u> probable immediately following shocks. (After
Church & Getty, 1972.)

response and to engage in other behavior. This is a "passive avoidance" theory of punishment.

These two versions of instrumental competing response theories may, for shorthand, be referred to as the escape and the avoidance theories of punishment, respectively. Miller (Miller & Dollard, 1941) has been closely linked with the former, while Dinsmoor (1954, 1955) and Mowrer (1950) have been linked with the latter.

Fowler and Miller (1963) have provided some data that are consistent with the instrumental analysis of punishment. They first trained three groups of rats to run down a runway for food. Whenever the rat took food in the goalbox, the experimenters presented the punisher but did so in a special way. One group was shocked only on the forepaws, while the second group was shocked only on the hind paws; the third group was not shocked at all. Their results are shown in Figure 7. In the "forepaw" group, the usual response-suppressing effect of a punisher was observed. But in the hindpaw group, the opposite -- a facilitation of running -- was observed. They said of their experiments that "hindpaw shock tends to elicit a response of lurching forward, and forepaw shock one of lurching back. Seemingly compatible with approach to the goal and reinforced by escape from pain... and conversely, the response of lurching back, also reinforced by escape from pain..." will modulate running in the alley (p.804). Clearly, Fowler and Miller's results indicate that a punished act can be facilitated or inhibited depending upon the nature of the response elicited by the punishment. The special treatment they give these responses is to infer that they are strengthened by virtue of their association with shock termination, that is, by escape.

In the Fowler and Miller study, we see a clear example of careful attention to the competing response. But in fact this is not usual; the "competing" response more typically has the status of an unobserved, hypothetical construct! That is, the competing response is not specified but is inferred from the occurrence of a decrease in the punished response. This yields a circular explanation of punishment. Unless the incompatible response is specified, such theories are simply not falsifiable. For this reason, Dunham (1971) has strongly urged a new general strategy in the study of punishment.

The new strategy involves the continuous measurement of all the activities in the animal's repertoire. Under such circumstances one can observe whether some other response comes to replace the punished response and whether this new response stands in the proper

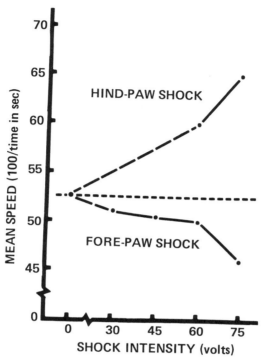

Figure 9.7   Speeds of running to food by rats under conditions of
punishment.   The two groups differed with respect to
how the shocks were delivered in the goal box; for one
group only the hind paws were shocked whereas for the
other only the fore paws were shocked.   Fore paw shocks
tended to elicit forward lurches, the amplitude of
these increasing with shock intensity.   Thus, the figure suggests that the degree of compatibility between
the running behavior and the punishment elicited response is an important determiner of asymptotic behavior in the face of punishment.   (After Fowler &
Miller, 1963.)

relationships to shock or fear termination to be consistent with
theory. Dunham (1971, 1972) has carried out some work in this
vein. Unfortunately, his results are not consistent with any of
the competing response theories. Although the punished response
decreased in frequency and other responses increased in frequency,
the temporal ordering of these changes did not meet the requirements of theory. He found that the punished response suppressed,
and then the new response emerged.

We have looked at three major classes of theories of punishment, the Negative Law of Effect, CER, and Competing Response.
None of them seems wholly satisfactory -- even verging on being
untestable at times. But the task of the theory builder is even
more complicated than we have heretofore suggested. This is because, up to now, we have only concerned ourselves with simple
punishment phenomena. There are more complex phenomena that pose
special challenges to any theory. Let us consider selected
examples arising from consideration of the other factors listed
earlier that also determine the effects of punishment operations.

OTHER FACTORS

At the outset we listed three major factors that merited consideration: (1) the punisher, (2) response maintenance conditions,
and (3) previous experiences with the punisher. We now shall give
attention to these other factors. The more brief treatment that
they shall receive is not an indication of lesser importance but
of a smaller data base.

Response Maintenance Conditions

It makes intuitive sense that the response decrementing effect
of punishment would be inversely related to the response strength
and vigor of the to-be-punished response. For example, we already
know that increases in reward magnitude and motivation both yield
increased response vigor. Thus, we would expect that a given punisher would have lesser effect under conditions of large reward and
high motivation than under conditons of small reward and low motivation. This is exactly what has been reported. Capaldi and Levy
(1972) found that increases in reward magnitude led to reduced
effectiveness of a punisher of fixed severity. Similarly, Azrin,
Holz, and Hake (1963) found that increases in motivation (decreases
in body weight) led to reduced effectiveness of a fixed punisher.

One reward manipulation known to have dramatic effects upon resistance to extinction is partial reinforcement. A reasonable question to ask is whether partial reinforcement modulates resistance to punishment. This question has been thoroughly studied now (Brown & Wagner, 1964; Linden, 1974), and the results are quite clear. The experience of partial reinforcement during acquisition and maintenance of responding makes the animal more resistant to punishment when it is later introduced. Of special interest is that the inverse effect was also found; that is, punishment during acquisition training resulted in animals more resistant to omission of reward (extinction)! Finally, Linden (1974) has obtained data that suggests that these two persistence inducing treatments (partial reinforcement and punishment during acquisition) may be additive.

Theorists (e.g., Wagner, 1969) have used observations like those just discussed to suggest that the aftereffects of non-reward and punishment have elements in common. D'Amato (1970) has even suggested that extinction is "not a process sui generis but rather a special case of punishment" (p.353). In any case, the observation that punishment during acquisition results in latter increased resistance to extinction is our first hint that punishment can have "facilitating" effects upon behavior.

History With Respect to the Punisher

Effectiveness of a punisher is not determined solely by the present conditions of punishment but by prior experiences with the punisher. These prior experiences may have been under punishment contingencies or under escape/avoidance training contingencies. We will consider both cases.

Sequential Punishment Effects. In our society, there are strong prohibitions against cruelty. Therefore, when one undertakes to apply punishment -- either in the laboratory or the home -- a common practice is to first introduce the punisher at its minimum intensity and then to gradually increase the intensity until the desired level of response decrementing is achieved. The goal of this practice is, of course, to minimize the severity of punisher used. The effect of this practice, however, is often quite the opposite! Graded introduction of a punisher results in adaptation-like outcomes with more intense punishers (and often more of them) ultimately required to achieve a given effect.

A clear example of this is provided by Miller (1960). Indeed, he studied this phenomenon as an example of training to resist pain. He compared a group of rats exposed to gradually increasing shock punishers in an alley goalbox to ones who experienced the maximum intensity from the outset of the punishment phase. Miller found that gradually increasing the punisher to its terminal value led to only a 50 percent reduction in speed of running down the alley, while sudden introduction yielded   80 to 100 percent reduction. These results are shown in Figure 8. Also shown is a special control group (labelled "gradual outside") which also experienced the series of increasing punisher intensities, but this experience was outside of the alley. When this group was exposed to the full strength punisher in the alley goalbox, it too showed a 90 percent reduction in speed. Thus, the reduced effectiveness is not habituation to shock per se because there was no transfer between apparatus units. Hence, Miller (1960) interpreted his results as indicating a learned behavioral tolerance of punishment specific to the task being punished.

The inverse of this behavioral tolerance phenomenon also exists. That is, if an animal has had prior experience with an intense punisher in a given task, the effectiveness of subsequent presentations of a weak punisher is enhanced. Raymond (1968) has provided an excellent demonstration of this phenomenon, and his results are shown in Figure 9. The figure shows the effects of intermittent 110-volt punishment (VI-2 min) upon the bar pressing for food (FR-22) by two groups of rats. The panel on the right shows that if the rat has had no prior experience with shocks, a 110-volt punisher had little effect. But in sharp contrast, the three panels on the left show that if the rats were first exposed to a severe punisher which completely suppressed behavior, the later introduction of the mild, usually ineffective 110-volt punisher results in nearly complete suppression. This occurs even after "full" recovery of the food getting behavior.

Kurtz and Walters (1962) and Hollis and Overmier (1973) have shown that this behavioral sensitization phenomenon will transfer across exposures in different apparatus units, and perhaps more importantly, across maturational/developmental stages of an organism's life (Brookshire, Littman, and Stewart, 1961). One thing that has not been shown, but is likely to be the case, is that these sequential effects can be found across qualitatively different punishers as well. Such a demonstration would open a new area of research and applications.

In addition to the obvious operational conclusions to be drawn from the Miller and the Raymond experiments, there are two additional reasons for attending to the discussed phenomena. One is methodological and one is theoretical.

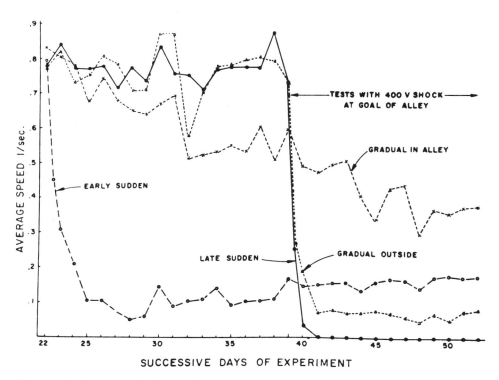

Figure 9.8 Speed of running to food by rats under various conditions of introduction of concurrent punishment: (a)
GRADUAL IN ALLEY rats were given 60-V shocks in the
goal box and this intensity was increased by 20 V per
day reaching 400 V by Day 40; (b) GRADUAL OUTSIDE OF
ALLEY rats received exactly the same sequences of
shocks except that the shocks were administered in a
chamber separate from the alley and its goal box until
Day 40 with all shocks thereafter in the boal box; (c)
EARLY SUDDEN rats always received 400-V shocks in the
goal box starting on Day 22; (d) LATE SUDDEN rats
always received 400-V shocks in the goal box starting
on Day 40. All groups were treated alike on Days 1-21
(i.e., food and no shock punishment) and on Days 40-52
(i.e., food plus 400-V shock). Note that the effectiveness of the punishment is a function of history
with shocks and whether these shocks were received as
part of the criterial task or elsewhere. (After
Miller, 1960.)

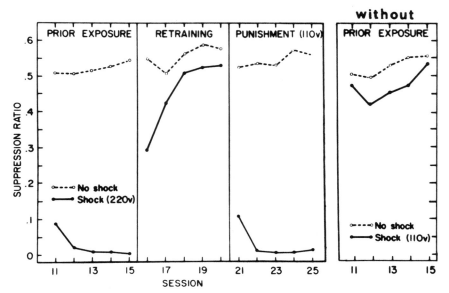

Figure 9.9   Effects of 110-V shock punishment of bar-pressing for
             food as a function of prior exposures to severe shocks
             (left panel) or without any prior exposure to shocks
             (right panel). Note that prior exposure to severe
             shocks results in transformation of an usually inef-
             fective shock punisher into an exceedingly effective
             one.   (Adapted from Church, 1969.)

On the methodological side, the existence of these sequential interactions of intensities implies that experiments seeking
to determine "punishment" thresholds using behavioral indices
and traditional psychophysical methods are likely to be fraught
with error. For example, the practice of determining the threshold
using an ascending series of shocks will underestimate the effects
of all but the first punisher (c.f. Fantino, 1973). This underestimate would occur even if the ascending series were then followed by a descending series in those same animals. In contrast,
a descending series, even when followed by an ascending series in
those animals, will overestimate the effectiveness of all punishers
in the series but the first. In so far as I can see, there is no
easy solution to this problem and its certain errors.

The theoretical issue raised again by these studies on the
sequential interactions speaks to the mechanism of punishment.
To the extent that the sequential effects cannot be attributed to
changes in sensory sensitives, as indeed Miller's (1960) results
cannot, then one seems pressed towards a learning analysis. Miller
proposed that the gradual introduction of shocks going from very
weak to intense allows for the animals to learn responses to shocks
and the anticipations of them (fractional anticipatory punishment
responses, r_p-s_p) that are different from those disruptive responses usually elicited by moderate and strong shock punishers.
Thus, when strong shocks are presented, they evoke the response
learned at the low intensity -- in Miller's case, running to the
reward. The opposite could be true of prior exposure to intense
shocks. The disruptive response patterns elicited by strong shocks
may become well established such that when weaker shocks are presented, the disruptive responses generalize to them yielding the
unexpected impairment reported by Raymond (1968) and others.

Thus, the observed pattern of results is consistent with those
theories of punishment which invoke "incompatible" overt responses
to explain the effects of punishment. This appeal to evocation .
of previously learned responses gains additional plausibility when
we explore the effects of the several punishment operations when
applied to previously trained escape or avoidance responses.

Punishment of Escape/Avoidance Behavior. The usual long-term
effect of strong punishers upon the extinction of escape/avoidance
behaviors is to suppress them if the punisher is administered at the
end of the response chain (Kamin, 1959; Imada, 1959). However, because such introduction does reinstate some of the training conditions, including some of the discriminative stimuli and, importantly,
the motivational conditions for responding, a temporary facilitation is commonly observed (Sidman, 1966). Furthermore, if the

conditions of punishment are arranged such that continued performance of the previously learned response will terminate the punisher,
then behavior may well be maintained almost <u>indefinitely</u>.

Gwinn (1949) was one of the first to demonstrate this latter
phenomenon. The apparatus he used was a circular runway shown in
Figure 10. First, Gwinn taught his rats to run around the circle
to escape from shock; then he modified the apparatus so that shock
was only present in the third (covered) section of the circular
runway. Thus, if rats did not run beyond the first two sections
of the alley, no shocks were received; however running further
<u>produced</u> shocks (i.e. punishment) which could, as in original train-
ing, be escaped by completing the·circle. Gwinn found that his rats
continued to run into the shock/punishment portion of the runway,
thus engaging in self-punitive behavior! When shocks were discontinued, the running also stopped. This surprising behavior is
often described as "vicious-circle behavior", perhaps partly as a
pun on the apparatus used!

There is no doubt about the reliability of the phenomenon.
There is, however, a rather large and contentious literature surrounding this vicious circle phenomenon with respect to the necessary and sufficient conditions for producing it (Eaton, 1972;
Dreyer & Renner, 1971). One critical feature, as we noted earlier,
is that the animal can escape from the punisher. If this is not
the case, as when punishment is given only in the goalbox, the
punishment decrements responding (Seligman & Campbell, 1965).
Nonetheless, it is now clear that punishment of behavior originally
established under conditions of aversive control may result in
marked facilitation.

Punishment as a Discriminative Stimulus

The effect of a punisher is not determined solely by the present conditions of punishment and their relationship to prior experiences with punishment, but by the punisher's relationship to
reward as well. When the punisher is also a discriminative stimulus
for rewards, some rather counter intuitive outcomes are again obtained.

Let us consider a dramatic example obtained in a multiphase
experiment (Holz & Azrin, 1961). Pigeons are trained in Phase A
to peck a key on a VI schedule for food reward. Then in Phase B,
they are subject to alternating periods of reward training and
extinction. Also during Phase B, the pigeons are punished with
moderate electric shocks for pecking when rewards are available,

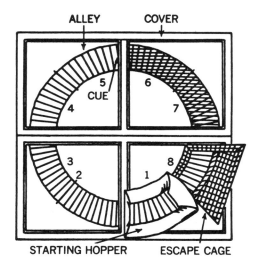

Figure 9.10   Diagram of Gwinn's apparatus for generating "self-
              punitive" behavior. The circular alley had an elec-
              trifiable grid floor throughout its full circumfer-
              ence although sections could be electrified indepen-
              dently. In the first phase all sections were elec-
              trified and rats were dropped into section 1 and had
              to run to section 8 and jump onto the safe platform
              (ESCAPE CAGE) there. In the second phase only sec-
              tions 6 and 7 were electrified but rats dropped into
              section 1 continued to run from section 1 to section
              8 more than 40 times. In a second phase control con-
              dition with no shock anywhere, control rats ran from
              section 1 to section 8 less than 10 times. (After
              Gwinn, 1949.)

but they are <u>not</u> punished during the extinction component. At
first, the punishment suppresses pecking, but over sessions, as to
be expected, the birds show partial recovery and continue to peck
for food -- albeit at a lower rate -- despite the punishment contingency. During the food extinction component, they of course
do not peck. Finally, in Phase C, food rewards are eliminated
altogether in total extinction. One might expect that the addition
of punishment would serve to decrement responding even further,
 but this is not what Holz and Azrin found; they found that introduction of the punisher in this last phase <u>facilitated</u> responding!
Their results are shown in Figure 11 for each phase. Holz and
Azrin (1961) attributed the facilitation observed during punishment in Phase C to the discriminative stimulus properties of the
punisher developed in Phase B.

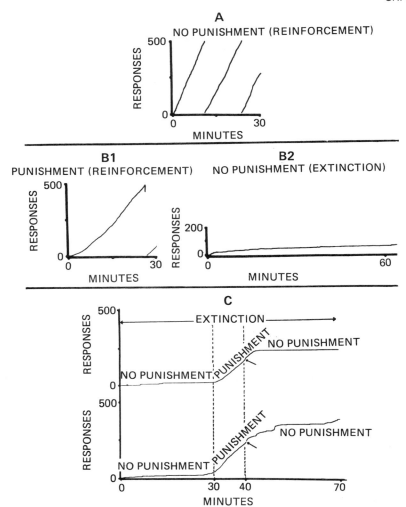

Figure 9.11   Demonstration that shock punishments can, with appropriate training, function as discriminative stimuli
for responding. This was achieved in a three-stage
experiment: A. training of pigeons to peck a key for
reward; B. a mixed schedule in one component of which
(B1) responding produced both food and punishment
while in the other (B2) responding produced neither
food nor punishment; and finally C. food reward extinction session in which periods of no punishment
alternated with periods of punishment. Note that during this extinction, responding was highest during
periods of punishment. (After Holz & Azrin, 1961.)

This is not a unique, aberrant observation; others have reported similar results. Indeed, it can even be shown that, after training such as that in Phase B, the presentation of a classical CS previously paired with shocks can facilitate responding (Rosellini & Terris, 1976), indicating that the increase is not a direct effect of shocks per se.

That acquisition of choice behavior might be facilitated by punishment was demonstrated very early. For example, Warden and Aylesworth (1927) demonstrated that punishment for wrong responses speeded learning. This result is not at all surprising; we expect such punishment to help stamp out errors. But Meunzinger (1934) has shown that the moderately intense punishment for making the correct response can also facilitate acquisition of choice behavior! This result is shown in Figure 12; note the positive contribution of shock for the correct response is almost as large as that for the wrong response. This is certainly a most counterintuitive effect, but a generally reliable one.

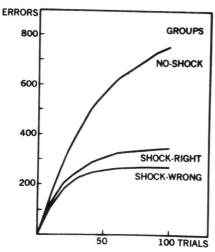

Figure 9.12 Cumulative error curves by rats learning a T-maze for food rewards under three different conditions: food reward only for correct responses (NO SHOCK); food reward plus mild electric shock punishment for correct responses (SHOCK-RIGHT); and food reward for correct responses plus shock punishment for incorrect responses (SHOCK-WRONG). Note that both punished groups learned faster than the reward-only group. While this was expected in the case of the SHOCK-WRONG group, it was unexpected in the case of the SHOCK-RIGHT group. (After Muenzinger, 1934.)

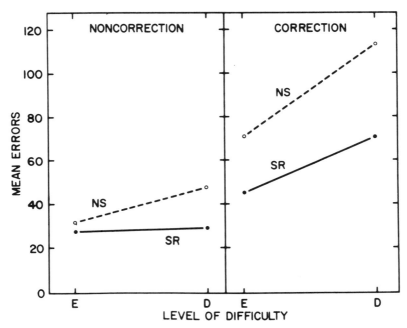

Figure 9.13   Contrast of the rates of learning in a T-maze task
               (indexed by errors to criterion), by groups either
               rewarded for correct choices (NS) or both rewarded and
               shocked (SR), as a function of level of difficulty of
               the discrimination task and the method of training.
               Groups were trained on either an easy (E) or a diffi-
               cult (D) visual discrimination task using either the
               non-correction or the correction technique. Note
               that mild shock for correct choices (shock-right, SR)
               facilitates learning under all conditions but more so
               when the problem is difficult and this effect is
               amplified by use of the correction technique. (After
               Fowler, Spelt, & Wischner, 1967.)

As is usually the case with unexpected phenomena, there have
been extensive efforts to identify those factors contributing to
it. It seems that facilitation of learning of choice behavior is
greatest with difficult discriminations under correction procedures
(Fowler, Spelt, & Wischner, 1967), with almost no facilitation at
all observed on easy discriminations under non-correction choice
procedures. This interaction is illustrated in Figure 13.

How is it that punishing the correct response facilitates
learning? A number of speculations have been offered. Meunzinger

thought that the punishment made the animal run slower and attend more carefully to the apparatus stimuli. Others have suggested that the shocks presented after the correct choice come to function as secondary reinforcers, but no experiments have demonstrated that such shocks could secondarily reinforce new behavior. Even more complex "explanations" are to be found, yet none have achieved broad acceptance (e.g., Fowler, Goodman, & DeVito, 1977).

Whatever the final explanations, it is now clear that punishment does _not_ always suppress behavior that leads to it. One goal of these last sections has been to illustrate just that point.

Punishment has been insufficiently studied. Our ignorance here is appalling. Part of the lack of attention can be traced to the early assertions that punishment was ineffective in producing permanent behavioral changes. We presently know this to be a myth. New facts are now needed that will enable us to develop a unifying conceptualization of punishment. We have a long way to go.

REFERENCES

Annau, Z., & Kamin, L. J. The conditioned emotional response as a function of intensity of the US. JOURNAL OF COMPARATIVE AND PHYSIOLOGICAL PSYCHOLOGY, 1961, 54, 428-432.

Azrin, N. H., & Holz, W. C. Punishment during fixed interval reinforcement. JOURNAL OF THE EXPERIMENTAL ANALYSIS OF BEHAVIOR, 1961, 4, 343-347.

Azrin, N. H., Holz, W. C., & Hake, D. F. Fixed-ratio punishment. JOURNAL OF THE EXPERIMENTAL ANALYSIS OF BEHAVIOR, 1963, 6, 141-148.

Baron, A. Delayed punishment of a runway response. JOURNAL OF COMPARATIVE AND PHYSIOLOGICAL PSYCHOLOGY, 1965, 60, 131-134.

Boe, E. E. Bibliography on punishment. In B. A. Campbell & R. M. Church (Eds.), PUNISHMENT AND AVERSIVE BEHAVIOR. New York: Appleton-Century-Crofts, 1969.

Boe, E. E., & Church, R. M. Permanent effect of punishment during extinction. JOURNAL OF COMPARATIVE AND PHYSIOLOGICAL PSYCHOLOGY,, 1967, 63, 486-492.

Brookshire, K. H., Littman, R. A., & Stewart, C. N. Residua of shock-trauma in the white rat: a three factor theory. PSYCHOLOGICAL MONOGRAPHS, 1961, 75, No. 10 (Whole No. 514).

Brown, R. T., & Wagner, A. R. Resistence to punishment and extinc-
    tion following training with shock or non-reinforcement.
    JOURNAL OF EXPERIMENTAL PSYCHOLOGY, 1964, 68, 503-507.

Camp, D. S., Raymond, G. A., & Church, R. M. Temporal relationship
    between response and punishment. JOURNAL OF EXPERIMENTAL
    PSYCHOLOGY, 1967, 74, 114-123.

Capaldi, E. J., & Levy, K. J. Stimulus control of punished reac-
    tions: Sequence of punishment trials and magnitude of rein-
    forcement trials. LEARNING AND MOTIVATION, 1972, 3, 1-19.

Carlsmith, J. M. The effect of punishment on avoidance responses:
    The use of different stimuli for training and punishment.
    Paper presented at Eastern Psychological Association, Phila-
    delphia, 1961.

Church, R. M. The varied effects of punishment on behavior.
    PSYCHOLOGICAL REVIEW, 1963, 70, 369-402.

Church, R. M. Response suppression. In B. A. Campbell & R. M.
    Church (Eds.), PUNISHMENT AND AVERSIVE BEHAVIOR. New York:
    Appleton-Century-Crofts, 1969.

Church, R. M., & Getty, D. J. Some consequences of the reaction to
    an aversive event. PSYCHOLOGICAL BULLETIN, 1972, 78, 21-27.

Church, R. M., Raymond, G. A., & Beauchamp, R. O. Response suppres-
    sion as a function of intensity and duration of a punishment.
    JOURNAL OF COMPARATIVE AND PHYSIOLOGICAL PSYCHOLOGY, 1967, 63,
    39-44.

Church, R. M., Wooton, G. L., & Matthews, T. J. Discriminative
    punishment and the conditioned emotional response. LEARNING
    AND MOTIVATION, 1970, 1, 1-17.

Czeh, R. S. Response strength as a function of the magnitude of
    the incentive and consummatory time in the goal box. Unpub-
    lished Master's Thesis, State University of Iowa, 1954.
    (Cited in Spence, 1956)

D'Amato, M. R. EXPERIMENTAL PSYCHOLOGY: METHODOLOGY, PSYCHOPHYSICS,
    AND LEARNING. New York: McGraw-Hill, 1970.

Davis, H. Conditioned suppression: A survey of the literature.
    PSYCHONOMIC MONOGRAPH SUPPLEMENTS, 1968, 2, No. 14 (Whole
    No. 30).

Dinsmoor, J. A. Punishment: I. The avoidance hypothesis. PSYCHO-
    LOGICAL REVIEW, 954, 61, 34-46.

Dinsmoor, J. A. Punishment: II. An interpretation of empirical
    findings. PSYCHOLOGICAL REVIEW, 1955, 62, 96-105.

Dreyer, P., & Renner, K. E.   Self-punitive behavior - masochism or confusion.   PSYCHOLOGICAL REVIEW, 1971, 78, 333-337.

Dunham, P. J.   Punishment: Method and theory.   PSYCHOLOGICAL RE-VIEW, 1971, 78, 58-70.

Dunham, P. J.   Some effects of punishment upon unpunished responding. JOURNAL OF THE EXPERIMENTAL ANALYSIS OF BEHAVIOR, 1972, 17, 443-450.

Eaton, N. K.   Self-punitive locomotor behavior as a function of fixed or varied shock zone location.   Unpublished Doctoral dissertation, University of Iowa, 1972.

Estes, W. K.   An experimental study of punishment.   PSYCHOLOGICAL MONOGRAPHS, 1944, 57, No. 3 (Whole No. 263).

Estes, W. K., & Skinner, B. F.   Some quantitative properties of anxiety.   JOURNAL OF EXPERIMENTAL PSYCHOLOGY, 1941, 29, 390-400.

Fantino, E.   Aversive control.   In J. A. Nevin (Ed.), THE STUDY OF BEHAVIOR: LEARNING, MOTIVATION, EMOTION, AND INSTINCT. Glenview, Ill.: Scott, Foresman, & Co., 1973.

Fowler, H., Goodman, J. H., & DeVito, P. L.   Across reinforcement blocking effects in a mediational test of the CS's general signaling property.   LEARNING AND MOTIVATION, 1977, 8, 507-519.

Fowler, H., & Miller, N. E.   Facilitation and inhibition of runway performance by hind- and forepaw shock of various intensities. JOURNAL OF COMPARATIVE AND PHYSIOLOGICAL PSYCHOLOGY, 1963, 56, 801-805.

Fowler, H., Spelt, P. F., & Wischner, G. J.   Discrimination perfor-mance as affected by training procedure, problem difficulty and shock for the correct response.   JOURNAL OF EXPERIMENTAL PSYCHOLOGY, 1967, 75, 432-436.

Frankel, F. D.   The role of the response-punishment contingency in the suppression of a positively-reinforced operant. LEARNING AND MOTIVATION, 1975, 6, 385-403.

Guthrie, E. R.   Reward and punishment.   PSYCHOLOGICAL REVIEW, 1934, 41, 450-460.

Gwinn, G. T.   The effects of punishment on acts motivated by fear. JOURNAL OF EXPERIMENTAL PSYCHOLOGY, 1949, 39, 260-269.

Hollis, K. L., & Overmier, J. B.   Effect of inescapable shock on efficacy of punishment of appetitive instrumental responding by dogs.   PSYCHOLOGICAL REPORTS, 1973, 33, 903-906.

Holz, W. C., & Azrin, N. H. Discriminative properties of punish-
    ment. JOURNAL OF THE EXPERIMENTAL ANALYSIS OF BEHAVIOR,
    1961, 4, 225-232.

Hull, C. L. PRINCIPLES OF BEHAVIOR. New York: Appleton-Century-
    Crofts, 1943.

Imada, H. The effects of punishment on avoidance behavior. JAPANESE
    PSYCHOLOGICAL RESEARCH, 1959, 1, 27-38.

Kamin, L. J. The delay-of-punishment gradient. JOURNAL OF COMPARA-
    TIVE AND PHYSIOLOGICAL PSYCHOLOGY, 1959, 52, 434-437.

Kurtz, K. H., & Walters, G. C. The effects of prior fear experiences
    on approach-avoidance conflict. JOURNAL OF COMPARATIVE AND
    PHYSIOLOGICAL PSYCHOLOGY, 1962, 55, 1075-1078.

Linden, D. R. Transfer of approach responding between punishment,
    frustrative nonreward, and the combination of punishment and
    nonreward. LEARNING AND MOTIVATION, 1974, 5, 498-510.

Logan, F. A. The negative incentive value of punishment. In B. A.
    Campbell & R. M. Church (Eds.), PUNISHMENT AND AVERSIVE BEHA-
    VIOR. New York: Appleton-Century-Crofts, 1969.

Masserman, J. H. BEHAVIOR AND NEUROSIS. Chicago: University of
    Chicago, Press, 1943.

Miller, N. E. Learning resistance to pain and fear: Effects of
    overlearning, exposure and rewarded exposure in context.
    JOURNAL OF EXPERIMENTAL PSYCHOLOGY, 1960, 60, 137-145.

Miller, N. E., & Dollard, J. C. SOCIAL LEARNING AND IMITATION.
    New Haven: Yale University Press, 1941.

Mowrer, O. H. LEARNING THEORY AND PERSONALITY DYNAMICS. New York:
    Ronald Press, 1950.

Muenzinger, K. F. Motivation in learning: I. Electric shock for
    correct responses in the visual discrimination habit. JOURNAL
    OF COMPARATIVE PSYCHOLOGY, 1934, 17, 267-277.

Rachlin, H., & Herrnstein, R. J. Hedonism revisited: On the nega-
    tive law of effect. In B. A. Campbell & R. M. Church (Eds.),
    PUNISHMENT AND AVERSIVE BEHAVIOR. New York: Appleton-
    Century-Crofts, 1969.

Raymond, G. A. Accentuation and attenuation of punishment by prior
    exposure to aversive stimulation. Unpublished Doctoral
    dissertation, Brown University, 1968.

Rosellini, R. A., & Terris, W. Fear as a discriminative stimulus
    for an appetitive instrumental response. LEARNING AND
    MOTIVATION, 1976, 7, 327-339.

Seligman, M. E. P., & Campbell, B. A. Effect of intensity and duration of punishment on extinction of an avoidance response. JOURNAL OF COMPARATIVE AND PHYSIOLOGICAL PSYCHOLOGY, 1965, 59, 295-297.

Sidman, M. Avoidance behavior. In W. K. Honig (Ed.), OPERANT BEHAVIOR: AREAS OF RESEARCH AND APPLICATION. New York: Appleton-Century-Crofts, 1966.

Skinner, B. F. BEHAVIOR OF ORGANISMS. New York: Appleton-Century-Crofts, 1938.

Spence, K. W. BEHAVIOR THEORY AND CONDITIONING. New Haven: Yale University Press, 1956.

Thorndike, E. L. EDUCATIONAL PSYCHOLOGY. II. THE PSYCHOLOGY OF LEARNING. New York: Teachers College, Columbia University, 1913.

Thorndike, E. L. FUNDAMENTALS OF LEARNING. New York: Teachers College, Columbia University, 1932.

Tolman, E. C. PURPOSIVE BEHAVIOR IN ANIMALS AND MEN. New York: Century, 1932.

Wagner, A. R. Frustrative nonreward: A variety of punishment? In B. A. Campbell & R. M. Church (Eds.), PUNISHMENT AND AVERSIVE BEHAVIOR. New York: Appleton-Century-Crofts, 1969.

Walters, G. C., & Grusec, J. E. PUNISHMENT. San Francisco: Freeman & Co., 1977.

Warden, C. J., & Aylesworth, M. The relative value of reward and punishment in the formation of a visual discrimination habit in the white rat. JOURNAL OF COMPARATIVE PSYCHOLOGY, 1927, 7, 117-127.

10. AVOIDANCE LEARNING

J. B. Overmier

University of Minnesota
Minneapolis, Minnesota, USA

Review of the experimental and theoretical analysis of avoidance behaviors can provide fascinating insights into the development of the psychology of learning in America. This is because problems in the analysis of avoidance behaviors often stimulated the development of new experimental designs and new theoretical concepts which in turn had substantial impact upon how we interpreted all other learning phenomena. Most simply, the phenomenon of avoidance learning was always--and continues to be--a source of difficulties. The difficulties are of three types: (a) paradigmatic, (b) methodological, and (c) theoretical. We shall consider each in turn.

THE PARADIGM

The paradigmatic problem posed by avoidance behaviors was one of recognizing that the conditions which produced such behaviors constituted a unique paradigm to be distinguished from those of classical conditioning and instrumental training. Once such recognition took place, the special properties of the resulting avoidance learning were also discriminated. But such recognition was not immediate. The avoidance paradigm was used for about 20 years before its distinctive features were recognized--at least in America. It was most often confused with the classical conditioning paradigm. Indeed, the avoidance paradigm can be arrived at through a very slight modification of classical conditioning operations which use primary aversive events like electric shocks as the US.

313

Consider a dog standing on a platform with two electric
shocking electrodes (the anode and the cathode) attached to the
right forepaw approximately 1 cm apart. In the course of classical conditioning each CS will be followed inexorably by the electric shock US to the leg which elicits flexion; we can anticipate
that the dog as a result of this conditioning will come to flex
its leg during the isolated action of the CS, such flexions of
course having no effect upon US delivery. But suppose in the
course of such an experiment one had some difficulty with smeared
electrode paste completing the circuit and hence allowing the
electric shocks to "by-pass" the dog. To overcome this problem
one might separate the electrodes further and fix only the cathode
to the leg while letting the footpad rest on the anode. Under
these circumstances the dog will also learn to flex its leg during the CS; but note now that such flexions have an important consequence, viz. the electric shock to the foot is prevented because
the circuit is open. That is, with these modified circumstances,
the shock is avoided. Bekhterev and his coworkers were likely the
first to use the avoidance paradigm, possibly through just this
sort of "accident", but without clear recognition of the importance
for behavior theory of the procedural change. Certainly, this
importance was long unrecognized in America only to become a subject of argument in the late 1930's (Mowrer, 1947).

Even after avoidance training procedures were recognized as
distinct from classical conditioning, confusions remained. These
confusions were between the avoidance paradigm and the alternative
escape paradigm, on one hand, and the punishment paradigm on the
other. Avoidance and escape are often confused because in both
cases the aversive event is not presented (or continued) after
the experimental subject makes the requisite response. But note
that these differ with respect to (a) the conditions prevailing
prior to the response and (b) what occurs at the moment of the response. In the case of escape, the primary aversive stimulus is
acting on the subject prior to the response; upon occurrence of
the response there is a dramatic change in the situation because
the aversive stimulus ceases to act. Under the avoidance paradigm, however, the aversive stimulus is not acting prior to the
response, nor is there necessarily a dramatic stimulus change upon
occurrence of the response. Here, it is just that the not-acting
aversive stimulus will not be forthcoming--it is avoided altogether.
This distinction is often blurred by the concatenation of contingencies experimenters choose to impose. It is not uncommon for
one to arrange that, after a trial signal has occurred, an electric
shock be delivered and continue until the subject responds appropriately, but with the supplemental contingency that should the
animal respond before the shock is delivered the trial will be

terminated. This procedure has in fact often been used and at times unwittingly called escape training (e.g., May, 1948). We can see, however, that it is a compounding of the contingencies characterizing the two different paradigms of escape and avoidance. Escape contingencies are not a defining feature of the avoidance paradigm and vice versa.

Finally, we should distinguish paradigmatically between punishment, and avoidance, In the instrumental training paradigm, including punishment, the occurrence of the requisite response results in the presentation of the arranged outcome event. In the avoidance paradigm, the occurrence of the requisite response results in the omission of the arranged outcome event. Thus, the two paradigms are quite opposite.[1]

While learning under avoidance contingencies seems sensible enough and its adaptive value obvious, scientific accounts of the behavior strengthening mechanisms have proved elusive. This is because not only have the optimum procedures proved difficult to specify, but even the necessary and sufficient conditions have proved difficult to identify.

METHODOLOGY

There are two major classes of avoidance training techniques: signalled discrete-trial avoidance training and unsignalled, temporally paced avoidance training. Both techniques can produce avoidance learning over the range of vertebrate species. These two techniques are illustrated in Figure 1.

The first is the classical technique derived by Bekhterev which we have already described as involving signal-shock pairing distributed in time; but these pairings are not independent of the subject's behavior. When the subject makes the response during the interval between signal (typically called CS) onset and shock (typically called US) onset, the shock is omitted.

The second is called free-operant avoidance or Sidman avoidance, after an early proponent of the procedure. Sidman avoidance also involves two contingencies. First, shocks are arranged to occur according to a fixed schedule, say every 3-sec; this is called the shock-shock interval (S-S). Second, when the subject makes the

[1]We need recall, however, that Mowrer has tried to show that avoidance learning and punishment training are dependent upon a common underlying associative mechanism. To establish parallels between them, Mowrer redefined the requisite response under punishment conditions as being the non-occurrence of a behavior.

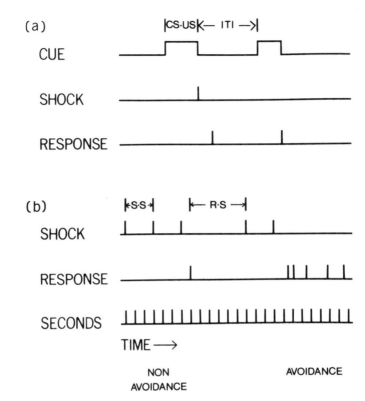

Figure 10.1   Illustration of the contingencies under two avoidance
               training procedures (a) traditional ("classical"),
               discrete-trial avoidance training and (b) Sidman,
               temporally-paced free operant avoidance training.
               The left side of the figure presents the conditions
               prior to avoidance learning while the right side rep-
               resents those after learning. See the text for full
               description of methods a and b.

requisite response, it is given a reprieve from shocks for a period, say 6-sec; this is called the response-shock interval (R-S). Thus, in the absence of the requisite behavior, shocks are delivered according to the S-S schedule but the occurrence of response switches the delivery of shocks to the R-S schedule. So long as the subject reliably responds with an inter-response interval less than the R-S interval, no shocks will be delivered.

We have already noted that both techniques can yield avoidance learning. Sometimes surprisingly powerful behavioral control results with avoidance training. At other times, success in controlling behavior by avoidance methodologies is quite surprisingly elusive. Learning under avoidance contingencies is critically dependent upon a variety of factors and their interactions. In addition to the obvious factors, including CS properties, US properties, CS-US or S-S to R-S interval relations, there are some ill-defined factors like apparatus configuration, topography of the required response, and species-specific organismic characteristics which modulate success in avoidance learning. Perfectly "reasonable" tasks, in the sense that learning would proceed rapidly under instrumental reward contingencies, occasionally seem almost impossible to learn under avoidance contingencies. In sharp contrast, however, are reports that other avoidance tasks are learned within a trial or two and, once learned, are nearly inextinguishable. Such a Janus-faced paradigm has provided many stumbling blocks to adequate theoretical conceptualization of avoidance learning. Before turning to theory, we first need to know a bit more about this complex behavioral phenomenon and the data surrounding it.

RESPONSE TOPOGRAPHY

Solomon and Wynne (1953), following Warner's (1932) lead, trained dogs to avoid electric shocks to the feet by jumping over a low barrier separating two compartments. This so-called shuttle-box avoidance training has become one standard technique for avoidance training. In this particular experiment, the dogs were trained using the signalled discrete trial technique. The rapid acquisition observed is reflected in the representative behavior of one dog shown in Figure 2. After only seven trials, the dog was responding promptly and avoiding rapidly.

In contrast, D'Amato and Schiff (1964) tried to train rats to press a lever to avoid footshocks also using a signalled discrete trial procedure. After literally thousands of trials, the rats were averaging about 10 percent avoidance responses--a pitiful showing indeed!

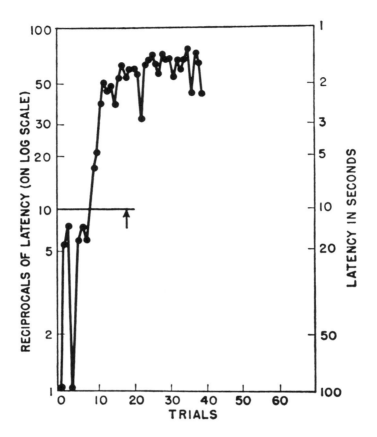

Figure 10.2   Representative avoidance learning data obtained from
              one dog in a two-way shuttlebox. The horizontal line
              at 10 seconds represents the CS-US interval used in
              this training; responses with latencies shorter than
              this CS-US interval prevented the occurrence of shock
              on that trial. The arrow represents the point in the
              experiment when the shock source was disconnected so
              that additional shock presentations were not possible,
              usually described as "extinction" conditions. (After
              Solomon & Wynne, 1953.)

These two experiments, of course, confound response topography and species. It is possible that rats simply cannot learn
to avoid. But this is not so. Riess (1971) compared avoidance
learning of shuttle and bar-press responses in rats using the
Sidman technique. The data from this comparison are shown in
Figure 3. The figure makes clear that learning of the shuttle response proceded very much more rapidly than that of the bar-press
response. Even after nine sessions of training the animals pressing
the bar were receiving about five shocks per minute, although they
were still improving. By way of contrast, the animals shuttling
over a barrier to avoid had reached asymptotic performance by
the third session and were only occassionally receiving shocks.

Given the poor learning shown by rats in the bar-pressing
task, one cannot help but be surprised that the procedure is used
so routinely for the parametric analysis of avoidance phenomena.
One reason may lie in just the fact that the animal does continue
to come into contact with the temporal contingencies in the bar-
pressing task and is thus sensitive to changes in such contingencies.
That the Sidman avoidance behavior is sensitive to the parameters
in effect is readily illustrated by the performance of one rat presented in Figure 4. Here we see that behavior is a function of
both the S-S interval, the R-S interval, and their interaction. In
general, the response rate is very low when the R-S interval is
less than the S-S interval, reaches a maximum when the R-S and S-S
intervals are about equal, and then decreases as the R-S interval
gets increasingly longer than the S-S interval (Sidman, 1953). For
a given R-S interval, response rate seems to be inversely related
to the S-S interval (Leaf, 1965).

There is a special problem in evaluating learning and performance in Sidman avoidance tasks. The problem here is how to assess
the efficacy of the avoidance behavior. Do we use response rate
as the index of avoidance learning? Surely not. After all, at
long R-S or S-S intervals rapid responding is not necessary to a
avoid all the shocks. A common measure is number of shocks received,
but this too seems inappropriate as the number of shocks scheduled
also decreases with increasing R-S and S-S intervals. Clearly, some
kind of relative index is called for although none has yet come into
common usage. Hurwitz, Harzem, and Kulig (1972) have tested the
applicability of several such relative indices but were unable to
show that one was always superior.

The message is that indexing avoidance efficiency is not a
simple matter in Sidman tasks. This indexing of mastery also arises
in classical avoidance tasks, for that matter (See Bitterman, 1965),
because as the CS-US interval increases the likelihood of "accidental"

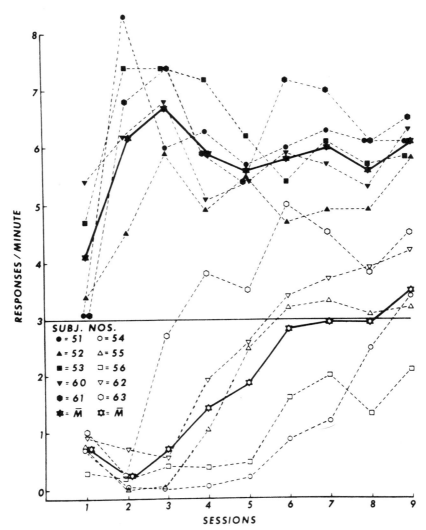

Figure 10.3 Avoidance response learning by 10 rats: Five were
trained on a lever-pressing task (open symbols) while
five were trained in a two-way shuttlebox (closed
symbols). The same temporally paced avoidance
schedule was used in both (S-S = 5 sec, R-S = 20 sec),
and a minimum response rate of three responses per
min, properly spaced, was required to avoid all
shocks. The heavy lines are group means. Note that
the shuttle response was much more readily learned
than the lever-pressing response. (After Riess, 1971)

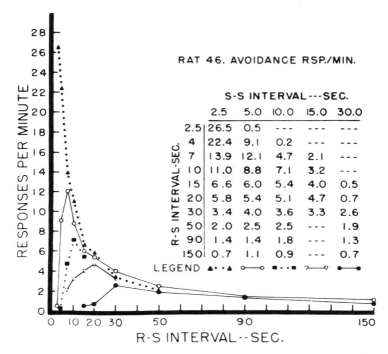

RAT 46. AVOIDANCE RSP./MIN.

R-S INTERVAL-SEC.	S-S INTERVAL---SEC.				
	2.5	5.0	10.0	15.0	30.0
2.5	26.5	0.5	---	---	---
4	22.4	9.1	0.2	---	---
7	13.9	12.1	4.7	2.1	---
10	11.0	8.8	7.1	3.2	---
15	6.6	6.0	5.4	4.0	0.5
20	5.8	5.4	5.1	4.7	0.7
30	3.4	4.0	3.6	3.3	2.6
50	2.0	2.5	2.5	---	1.9
90	1.4	1.4	1.8	---	1.3
150	0.7	1.1	0.9	---	0.7

LEGEND ▲··▲ ○—○ ■··■ ▽—▽ ●—●

Figure 10.4   Performance on temporally-paced, Sidman avoidance as a function of the temporal parameters defining the contingencies.   The curves in the figure represent different S-S intervals.   (After Sidman, 1966.)

or spontaneous responses leading to avoidance increases.   This then confounds comparisons of relative avoidance efficacy when different CS-US intervals are to be compared.

Given that response topography is shown to be so powerful an influence, the obvious question is "Why?"   Basically, two answers have been suggested.   One is that shocks elicit behaviors that are incompatible with the required response.   The usual candidate is "freezing" or immobility.   Blanchard and Blanchard (1969) have shown that free shocks do result in an overall reduction in activity including exploratory movements.   A second is that occurrences of some response are simply less discriminable than others.   It is argued that the less discriminable a response the harder it should be to learn.   And, bar-pressing seems to provide less stimulus change (both proprioceptive and exteroceptive) than does shuttling, thus perhaps accounting for the difference in performance between the two.

Both answers encounter some difficulties. First, if freezing is elicited by the shocks, why does it <u>differentially</u> impair pressing and running? Second, "freezing" is actually only rarely measured; it is <u>inferred</u> from the observations of avoidance failure. Such circularity is highly unsatisfactory. Third, if "freezing" is elicited by the shocks, why is the pattern of behavior in barpressing avoidance tasks usually <u>bursts of responding after shocks</u> with the frequency declining as a function of time since last shock (Church & Getty, 1972)?

With respect to discriminability, it is true that increasing the amount of stimulus change (feedback) contingent upon the occurrence of the avoidance response facilitates avoidance learning (Bower, Starr, & Lazarovitz, 1965). But which responses would be expected to produce the minimum response feedback? Immobility itself is surely a likely candidate. Yet, immobility seems readily learned as an avoidance response (Bolles & Riley, 1973). Bolles and Riley further demonstrated that the freezing shown by their avoidance groups was a function of the avoidance contingencies; groups which got no further shocks after initial acquisition (i.e., placed on extinction) or got the same density of shocks as the avoidance group but independently of their behavior (1/900 sec noncontingently) both showed less freezing than the avoidance group.

Bolles (1970) has proposed a hypothesis which transcends some of these difficulties. It is the <u>species specific defense reaction</u> hypothesis (SSDR). The SSDR hypothesis seeks to specify the set of learnable avoidance responses. The hypothesis is that given aversive stimuli elicit a certain narrow range of defensive behavior unique to the species. In addition, the hierarchy among these behaviors is modulated by the situational cues in the presence of which the aversive stimulus occurs. According to the hypothesis, only responses from the set of SSDRs is learnable as an avoidance response, with the rate of learning being a function of the response's ranking within the hierarchy.

The SSDR hypothesis is intuitively appealing. But it does not in fact provide us with a list of SSDRs or insure us a way of making up such a list. One cannot simply assault the animal in one situation and from the hierarchy of behaviors generated there, infer what the hierarchy will be in the avoidance situation. This is because any change in the situation, e.g. providing a manipulandum, changes the stimulus features of the situation and thus, possibly, the response set and the hierarchical arrangement of responses. Such a failing makes the SSDR notion circular: if a response is learned it must have been a member of the SSDR set; if not, it was not an SSDR. When we learn how to independently assess the SSDR set, we will be in a position to test this hypothesis.

TASK DEMANDS

It is possible to alter the task demands without altering the experimental chamber or the required response topography. When we do this, we find that task demands turn out to be a very potent variable as well (e.g., Maier, Albin, & Testa, 1973)--nearly as potent as response topography. One example of the importance of task demands can be found in the shuttlebox apparatus wherein the animal must jump a barrier between two chambers as the avoidance response. The experiment can be conducted so that the animal always jumps in the same direction (one-way) or the animal can be required to sometimes jump from side A to side B and on other occasions jump from B to A (two-way).

Learning does not proceed equally well under these two different task demands. Theios Lynch, and Lowe (1966) have provided us with a direct comparison of these two procedures. Figure 5 presents their results and illustrates that the one-way task is learned much more easily than the two-way task under a variety of shock conditions. Of equal importance is the observation that the shock intensity parameter interacts with the task demands. Increases in shock US intensity tend to facilitate learning in the one-way task. In the two-way task, however, increases in shock US intensity result in increasingly poorer learning! This finding is especially difficult for theories of avoidance learning which emphasize the classical conditioning contingencies embedded within the avoidance paradigm. This is because the efficacy of classical conditioning

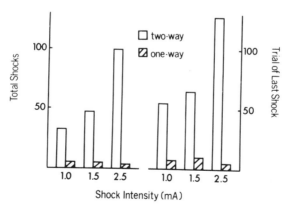

Figure 10.5 Comparison of indices of rates of avoidance learning in one-way and two-way shuttlebox tasks as a function of avoidance shock intensity. Note that two-way avoidance learning is impaired by increments in shock intensity. (Data from Theios, Lynch, & Lowe, 1966.)

is usually monotonic with US intensity. A number of "explanations" of this "unfortunate" interaction have been proposed. For example, Olton (1973) proposed that the two-way task produces approach-avoidance conflict which increases as US intensity increases. To date, no explanation has proved fully satisfactory.

EXTINCTION

Resistance to extinction was at one time touted as an index of the strength of learning (Skinner, 1938). Indexing resistance to extinction, however, is itself problematic when different response rates exist just prior to the extinction treatment. (Anderson, 1963, has suggested "relative rate of decline" as one solution, although it implausibly assumes a linear ratio scale relationship between performance and habit strength). It is with respect to extinction that some apochryphal notions about avoidance have come into being. An experiment by Solomon, Kamin, and Wynne (1953) played a major role here; after training dogs to avoid shock in a shuttlebox, they reported that the dogs seemed immune to a number of extinction procedures. As a result, current textbooks have come to perpetuate a notion of "partial irreversibility" of avoidance responding (Hulse, Deese, & Egeth, 1975).

Avoidance responses do extinguish! The rate of extinction is a function of the methodology employed. In part, the issue is one of confusion as to what is the "proper" extinction methodology. The typical appetitive extinction procedure has been to cease delivering, say, food rewards during extinction. The common analog in the avoidance experiments has been to cease delivering shocks. But note the asymmetry here: in the appetitive case the contingent event which supports behavior (delivery of food) is omitted, while in the avoidance case the contingent event associated with learning (omission of shocks) is continued! In addition, there is less change between the conditions of acquisition and extinction for the avoidance animals, leading to less stimulus generalization decrement between phases. Is it any wonder that avoidance behavior persists?

An alternative definition of extinction of avoidance is to simply negate the contingency between responding and omission of shocks (either with or without a change in frequency of shocks). Uhl and Eichbauer (1975) have compared extinction of both instrumental reward and avoidance under the conditions where the food or shock is not delivered (EXT) to extinction under conditions where the reinforcer is delivered (FREE VC). Their data are shown in Figure 6. If avoidance EXT is the parallel to appetitive FREE VC, then appetitive responses are to be considered the more resistant to extinction. But the opposite conclusion would follow from the comparison of avoidance FREE VC and appetitive EXT!

Relative persistence then is a function of ones decision about what constitutes the proper extinction methodology. In both cases, however, we should note that extinction of avoidance responding does occur.

We have now seen a variety of avoidance phenomena, enough to see some of its complexities. How then have theoreticians dealt with avoidance behavior?

THEORIES

The paradox of avoidance learning for the behaviorists was made clear by Mowrer who asked "How can omission of an event result in the stregthening of behavior?" After all, an infinitely large number of events fail to follow every response. The problem that faces the theorist is to account for the origin of the to-be-learned response and its subsequent strengthening.

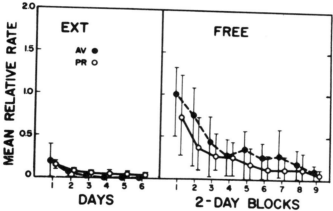

Figure 10.6 Rates of responding during extinction relative to the asymptotic pre-extinction response rates. Rats were trained either to respond for food (PR) or to avoid shocks (AV) and then extinguished under two different conditions: simple extinction in which all further food or shocks were precluded (EXT) or with continued delivery of food to group PR and shocks to group AV but with such deliveries being independent of responding (FREE). (After Uhl & Eichbauer, 1975.)

Two classes of theory have been proposed: single-process and two-process. The early theoretical efforts were of the single-process variety. These were generally completely displaced, until very recently, by two-process theory. There are several versions of two-process theory of avoidance, but they have sufficient commonality that, for our purposes here, we may treat them as a single theory for discussion.

Single-Process Theories

Historically, there have been two primary mechanisms of association suggested: contiguity and reinforcement. And, each has been offered as a sufficient basis for explaining avoidance. Let us consider each and look at the data bearing on their adequacy.

Contiguity explanations of avoidance (e.g., Hull, 1929, Sheffield, 1948) are essentially Pavlovian Theories. The basic notions are that:

(a)   the aversive US elicits the to-be-learned response,
(b)   the elicitation of this response in the presence of warning signal (CS) results in the CS acquiring the power to elicit the response, and
(c)   the experimenter arranges for the occurrence of the response to prevent the US, but this contingency is not critical to the emergence of the response.

From this summary of contiguity theory, we can derive several expectations about experiments in avoidance learning. First, only responses elicited by the US should get learned as avoidance responses. This is related to the issue of similarily of CR and UCR discussed earlier (Chapter 2). Second, the avoidance contingency per se, is irrelevant. Third, although the avoidance contingency per se is thought to be irrelevant, it has a correlated consequence that as the animal's performance improves, the US is omitted. This consequence should constitute partial reinforcement conditions during acquisition and extinction conditions when the organism comes to avoid on every trial.

Does the avoidance response closely resemble the US-elicited response as required by contiguity theory? It is easy to find cases where this is the case; the leg-flexion paradigm we considered at the outset of our treatment of avoidance is but one example. Of interest, however, are those cases in which a US-elicited response is specified as the avoidance response but is not learned (e.g., toe twitch, Turner & Solomon, 1962). Such examples are especially important when there is evidence that the response can be classically conditioned (e.g., Schlosberg, 1934, 1936).

Experiments critical to this view are those in which an avoidance response is learned and it is distinctively different from the UR elicited response. An interesting example here is an experiment by Fonberg (1962). She taught dogs to flex their leg to avoid an air puff to the ear; yet, the UR here was head shaking. The avoidance training of the heart rate response by DiCara and Miller (1968) is another intriguing example because the UR to their shock US was a heart rate increase, but the Pavlovian CR is a decrease in rate. Yet, they were able to train either an increase or decrease as an avoidance response. From these experiments, we may conclude that identity of UR and avoidance response is neither a necessary nor a sufficient condition for avoidance learning.

The second conclusion we derived from contiguity theory was that the avoidance contingency _per se_ was of no importance. If avoidance training were just classical conditioning in which the experimenters arranges for the classical CR to prevent the US, then learning should lead to partial reinforcement and pure classical conditioning should reliably lead to better conditioning than avoidance contingencies. As a first approximation to assessing this, one might simply compare response acquisition under avoidance and classical conditioning contingencies. A number of experiments have made such comparisons. The results of one comparison are presented in Figure 7. It is clear in this case that the avoidance procedure is markedly superior.

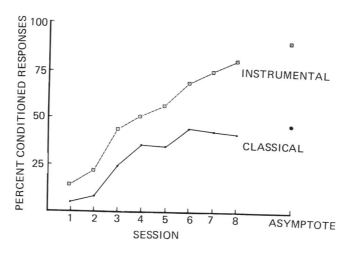

Figure 10.7   A comparison of the rate of learning a leg flexion response for groups under avoidance training procedures (INSTRUMENTAL) and classical conditioning procedures (CLASSICAL) with trials per session equated. (After Wahlsten & Cole, 1972.)

There are some problems however with such simple contrasts (c.f. Moore & Gormezano, 1961). One major problem is that the two groups differ in the number and distribution of USs delivered. A possible solution to this problem has been to yoke the two treatments in terms of the CS and US deliveries with the behavior of the avoidance animals determining the schedule of US deliveries for both treatment contingencies. This, then, equates the classical and avoidance animals in terms of number and distribution of USs, allowing them to differ only in the response contingency. Under these conditions, avoidance groups continue to be superior (e.g., Woodard & Bitterman, 1973).

The final derivation we made from contiguity theory is that omission of US in classical conditioning results in extinction. To the extent that avoidance behavior is dependent upon CS-UCR contingency, omission of the US should lead to prompt weakening of response strength. But the experiments just presented showed that such omissions did not weaken the response tendency. Furthermore, our earlier Figure 1, from Solomon and Wynne (1953), shows an instance of improving performance with omissions of shocks--an observation totally inconsistent with the contiguity analysis.

The major single-process alternative to the contiguity analysis was Hull's (1943, p. 73-74) reinforcement analysis. While he agreed that contiguity was important in learning, responses were thought to be strengthened only by reinforcement. In this single-process theory, the only source of such reinforcement was the termination of the aversive shock US. According to the view, strong responses tend to be evoked anticipatorily by the stimulus conditions, the latency being inversely related to habit strength. Viewed in this way, avoidance responses were little more than "extra-short" latency escape responses.

Again, we may derive certain expectations from the theory. For example, only responses that occur during the US and are effective in terminating the US should be learned as avoidance responses; that is, escape responding is a critical feature of the reinforcement analysis of avoidance learning. Second again, the avoidance contingency per se is seen as "irrelevant", yet it is clear that omission of the US necessarily prevents the response from terminating the US and thus would eliminate any further strengthening of the response through the escape contingency; indeed, the response should weaken because reinforcement would no longer be following the response.

We have already seen that the vigor of avoidance responding is maintained and can even increase in spite of the omission of the US. This fact is inconsistent with the single-process reinforcement analysis. But the focus of the testing of the reinforcement analysis has been the relationship between the avoidance and the escape responses.

Three kinds of data help us assess whether the avoidance response emerges from the escape response and is simply an anticipatory form of the latter. These are derived from experiments designed to assess whether or not:

a) an escape contingency is critical to avoidance learning,
b) prior escape training modulates avoidance acquisition, and
c) there is identity of the escape and avoidance responses.

A summary of the data bearing on these questions is generally negative for the reinforcement view. Although the existence of an escape contingency often aids avoidance learning, this is not always the case. Furthermore, the presence of such an escape contingency is neither necessary nor sufficient for avoidance learning to proceed. For example, Bolles, Stokes, and Younger (1966) have shown that whether or not escape contingency facilitates avoidance learning is a function of the response task. They studied avoidance learning in a shuttlebox and a wheel turning task; an escape contingency only significantly aided the former. But more to the present point is that avoidance learning in both tasks can proceed in the absence of an escape contingency. Indeed, other experiments suggest that the presence of an escape contingency can sometimes interfere with avoidance learning (Hurwitz, 1964).

Most difficult for the Hullian reinforcement theory are the data showing that it is possible to establish avoidance behavior while requiring one response to escape shock and a second, different response to avoid shock. Mowrer (Mowrer & Lamoreaux, 1946) was the first to show this, and Bolles (1969) has extended Mowrer's early observations. Figure 8 summarizes Bolles' data. Although rats show differential efficacy of avoidance learning as a function of task, they can learn any of them and do so whether or not the escape and avoidance responses are topographically congruent. Indeed, only for the turning response is such congruence helpful.

There is a final signal-factor theory of avoidance to consider briefly. (I refer to it as single "factor" because its authors do not specify the mechanism of the learning.) This theory proposes shock frequency reduction achieved by the animal as the factor that explains avoidance. Indeed, this factor seems by definition of avoidance to be the one factor common to all avoidance training procedures. Herrnstein (1969) has been the major proponent; it is not proposed as a cognitive theory. In contrast to the beliefs

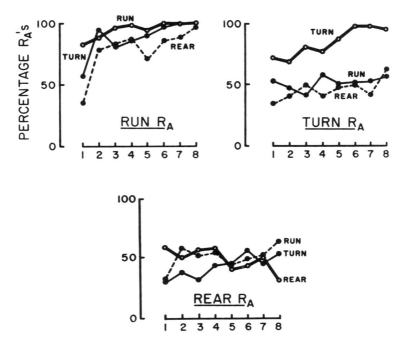

Figure 10.8   The learning of three different avoidance responses
               as a function of type of escape response. Nine dif-
               ferent groups of rats were required to run forward in
               a wheel (RUN), turn around (TURN), or rear onto their
               hind legs (REAR) to avoid shock; subgroups of these
               had to run, turn, or rear to escape the shocks if
               they failed to avoid. Thus, for example RUN-RUN is a
               case where the avoidance and escape responses are
               topographically identical while in TURN-REAR they are
               different. Note that all groups learned to avoid to
               some degree and that in only one case (TURN-TURN) was
               congruence facilitatory. (After Bolles, 1969.)

of its author, most interpreters of the literature believe it to be a "covert" two-process theory (e.g., Mackintosh, 1974). Nonetheless, we shall treat the shock-frequency-reduction hypothesis here as a single-factor theory where its <u>author</u> would argue it belongs.

The primary demonstration taken as the foundation for this theory is that rats will learn to press a lever even if it does not insure shock omission in the next few seconds so long as, on average, the shock frequency following the occurrence of a response is lower than the scheduled shock frequency in the absence of such responding. Herrnstein and Hineline (1966), using the contingencies illustrated in Figure 9, demonstrated just this. Rats would learn to press a lever to reduce the probability of shock, although they learned very slowly.

Several observations in the literature pose problems for the shock frequency reduction theory. For example, Hineline (1970) has shown that a schedule which does not result in reduction in shock frequency, but only delay of shock occurrence will result in good learning. More importantly, it has been shown by others (Gardner & Lewis, 1976) that the delay to the next shock is the critical feature maintaining such performance. Their experimental treatments and the resulting data are shown in Figure 10. When the delay is reduced by responding, behavior is suppressed; when it is increased, behavior is reinforced--this is with total shock frequency held constant.

Perhaps the most problematic data for Herrnstein's shock frequency reduction theory is that, given some minimum delay between the occurrence of the response and the next shock, rats will continue to make "avoidance" responses even when the overall frequency of shocks <u>increases</u> contingent upon those responses. Gardner and Lewis (1976) have shown that such "avoidance" responding will be maintained even when a doubling of the shock density follows the response! These data surely must count heavily against any shock-frequency-reduction view that does not incorporate supplementary principles taking into account the delay factor. What then is the alternative to these inadequate single-process views of avoidance?

Two-Process Theory

Two-process theory was Mowrer's (1947) solution to the paradox of avoidance learning. Basically, the theory asserts that a mediating response "fear" is classically conditioned by virtue of the CS-US pairings embedded within the early trials of the classical

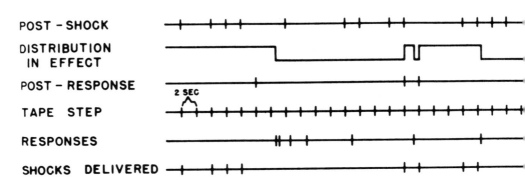

Figure 10.9   Representation of the experimental "avoidance" con-
               tingencies used by Herrnstein and Hineline (1966).
               Their tape advanced at regular 2-second intervals.
               Deflections on the lines marked "post-shock" and
               "post-response" indicate scheduled shocks for the two
               possible distributions of shocks. The "distribution
               in effect" line shows which of the schedules controls
               the delivery of shocks. The distribution in effect
               is determined by responses and shock deliveries: Con-
               trol is changed from the low density, post-response
               distribution to the high-density, post-shock distri-
               bution after each shock. Control is changed from the
               high density, post-shock distribution to the low den-
               sity, post-response distribution only by the first
               response after receiving a shock; additional responses
               are without consequence. Note that the contingencies
               are asymmetrical. Hypothetical response patterns and
               the resulting shocks experienced (i.e., delivered)
               are shown in the bottom two lines. (After Herrnstein
               & Hineline, 1966.)

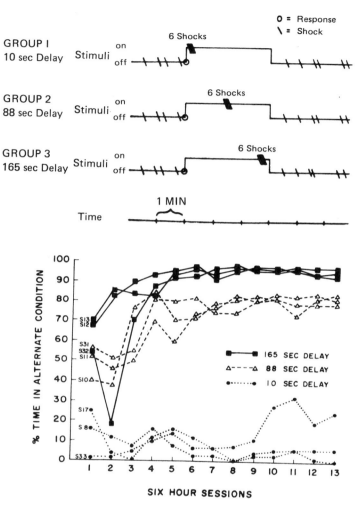

Figure 10.10   Experimental treatments (top) and data obtained
(bottom) by Gardner and Lewis (1976) showing that
animals choose a stimulus condition in which time-
until-next-shock is long (groups 2 and 3) relative
to the baseline VT 30-sec. shock schedule but do not
choose one in which the time-until-next-shock is
short (group 1). Responding by subjects turned on a
stimulus which was correlated with a shift in the
distribution of shocks for the next three minutes.
Note that the overall average density of shocks per
3-minute period is unchanged under all conditions.
(After Gardner & Lewis, 1976.)

avoidance paradigm. This conditioned fear state is assumed to have some of the response eliciting properties of the US; that is, the fear state is also aversive. The theory, then, suggests that any response that leads to the termination of the stimuli eliciting the aversive fear state will be reinforced consonant with Law of Effect principles. To recapitulate, the two-processes are classical conditioning of fear based upon contiguity of CS and US and instrumental motor response learning of the avoidance response with fear reduction as the reinforcer. These two processes are separate and independent processes which can lead to avoidance behavior through their sequential and interdependent action. Avoidance response learning can only proceed after fear has become conditioned.

The theory in its simplest form has features we may test. These are:
   a) that CS-US pairings condition a state which has aversive
      motivational properties,
   b) that these properties are classically conditioned through
      contiguity of CS and US, and
   c) that instrumental responding can be strengthened by termi-
      nation of the stimuli (usually a CS) which elicit the moti-
      vational state.
We shall briefly consider experiments which bear upon each of these.

Test of the question as to whether CS-US pairings condition a state which has motivational properties appear under the rubric of "acquired drive" experiments. The research here has been strongly confirmatory. Our traditional conceptualization of motivational variables has been that they energize any ongoing behavior. Brown, Kalish, and Farber (1951) gave us a test of whether CS-US pairings would endow the CS with behavior-facilitating properties. They placed rats in an apparatus sensitive to the activity of the rat. Here, they elicited startle reactions from individual rats from two treatment groups during CS presentation. For the experimental group the CS was paired with shocks; the control group did not receive CS and US pairings. Their data revealed the magnitude of the startle response to be greater as a function of the increasing number of CS—US pairings across days. This is consistent with the principle of conditioned motivational properties. Parallel data have been obtained by others; for example, Bull and Overmier (1968) showed the facilitating effects of a CS paired with shock when the CS was superimposed upon ongoing avoidance responding; a control CS had no such effect. These data, then, are consistent with the expectations based upon a principle of classical conditioning of a motivational state.

The second principle is that the conditioned motivational reaction is strengthened by contiguity of CS and US and not by drive reduction reinforcement. It is this principle which makes the theory a two-process one. Indeed, Miller (1948) agreed with Mowrer in substantial degree on all aspects except how "fear" was conditioned, Miller preferring to rely on shock termination as the reinforcer of fear. Thus, Miller's position might have been characterized as a two-factor, single-process theory.

If fear conditioning were strengthened by the drive reduction at US termination, as Miller suggested (1951), then for fixed CS—US interval, fear conditioning might be expected to be substantially poorer for long US duration because with longer US durations, reinforcement at US termination being necessarily more delayed with long USs. Delayed reinforcement yields poorer learning. In contrast, if fear were strengthened by contiguity, as Mowrer (1947) suggested, then the important parameter should prove to be CS-US onset interval. A number of experiments have compared classical conditioning using fixed CS-US onsets interval but varying the CS onset to US termination interval (Bitterman, Reed, & Krauskopf, 1952; Overmier, 1966 a, b). Longer US durations did not lead to poorer conditioning. This pattern of results is not consistent with the reinforcement view, but rather supports the contiguity view of how fear is learned.

The third major principle was that termination of stimuli, which elicit fear as a result of prior pairings with shock, can reinforce the learning of behavior. This has most commonly been tested by asking whether variation in CS termination contingencies modulated avoidance learning.

But before turning to those experiments, we might look a moment at a slightly more direct test. Brown and Jacobs (1949) asked whether the termination of a CS which had previously been paired with shock could serve as a reward for rats running down a runway. Theirs, then was a two phase experiment separating the fear conditioning and the response learning processes thought to underlie avoidance behavior. The results of their second phase in which the response terminated either a CS which had been paired with shock (Experimental Group) or one that had not (Control Group) are shown in Figure 11. Clearly the response contingent termination of the CS in the Experimental Group resulted in response learning. The results support the third principle.

Most tests of two-process theory focus on this last principle. Experiments manipulating the CS termination contingency within the avoidance paradigm also provide a test of this third reinforcement principle, but they are somewhat more indirect tests. This is because, say, delay of termination of the CS after a response confounds

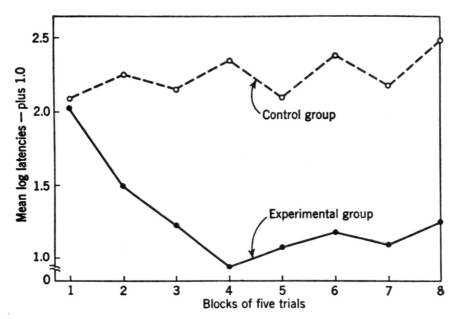

Figure 10.11   Latencies of responding to terminate a stimulus over
                a series of trials.   For the Experimental group the
                stimulus had previously been established on a clas-
                sical CS+ for electric shocks.   The Control group
                had no prior pairings of CS and shocks.   Note that
                during the test trials presented here no shocks were
                ever presented.   That the Experimental group learned
                the response for which the only contingent event was
                termination of the CS was taken as evidence that the
                CS+ had acquired aversive, motivational properties.
                (After Brown & Jacobs, 1949.)

the hypothesized delay of fear reduction with reduction in response feedback. Response feedback, we noted earlier, has been shown to be important in avoidance learning. Thus, such an experiment would have two factors that should reduce the efficacy of avoidance learning in the group receiving delay of CS termination.

In general, experiments testing the effects of delay of CS termination have confirmed that CS-termination contingent upon the occurrence of the avoidance response does facilitate avoidance learning. For example, Kamin (1956) manipulated factorially the termination-of-CS contingency and the avoidance-of-US contingency. Kamin found that the CS-termination contingency by itself produced response learning and that when this factor was added to the avoidance contingency (e.g., normal procedure), learning was maximal. These effects are shown in Figure 12. Bolles, Stokes and Younger (1966) did the same plus they added a third factor to the factorial of escape-from-the-US. Their results are consistent with Kamin's in that they show both termination of CS and the avoidance contingencies to be important, with some indication that the avoidance contingency may be more important. But, we must note that in both experiments some learning occurred when there was not prompt termination of CS following the response.

If avoidance learning proceeds reasonably well in the absence of a CS termination contingency, must we now infer that it is omission of the US that is reinforcing avoidance behavior? This is the very position that Mowrer struggled so vigorously against. But it is clear that there exist a host of experiments that show avoidance learning to proceed reasonably well in the absence of a CS-termination contingency. These include experiments using trace conditioning and those in which there is no explicit CS at all like Sidman avoidance.

One alternative has been to "invent" sources of stimuli. That is to say, theorists have inferred that the presence or absence of response correlated stimuli (either internal or external) may be the important cues. One view has been to assign the role of CS to these response related stimuli (c.f. Schoenfeld, 1950; Dinsmoor, 1954). This leads to conceptualization of the absence of those stimuli produced by the avoidance response as the fear eliciting cue.

A second view has been to focus upon the events closely consequent upon the occurrence of the avoidance response. By virtue of the avoidance contingency, the response "feedback" stimuli are rarely or never paired with shock. Since this feedback is rarely or never followed by shock but the "background" stimuli have been associated with shocks, we have the necessary and sufficient condition to establish the feedback stimuli as a conditioned CS-.

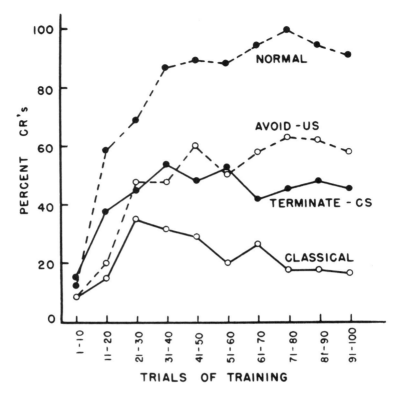

Figure 10.12   Percentage of crossings <u>during</u> CS-US interval ("CRs")
in a modified Miller-Mowrer shuttlebox by four groups
of rats over a series of learning trials. The groups
were identical with respect to trial initiation but
differed in the changes that occurred contingent
upon the response. Each trial began with the onset
of a buzzer CS and a shock US was scheduled to come
on 5 sec. later. For the CLASSICAL group, these
pairings were independent of behavior. For the TER-
MINATE-CS group, CRs terminated the CS but the shocks
were delivered as scheduled. For the AVOID-US group,
CRs prevented the delivery of the scheduled shock but
did not terminate the CS. Finally, for the NORMAL
group, CRs both terminated the CS and prevented the
delivery of the scheduled shock; this is standard
avoidance training procedure. (After Kamin, 1956.)

Thus, this view is to characterize the response feedback stimuli as a classically conditioned CS- which serves to depress the fear level elicited by those stimuli having a positive correlation with shock. If one accepts response feedback stimuli as a CS-, these can inhibit the fear elicited by either an explicit or implicit CS+ for shock yielding fear reduction reinforcement. Thus, we are no longer dependent upon only change in the CS+ for reinforcement.

Is there any reason to favor such a view? We can see in Figure 13, which presents the results of an experiment by Bull and Overmier (1968), that a CS- when superimposed upon ongoing avoidance behavior suppresses the avoidance responding presumedly by inhibiting the underlying fear which motivates the behavior. Furthermore, Weisman and Litner (1969) have shown that if a CS- is presented contingent upon the occurrence of free operant avoidance responses specified changes in the pattern of avoidance responding can be reinforced. Figure 14 shows that they were able to establish either increases (drh) or decreases (drl) in responding by using a CS- as the reinforcer for the change.

Morris (1974) has sought to show that external feedback stimuli (and by inference, internal ones, too) can acquire their CS- properties through classical conditioning contingencies embedded within the avoidance training. More important to the present view, however, would be the demonstration that feedback stimuli as a result of the embedded contingencies acquire the power to reinforce avoidance responding. Morris (1975) has done just that.

One might fairly ask here whether CS+ termination and inhibition of fear by CS- are not redundant and an inseperable identity. The answer, I think is "no." For example, Cicala and Owen (1976) have provided an experiment which suggests the two acquire their capacity to reinforce behavior rather independently through separate contingencies. In their experiment they first trained three groups of animals to avoid; each trial began with a CS+ onset. For one group (T), the CS+ terminated whenever an avoidance response occurred; for the second group (F), a feedback stimulus (potentially a CS-) was presented contingent upon the avoidance response, and after 10 seconds both CS+ and feedback stimulus (CS-) went off; the third group (TF) had both CS termination and CS- feedback. After all three groups had reached avoidance criterion, three trials were given in which the US was presented even though the animal avoided. Avoidance training then continued as before. The effect of the shock trials upon the T and TF groups was minimal; but, the effect upon the F was dramatic; the data are presented in Figure 15. The extra shocks markedly reduced the reinforcing properties of the feedback stimulus; one may presume that the pairing of the feedback stimulus with shocks

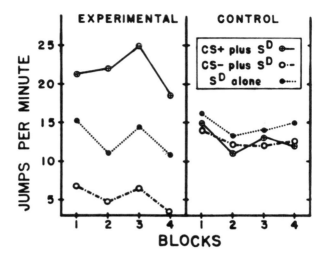

Figure 10.13   Effects of tone CSs upon responding during visual
cue for avoidance for two groups of dogs. After instrumental avoidance training in a shuttlebox, the
two groups received classical conditioning in a
small chamber: For the EXPERIMENTAL group the CS+
was paired with shocks while the CS- signalled a
period free from such shocks; the CONTROL group received an equal number of presentations of tones and
of shocks; the CONTROL group received an equal number of presentations of tones and of shocks but
these were programmed completely independently of
each other (TRC). Finally, the tone CSs were compounded with the cue for avoidance (S^D) to assess
the effects of the classical conditioning. (After
Bull & Overmier, 1968.)

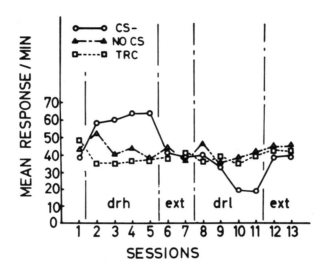

Figure 10.14   The effects upon rates of avoidance responding of
               response contingent presentations of a classical con-
               ditioned inhibitory stimulus (CS-) or a neutral stim-
               ulus (TRC). The response contingencies were that the
               tone (CS- or TRC history) was present upon increases
               in avoidance response rate (panel drh), then not pre-
               sented (panel ext), then presented contingent upon
               decreases in avoidance response rate (panel drl).
               Note that response contingent presentations of only
               the inhibitory stimulus served to reinforce changes
               in the rates of avoidance responding. (After Weis-
               man & Litner, 1969.)

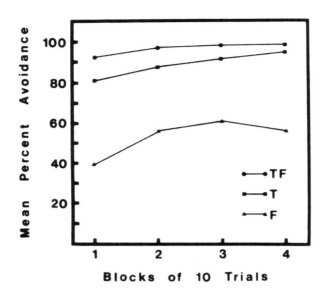

Figure 10.15   Avoidance performance following three punished trials
               as a function of the kind of prior training history:
               Group T had previously experienced termination of
               the CS after each avoidance response; group F had
               experienced presentation of a feedback stimulus after
               each such response; and group TF had experienced both
               termination of the CS and presentation of the feed-
               back stimulus.   Punishment for group F, which re-
               sulted in pairing of the feedback stimulus with shock
               thus diminishing its status as an inhibitory CS- and
               removing that group's source of "reinforcement" for
               avoidance, had the greatest debilitating effect upon
               avoidance performance.   (After Cicala & Owen, 1976.)

on those trials must have reduced CS-'s fear-inhibitory power. In contrast, the extra shock could only enhance the CS+ fear eliciting powers such that later CS+ terminations were enhanced as a reinforcer for the avoidance behavior. While this experiment is not definitive, it clearly supports the idea that the events of CS+ termination and CS- presentation accrue their reinforcing power through different mechanisms because the identical operation on both events (presentation of 3 shocks) had quite different effects.

This positive reinforcement through fear inhibition view is quite useful in dealing with some difficult data. It tells us why responses generating substantial intrinsic feedback are acquired more readily than those with poor feedback. It tells us why supplementary feedback, either through CS-termination or by added cue, facilitates most those responses with poor intrinsic feedback. It tells us that long delays to the next shock or long intertrial intervals should facilitate avoidance learning because these are ideal conditions for establishing a CS- (Moscovitch & LoLordo, 1968). Indeed it gives considerable supplementary power to two-process theory. The preceding is not to discount reinforcement through fear reduction achieved by CS+ termination, if such is available in the paradigm. But if it is not, avoidance learning still should proceed according to the supplementary two-process principles derived here.

The idea of a classically conditioned state of fear (and the parallel idea of incentive motivation, as discussed earlier) mediating the learning of instrumental motor acts gave learning theorists new purchase on complex learning phenomena and served as a powerful heuristic in the search for new phenomena (e.g., Trapold & Overmier, 1972).

REFERENCES

Anderson, N. H. Comparison of different populations: Resistance to extinction and transfer. PSYCHOLOGICAL REVIEW, 1963, 70, 162-179.

Bitterman, M. E. The CS-US interval in classical and avoidance conditioning. In W. F. Prokasy (Ed.), CLASSICAL CONDITIONING. New York: Appleton-Century-Crofts, 1965.

Bitterman, M. E., Reed, P. C., & Krauskopf, J. The effect of the duration of the unconditioned stimulus upon conditioning and extinction. AMERICAN JOURNAL OF PSYCHOLOGY, 1952, 65, 256-262.

Blanchard, R. J., & Blanchard, D. C.   Crouching as an index of fear.
    JOURNAL OF COMPARATIVE AND PHYSIOLOGICAL PSYCHOLOGY, 1969, 67,
    370-375.

Bolles, R. C.   Avoidance and escape learning: Simultaneous acquisi-
    tion of different responses. JOURNAL OF COMPARATIVE AND PHYSIO-
    LOGICAL PSYCHOLOGY, 1969, 68, 355-358.

Bolles, R. C.   Species specific defense reactions and avoidance
    learning. PSYCHOLOGICAL REVIEW, 1970, 71, 32-48.

Bolles, R. C., & Riley, A. L.   Freezing as an avoidance response:
    Another look at the operant-respondent distinction. LEARNING
    AND MOTIVATION, 1973, 4, 268-275.

Bolles, R. C., Stokes, L. W., & Younger, M. S.   Does CS termination
    reinforce avoidance behavior. JOURNAL OF COMPARATIVE PHYSIO-
    LOGICAL PSYCHOLOGY, 1966, 62, 201-207.

Bower, G., Starr, R., & Lazarovitz, L.   Amount of response-produced
    change in the CS and avoidance learning. JOURNAL OF COMPARA-
    TIVE AND PHYSIOLOGICAL PSYCHOLOGY, 1965, 59, 13-17.

Brown, J. S., & Jacobs, A.   The role of fear in the motivation and
    acquisition of responses. JOURNAL OF EXPERIMENTAL PSYCHOLOGY,
    1949, 39, 747-759.

Brown, J. S., Kalish, H. I., & Farber, I. E.   Conditioned fear as
    revealed by the magnitude of startle response to an auditory
    stimulus. JOURNAL OF EXPERIMENTAL PSYCHOLOGY, 1951, 41, 317-
    328.

Bull, J. A. III, & Overmier, J. B.   The additive and subtractive
    properties of excitation and inhibition. JOURNAL OF COMPARATIVE
    AND PHYSIOLOGICAL PSYCHOLOGY, 1968, 66, 511-514.

Church, R. M., & Getty, D. J.   Some consequences of the reaction to
    an aversive event. PSYCHOLOGICAL BULLETIN, 1972, 78, 21-27.

Cicala, G. A., & Owen, J. W.   Warning signal termination and a feed-
    back signal may not serve the same function. LEARNING AND
    MOTIVATION, 1976, 7, 356-367.

D'Amato, M. R., & Schiff, D.   Long-term discriminated avoidance
    performance in the rat. JOURNAL OF COMPARATIVE AND PHYSIOLOG-
    ICAL PSYCHOLOGY, 1964, 57, 123-126.

DiCara, L. V., & Miller, N. E.   Changes in heart rate instrumentally
    learned by curarized rats as avoidance response. JOURNAL OF
    COMPARATIVE AND PHYSIOLOGICAL PSYCHOLOGY, 1968, 65, 8-12.

Dinsmoor, J. A.   Punishment: I. The avoidance hypothesis.
    PSYCHOLOGICAL REVIEW, 1954, 61, 34-46.

Fonberg, E. Transfer of the conditioned avoidance reaction to the unconditioned noxious stimuli. ACTA BIOLOGIAE EXPERIMENTALIS (Warsaw), 1962, 22, 251-258.

Gardner, E. T., & Lewis, P. Negative reinforcement with shock frequency increase. JOURNAL OF THE EXPERIMENTAL ANALYSIS OF BEHAVIOR, 1976, 25, 1-14.

Herrnstein, R. J. Method and thoery in the study of avoidance. PSYCHOLOGICAL REVIEW, 1969, 76, 49-69.

Herrnstein, R. J., & Hineline, P. N. Negative reinforcement as shock frequency reduction. JOURNAL OF THE EXPERIMENTAL ANALYSIS OF BEHAVIOR, 1966, 9, 421-430.

Hineline, P. N. Negative reinforcement without shock reduction. JOURNAL OF THE EXPERIMENTAL ANALYSIS OF BEHAVIOR, 1970, 14, 259-268.

Hull, C. L. A functional interpretation of the conditioned reflex. PSYCHOLOGICAL REVIEW, 1929, 36, 498-511.

Hull, C. L. PRINCIPLES OF BEHAVIOR. New York: Appleton-Century-Croft, 1943.

Hulse, S. H., Deese, J., & Egeth, H. THE PSYCHOLOGY OF LEARNING. New York: McGraw-Hill, 1975.

Hurwitz, H. M. B. Method for discriminative avoidance training. SCIENCE, 1964, 145, 1070-1071.

Hurwitz, H. M. B., Harzem, P., & Kulig, B. Comparisons of two measures of free-operant avoidance under two conditions of response feedback. QUARTERLY JOURNAL OF EXPERIMENTAL PSYCHOLOGY, 1972, 24, 92-97.

Kamin, L. J. The effects of termination of the CS and avoidance of the US on avoidance learning. JOURNAL OF COMPARATIVE AND PHYSIOLOGICAL PSYCHOLOGY, 1956, 49, 420-424.

Leaf, R. C. Acquisition of Sidman avoidance responding as a function of S-S interval. JOURNAL OF COMPARATIVE AND PHYSIOLOGICAL PSYCHOLOGY, 1965, 59, 298-300.

Mackintosh, N. J. THE PSYCHOLOGY OF ANIMAL LEARNING. London: Academic Press, 1974.

Maier, S. F., Albin, R. W., & Testa, T. J. Failure to learn to escape in rats previously exposed to inescapable shock depends on the nature of escape response. JOURNAL OF COMPARATIVE AND PHYSIOLOGICAL PSYCHOLOGY, 1973, 85, 581-592.

May, M. A. Experimentally acquired drives. JOURNAL OF EXPERIMENTAL PSYCHOLOGY, 1948, 38, 66-77.

Miller, N. E. Studies of fear as an acquirable drive: I. Fear
    as motivation and fear reduction as reinforcement in the
    learning of new responses. JOURNAL OF EXPERIMENTAL PSYCHOLOGY,
    1948, 38, 89-101.

Miller, N. E. Comments on multiple-process conceptions of learning.
    PSYCHOLOGICAL REVIEW, 1951, 58, 375-381.

Moore, J. W., & Gormezzano, I. Yoked comparisons of instrumental
    and classical eyelid conditioning. JOURNAL OF EXPERIMENTAL
    PSYCHOLOGY, 1961, 62, 552-559.

Morris, R. G. M. Pavlovian conditioned inhibition of fear during
    shuttlebox avoidance behavior. LEARNING AND MOTIVATION, 1974,
    5, 424-447.

Morris, R. G. M. Preconditioning of reinforcing properties to an
    exteroceptive feedback stimulus. LEARNING AND MOTIVATION,
    1975, 6, 289-298.

Moscovitch, A., & LoLordo, V. M. Role of safety in the Pavlovian
    backward fear conditioning procedure. JOURNAL OF COMPARATIVE
    AND PHYSIOLOGICAL PSYCHOLOGY, 1968, 66, 673-678.

Mowrer, O. H. On the dual nature of learning: A reinterpretation
    of "conditioning" and "problem solving." HARVARD EDUCATIONAL
    REVIEW, 1947, 17, 102-148.

Mowrer, O. H., & Lamoreaux, R. R. Fear as an intervening variable
    in avoidance conditioning. JOURNAL OF COMPARATIVE PSYCHOLOGY,
    1946, 39, 29-50.

Olton, D. S. Shock-motivated avoidance and the analysis of behavior.
    PSYCHOLOGICAL BULLETIN, 1973, 79, 243-251.

Overmier, J. B. Differential transfer of control of avoidance re-
    sponses as a function of UCS duration. PSYCHONOMIC SCIENCE,
    1966, 5, 25-26. (a)

Overmier, J. B. Instrumental and cardiac indices of Pavlovian
    fear conditioning as a function of UCS duration. JOURNAL OF
    COMPARATIVE AND PHYSIOLOGICAL PSYCHOLOGY, 1966, 62, 15-20. (b)

Riess, D. Shuttleboxes, Skinner boxes, and Sidman avoidance in
    rats: Acquisition and terminal performance as a function of
    response topography. PSYCHONOMIC SCIENCE, 1971, 25, 283-286.

Schlosberg, H. Conditioned responses in the white rats. JOURNAL
    OF GENETIC PSYCHOLOGY, 1934, 45, 303-335.

Schlosberg, H. Conditioned responses in the white rat: II. Condi-
    tioned responses based upon shock to the foreleg. JOURNAL OF
    GENETIC PSYCHOLOGY, 1936, 49, 107-138.

Schoenfeld, W. N. An experimental approach to anxiety, escape, and avoidance behavior. In P. H. Hock & J. Zubin (Eds.), ANXIETY. New York: Grune & Stratton, 1950.

Sheffield, F. D. Avoidance training and the contiguity principle. JOURNAL OF COMPARATIVE AND PHYSIOLOGICAL PSYCHOLOGY, 1948, 41, 165-177.

Sidman, M. Avoidance conditioning with a brief shock and no extero-ceptive warning signal. SCIENCE, 1953, 118, 157-158.

Sidman, M. Avoidance behavior. In W. K. Honig (Ed.), OPERANT BEHA-VIOR: AREAS OF RESEARCH APPLICATION. New York: Appleton-Century-Crofts, 1966.

Skinner, B. F. THE BEHAVIOR OF ORGANISMS. New York: Appleton-Century-Crofts, 1938.

Solomon, R. L., Kamin, L. G., & Wynne, L. C. Traumatic avoidance learning: The outcomes of several extinction procedures with dogs. JOURNAL OF ABNORMAL AND SOCIAL PSYCHOLOGY, 1953, 48, 291-302.

Solomon, R. L., & Wynne, L. C. Traumatic avoidance learning: Acquisition in normal dogs. PSYCHOLOGICAL MONOGRAPHS, 1953, 67, No. 4 (Whole No. 354).

Theois, J., Lynch, A. D., & Lowe, W. F. Differential effects of shock intensity on one-way and shuttle avoidance conditioning. JOURNAL OF EXPERIMENTAL PSYCHOLOGY, 1966, 72, 294-299.

Trapold, M. A., & Overmier, J. B. The second learning process in in-strumental learning. In A. H. Black & W. F. Prokasy (Eds.), CLASSICAL CONDITIONING II: CURRENT RESEARCH AND THEORY. New York: Appleton-Century-Crofts, 1972.

Turner, L. H., & Solomon, R. L. Human traumatic avoidance learning: Theory and experiments on the operant-respondent distinction and failures to learn. PSYCHOLOGICAL MONOGRAPHS, 1962, 76 (Whole No. 559).

Uhl, C. N., & Eichbauer, E. A. Relative persistence of avoidance and positively reinforced behavior. LEARNING AND MOTIVATION, 1975, 6, 468-483.

Wahlston, D. L., & Cole, M. Classical and avoidance training of leg flexion in the dog. In A. H. Black & W. F. Prokasy (Eds.), CLASSICAL CONDITIONING II: CURRENT THEORY AND RESEARCH. New York: Appleton-Century-Crofts, 1972.

Warner, L. H. The association span of the white rat. JOURNAL OF GENETIC PSYCHOLOGY, 1932, 41, 57-90. (a)

Warner, L. H. An experimental search for the conditioned response. JOURNAL OF GENETIC PSYCHOLOGY, 1932, 41, 91-115. (b)

Weisman, R. G., & Litner, J. S.   Positive conditioned reinforcement of Sidman avoidance behavior in rats.   JOURNAL OF COMPARATIVE AND PHYSIOLOGICAL PSYCHOLOGY, 1969, 68, 597-603.

Woodard, W. T., & Bitterman, M. E.   Pavlovian analysis of avoidance conditioning in the goldfish (*Carassius auratus*).   JOURNAL OF COMPARATIVE AND PHYSIOLOGICAL PSYCHOLOGY, 1973, 82, 123-129.

11. THEORIES OF INSTRUMENTAL LEARNING

J. B. Overmier

University of Minnesota
Minneapolis, Minnesota, USA

We have now reviewed the learning that occurs under various applications of the three-term contingency of stimulus→response→ outcome stimuli, i.e., the instrumental reward, escape, punishment, and avoidance paradigms. Each has proved to produce far more complex phenomena than one might have naively expected. The questions that shall interest us here in a summary fashion are: "What is learned?" and "How?" These are <u>the</u> perennial questions in the psychology of learning.

Different theories have focused upon different aspects of the three-term contingency and have, as a result, offered different answers. While we cannot deal with all the variations, we can characterize the major hypotheses offered. First, we distinguish theories which seek to account for all learned instrumental behavior by reference to a single principle from those which argue that two (or more) of the proposed single processes operating in parallel are required to adequately account for instrumental behavior.

SINGLE PROCESS THEORIES

The single process theories, while reflecting the <u>associative</u> tradition of psychology, divide on what elements enter <u>into associa</u>-tion. Some argue that stimuli become associated with antecedent stimuli; these are called S-S theories. Others argue that it is responses that become associated with antecedent stimuli, particularly those stimuli in the presence of which the responses occurred; these are called S-R theories.

349

These S-S and S-R theories may in turn be subdivided on what are thought to be the necessary and sufficient conditions for the strengthening of associations. Some have argued that contiguity of elements is sufficient; others reflecting hedonistic principles argued that, in addition to contiguity, reinforcement is also necessary. Table 1 presents an idealized summarization of the varieties of single process theories and some of their proponents. That such a variety of answers to the "what" and "how" questions have been offered suggests that the resolution is no simple matter.

Let us briefly review the major arguments for and against some of these positions. Before doing so, we must distinguish between operations and hypothesized mechanisms. Antecedent and consequent environmental stimuli are manipulated, and responses are observed by all behaviorists. But the theoretical inferences drawn are about unobserved associations within the organism. Thus, both the S-R and S-S theorists are invoking an intervening construct or intervening variable that is itself not directly observed but inferred from the behavior of the animal. The degree of validation for such inferences is a function of the degree to which they enable one to account for the existing data and predict new phenomena. Let us now turn to the "theories" themselves.

One early class was S-R contiguity; both Watson and Guthrie offered such theories, although differing on details. This class of theory reflected the associationistic tradition while emphasizing the response as the terminal product of learning. The basic assumptions were (1) that if a response (R) occurs in the presence of stimulus conditions (S), an association between S and R comes into existence because of their contiguous occurrence; (2) this association results in an increased probability that R will occur the next time that stimulus is presented; and (3) the more often a given response occurs in the presence of a given stimulus, the stronger the association. The discoveries of Pavlov on classical conditioning of reflexes increased the plausibility of this S-R contiguity theory and became a keystone for the theory.

According to this view, the role of any response contingent outcome (reinforcer) is to either insure that R occurs (Watson, 1916) or to change the stimulus situation after R occurs spontaneously so that no other response can become associated with S (Guthrie's principle of postremity, 1935). The to-be-learned response increases in strength because the contingencies insure that it occurs on every trial. This view of instrumental learning has a degree of simple elegance that makes it very attractive as a base for further theorizing (cf., Estes, 1959). But its simplicity and adequacy are more apparent than real. It is with great difficulty that it accounts for why all responses that occur in a situa-

TABLE I

SINGLE PROCESS THEORIES CATEGORIZED BY WHAT IS LEARNED

AND THE NECESSARY CONDITIONS

	Contiguity	Continguity _plus_ Reinforcement
S—R	Watson Guthrie Estes	Thorndike Hull
S—S	Tolman Bindra	

TABLE II

TWO PROCESS THEORIES:

THE ASSUMED CONDITIONS FOR TYPES OF LEARNING

	HABITS	MOTIVATIONS
Spence	S-R contiguity	S-R reinforcement
Mowrer	S-R reinforcement	S-R contiguity

tion are not learned equally well -- that is, why some spontaneously occurring responses are not learned! Equally problematic
is why a response, once well learned, should ever extinguish.

Like Watson and Guthrie, Thorndike, too, argued that it was
response tendencies in the presence of stimuli that were learned;
but unlike them, Thorndike argued that it was the consequences of
those responses that caused the strengthening or weakening of such
S-R associations. This emphasis on response consequences enabled
Thorndike to explain why it was that only selected responses occurring in a stimulus situation were learned. Thorndike's attribu-
tion of a critical role to the hedonic consequences of responses
was called the Law of Effect. The Law of Effect may be considered
at two levels, as an empirical generalization or as a theoretical
assertion regarding the necessary and sufficient conditons for
learning. Here we shall be considering the latter because as an
empirical generalization it is generally quite sound.

Thorndike, and, later, Hull (1943), argued that all response
learning might be accounted for by S-R reinforcement principles.
Further, they argued that it is the dependence of learning upon
contingent reinforcement -- behavior's sensitivity to its consequences -- that accounts for learning's adaptive nature. Just as
features found to be useful in the history of a species are selected,
response found to be useful in the history of the individual are
selected. Thorndike observed that if reinforcement were contingent
upon the occurrence of one of two incompatible responses, the one
producing the reward would rapidly come to dominate, even if the
other were initially the more frequent! This he took as a contra-
diction of S-R contiguity theory.

In general, the experiments which are critical to all S-R reinforcement theories, including that of Hull (1943), are those
which show the importance of the response contingent reinforcing
event. We have reviewed several experiments which suggest such
importance (Chapter 5). These include demonstrations that the
greater the magnitude of reward the greater the amount of behavior
change and that delay of occurrence of the reinforcer impairs
learning. If the response-reinforcer contingency were not important, then delaying the reinforcer should have no effect, the argument goes. Another way to demonstrate the role of contingency in
response learning is to compare the performance of animals which
receive reinforcers contingent upon their own behavior with those
who recive the reinforcers independent of their own behavior but
"yoked" to those of the first animal. In the case of non-discrim-
inative training for naive animals, the data (Chapter 5) were generally clear and consistent with S-R reinforcement theory -- response independent reinforcers are not as effective as response

dependent ones in modulating behavior. In the discriminative case, however, we should recall that the data were substantially less clear in their consistency with a reinforcement view. For example, we may recall from Chapter 9 that the CER procedure produces nearly as much suppression of responding as does discriminative punishment.

Two obvious but critical assumptions of both S-R theories are that (1) it is responses that are acquired and extinguished, and (2) learning requires the occurrence of the response in the situation. Therefore, embarrassing to strict S-R theories are (1) those experiments in which one functionally equivalent but topographically dissimilar response substitutes for the originally trained response, and (2) those in which the strength of the trained behavior is associatively manipulated without occurrence of the S-R behavioral sequence.

Very early on, Tolman and his associates provided demonstrations which seem inconsistent with the S-R reinforcement theory. For example, he demonstrated place learning (Tolman, Ritchie, & Kalish, 1946) in which, after training rats to turn <u>right</u> in a T-maze for food reward, they rotated the apparatus 180° and carried out test trials; on these test trials it was observed that rats now turned <u>left</u>, hence going to the previously rewarded place by performing a response different than that reinforced! Tolman, following Hobhouse, placed great emphasis upon the goal directedness of behavior <u>and</u> the organism's flexibility in achieving this goal (fixed <u>end</u> product, not fixed responses) as illustrated by the demonstrations of place learning. Tolman accounted for this by hypothesizing that animals learn "what-leads-to-what" -- i.e., what stimuli are followed by what other stimuli and which stimuli signal goal objects (see Figure 2, in Chapter 6). The learning of these relationships was thought to be dependent only upon contiguity; thus Tolman may be characterized as an S-S contiguity theorist. (Sometimes Tolman is also referred to as an "expectancy" theorist.)

Tolman's latent learning experiments, presented in Chapter 6, are also problematic for S-R theorists because during the maze familiarization phase, the subjects seemed not to show learning; but after exposure to reward in the goalbox by the experimenter directly placing the animal there, the subjects on subsequent test trials ran quickly and directly to the place where the reward had been received. Latent learning has been demonstrated in a variety of ways (e.g., Tolman & Honzik, 1930; Seward, 1949). While several criticisms have been leveled at selected latent learning experiments (see Kimble, 1961), no one seems sufficiently broad to discredit all of them.

Further, there is another class of experiment that we encountered in the Chapter 2 discussion of classical conditioning which
seems to "survive" all such criticisms. This is the sensory preconditioning experiment in which as a result of prior bell-light pairings, a bell acquires the power to elicit a response conditioned to
the light (Brogden, 1939). In a sense, this is the ultimate latent
learning experiment representing in the bell-light association the
epitome of S-S contiguity learning.

The final coup-de-grace for the simple, single process S-R reinforcement view as exemplified by Hull's 1943 theory was the
Crespi (1942) experiment discussed extensively in Chapter 7. It was
a magnitude of reinforcement shift experiment in which animals when
faced with a changed level of reward, abruptly changed levels of performance as a function of the magnitude of the goal object -- even
showing "contrast effects" of overshooting the performance levels of
the unshifted subjects. The abruptness of the performance change
observed is not consistent with the traditional notions that the S-R
strengthening process is a gradual one. Clearly, then, the role of
reward magnitude is other than to determine the strength of the S-R
association. That similar abrupt performance shifts can be achieved
after exposing the subject to the new reward level to follow the
next response trial by placing the subject directly in the goal box
(e.g., Dyal, 1962) is consistent with the view that performance is a
function of the "expected" reinforcement level.

The major problem for S-S theory has been the translation of
associations into actions. Guthrie accused Tolman of leaving the
rat "buried in thought" at the choice-point of the maze (see also
MacCorquodale & Meehl, 1954). More contemporary S-S contiguity
analyses (Bindra, 1974, and Hearst & Jenkins, 1974) try to circumvent
this problem by assuming that just as rewards elicit approach behaviors and noxious events elicit withdrawal, signals for such reinforcer events acquire properties similar to those of reinforcer
events. These conditioned properties contribute attractiveness or
noxiousness to the signal stimuli themselves. The contributed or
enhanced hedonic qualities of the signals and their spatial locations
are in turn thought to directly elicit appropriate behaviors. Problems for this variant of S-S theory arise when the learned behaviors
are different than those elicited by the reinforcer stimuli. One
example of this would be learning as an avoidance response to approach a stimulus which signals impending shocks; this is learned
slowly -- but it is learned (e.g., Gallon, 1974).

Another attempt to account for behavior within an S-S framework (Mowrer, 1960) has been to place heavy emphasis upon response
feedback stimuli (kinesthetic or proprioceptive). In this theory,
Mowrer explicitly disavows S-R associations and argues that response
feedback stimuli come by virtue of contiguity to signal hedonically
powerful events. Actually, Mowrer prefers to class his view as an

S-S-r theory where r refers to the emotion elicited by the hedonic event; it is this emotion elicited by the hedonic event or by the conditioned image of it which serves to modulate the response. Thus, in an instrumental reward task, the occurrence of the appropriate response produces response feedback stimuli which signal reward and produce "hope"; hope then facilitates the behavior that initially pro· duced those feedback stimuli, which in turn generates more feedback stimuli to continue to signal reward. The analog here is to a cybernetic positive feedback loop. The opposite reasoning follows if the outcome event is aversive; here the emotion is "fear" which inhibits responses.

It has become increasingly popular to try to account for behavior using S-S contiguity principles and invoking response feedback as one of the sources of signal stimuli. There are two problems here. The first has to do with response selection in complex choice situations -- especially where there are many possible choices. If the secondary reinforcing effect is elicited by the response feedback, as hypothesized, then the subject must initiate the response before any differential effect can be generated. Thus, to find the "correct" response, the subject must sample each available response until the one which produces positive feedback is found. Since response probabilities per se are not themselves modulated by the contingencies, how can we account for increasingly more rapid correct choices with training? Even more difficult is the case wherein more than one response is rewarded, but the responses are rewarded with different magnitudes. In this case, the theory cannot account for consistent choice of the response associated with larger magnitude without invoking principles for response selection in addition to response feedback.

An especially knotty problem for any theory emphasizing response feedback is posed by the experiments by Taub and his associates (e.g., Taub & Berman, 1963). They have shown that animals can learn and maintain avoidance responses in the absences of both proprioceptive and exteroceptive feedback! While not denying the existence of response feedback or that it may have some role in learning, we should note that animals can learn to make skilled movements of a deafferented limb for reinforcement. This observation strikes at the heart of those theories that give a primary role to response feedback in acquisition and maintenance of responses.

If we reconsider what we have now said, it seems that none of the single process theories is entirely adequate. In general, traditional S-R theories are too restrictive with respect to behavior by not allowing for the generation of novel but appropriate S-R sequences. On the other hand, S-S theories have difficulty in accounting for the emergence of differentiated manipulative responses and choice behavior. In addition, there are demonstrated instances when the observed behavior is not consistent with the prior S-S experiences (e.g., Spence & Lippett, 1946).

One solution to these many problems has been to assume that two different associations are involved in the acquisition and control of behavior. Theories invoking this solution are the mediational theories of learning including the "two-process" theories of learning. While not all forms of mediational theory assert that different learning processes underlie the two parallel associations (e.g., Hull, 1952), it has been popular to do so (e.g., Mowrer, 1947). In either case, theorists have again and again felt compelled to invoke at least two associative links to account for the observed pattern of data; typically they appealed to classical conditioning mediators embedded within the instrumental contingencies.

TWO-PROCESS THEORIES

In the last chapter we gained some familiarity with Mowrer's (1947) theory of avoidance learning which extended Schlosberg's (1937) early distinction between "conditioning" and "success learning." Mowrer's theory had as its primary goal the resolution of the paradox of avoidance learning which is how the omission of a scheduled noxious event apparently reinforces response learning. Mowrer argued that instrumental responses were S-R associations strengthened through drive reduction, and his resolution of the paradox relied upon a hypothesized second, parallel, contiguity association--the classical conditioning of "fear"--as a drive.

Similarly, in Chapter 7, we were exposed to revisions of Hull's (1943) monistic theory (Hull, 1952; Spence, 1956). A revision was needed to explain latent learning phenomena (Chapter 6) and the prompt effects that shifts in reward magnitude had upon appetitive performance (Chapter 7). These experiments seemed to demonstrate that response-independent exposures to stimulus and reinforcer sequences could modulate responses. To account for this, Spence (1956) also hypothesized two parallel associations -- instrumental response learning and the classical conditioned anticipation of reward which he called "incentive motivation." To account for extinction of instrumental responses and the effects of partial reinforcement upon the rate of extinction, especially as a function of reward magnitude, Amsel (1958) extended Spence's theory by hypothesizing a classically conditioned anticipation of non-reward which he called "frustration" (see Chapter 8). Fear, frustration, and incentive motivation were similarly seen as independently learned mediational states with motivational properties that modulated instrumental motor response learning and performance.

These examples do not represent the first theoretical invocations of mediational states by behaviorists. As was noted in Chapter 1 and 5, Hull had introduced the concepts of "pure stimulus act" and "fractional anticipatory goal response" much earlier (1930, 1931). But the more contemporary mediational theories of Mowrer, Spence, Amsel and others differ from Hull's tentative steps in terms of their assumptions about the mediator -- for example, Hull did not attribute motivational properties to mediators.

Spence's and Mowrer's theories also differed substantially, primarily with respect to the assumed necessary and sufficient conditions for the establishment of the two parallel associations. These differences are illustrated in Table II. Spence's (1956) decision on the conditions for and types of association was a response to the problem that reinforcement magnitude only seemed to modulate discriminative Pavlovian conditioning and not discriminative instrumental running. Unlike Mowrer, Spence chose to resolve this by infering classical conditioning to be an S-R reinforcement sensitive process and instrumental response learning to be an S-R contiguity process. Still, Spence was only tentative in his suggestion that the two parallel associations were the products of different learning processes. We shall not focus upon the difference in detail between two process theories because the commonalitites and general orientation are far more important to current thinking and research.

The work of these pioneers has been amplified, expanded, and "adjusted" in a number of recent presentations of two-process theories. For example, while Mowrer and Spence each thought in terms of a single but different classical conditioned mediator, "fear" and "incentive motivation", respectively, Konorski, another early two-process theorist, argued that there were conditioned mediational states based upon both appetitive and aversive outcome events (Konorski, 1963, 1964). In addition, Konorski argued that these mediator states themselves could interact with one another to modulate instrumental performance (Konorski & Szwejkowska, 1956; Konorski, 1967), a conceptualization that has been heavily relied upon in other recent two-process analyses (Rescorla & Solomon, 1967; Overmier & Bull, 1970; Millenson & deVilliers, 1972; Gray, 1975), although the specific mechanisms hypothesized for this interaction have varied across theorists. Let us then examine the

principles common to the basic two-process theories. We specifically note that our list of principles characterizing these efforts
is strongly influenced by Mowrer's (1947) detailed presentation.
As we shall soon see, some of these principles have now been rejected in light of direct assessments.

Contemporary mediational theories argue that the behavior we
observe is the interactive product of two parallel but different
kinds of association, different in terms of what is associated and
how. One is the classical conditioning of a mediational state to
a stimulus through pairings with a reinforcer while the second is
the learning of instrumental habits. Thus, a first principle, obviously, is that there are two processes. Only Mowrer made a con-
certed effort to show empirically that the necessary and sufficient
conditions for the classical conditioning process were different
from those of the instrumental response learning processes. This
effort was stimulated, at least in part, by a technical problem in
the theory: If "fear" were reinforced by drive reduction and if
reduction in level of fear constituted drive reduction -- as it
must to support avoidance response learning -- then fear could
never extinguish. This follows because every instance of elicitation
of fear is eventually followed by its termination. So Mowrer set
himself the task of showing that the learning of fear was not
dependent upon drive reduction. This he did by manipulating the
duration of the shock UCS in classical conditioning. The reasoning was that if classical conditioning (including that of fear)
occurred by virtue of reinforcement at shock termination, then
longer UCSs should result in poorer conditioning because of the
concomitant increase in delay of reinforcement. However, Mowrer
and Solomon (1954) found that aversive classical conditioning was
not poorer when longer UCSs were used. Thus, Mowerer concluded that
it was the onset of the UCS that was critical for classical conditioning, not its termination. This was the basis for Mowrer's
assertation that the classical conditioning of fear is an S-R contiguity process. This distinction, between the S-R classical conditioning of the fear mediator as dependent upon only contiguity
and the instrumental S-R response learning process as dependent
upon drive reduction reinforcement, allowed Mowrer to explain the
importance of shock onset in the conditioning of fear while continuing to recognize the role of shock termination in escape response learning, wherein learning is poorer with delayed shock
termination (Fowler & Trapold, 1962).

It is perhaps worth noting that if both Mowrer and Spence had
chosen to infer that the classical conditioning of fear and incentive, respectively, were an S-S contiguity process, it would have
had few important consequences. Indeed, some interpreters of the

liteature (e.g., Trapold & Overmier, 1972) have chosen to treat the
theories so (primarily because the R in the motivation establishing
process is qualitatively different from the R in the S-R habit
process of their models, e.g., a former being anticipatory goal
responses like salivation while the latter is the instrumental motor
response, and stimulus-outcome pairing was the important operation).

In general, these two process theories gain their power from
the assumption that the two processes are independent though com-
plexly interacting in determining the animal's performance on a
given task. Now in the usual case, these two kinds of learning
would go on in parallel as the sequence of discriminative stimulus-
response-reinforcer unfolds. But the theories are explicit in suggesting that the two associations are potentially separable. Thus,
if special operations allow us to access one of the processes independently of the other, we ought to be able to alter performance by
influencing only one of the associations. The latent learning experiments, then, may be considered as one class of experiments where
such separation is achieved. In the latent learning experiment, the
probability and vigor of the locomotor response is altered by a
pairing operation (food in the goalbox) which is independent of
running off the locomotor sequence.

Closely coupled with the learning process distinction was an
effector response system distinction. Such restriction of type
of learning to a limited effector set has a long and venerable history (e.g., Miller & Konorski, 1928). Mowrer argued that learning
about stimulus-reinforcer sequences, i.e., classical conditioning,
was effected via the autonomic nervous system, while learning about
response-reinforcer sequences, i.e., instrumental training, was
effected via the skeletal-motor nervous system. In essence, Mowrer
equated fear motivation and peripheral autonomic arousal. Spence
did not explicitly make the nervous system distinction that Mowrer
did, but he too argued that the classically conditioned peripheral
responses like salivation (an autonomic response) and licking were
the source of the stimuli which produced incentive motivation. Thus,
for both theories, classically conditioned responses, substantially
autonomic in nature, were thought to be mediating events which modulated the performance of motor response habits.

The mechanism of modulation of habit performance by these conditioned responses was assumed to be primarily motivational, although Spence and Amsel did place some explanatory burden upon the
potential stimulus (cue) properties arising from feedback from the
mediating reactions. Both Mowrer and Spence thought of these "motivations" as rather general drive and activation sources -- sufficiently general to activate any response. Indeed, just this general drive assumption is relied upon by Amsel to explain the partial

reinforcement acquisition effect! By considering the case in which the two different mediating motivations might be simultaneously elicited, Konorski (1967) arrived at a different conclusion. He argued that the two different motivational states were reciprocally antagonistic. Such a conceptualization represented a significant increment in explanatory power for a wholly different class of experiment than theretofore considered -- experiments involving heterogeneous reinforcers. For example, the conditioned suppression experiment that we have considered before may be explained by Konorski's conceptualization. In it, we see the instrumental response as being partly determined by the incentive motivation elicited by the apparatus cues that signal food. Presentations of a CS which has been followed by shock would elicit fear motivation; because fear motivation is hypothesized to be inhibitory for incentive motivation, the incentive motivation for instrumental responding will be reduced and so the response rate would be predicted to decrease -- as in fact it does.

Two-process mediational theory would seem, then, to have substantial explanatory power. But as we have seen, the theories make a large number of assumptions. These assumptions are amenable to test. We shall consider tests of the following assumptions: (1) that a classically conditioned peripheral response mediates the instrumental responding; (2) that instrumental behavior can be modulated by classical conditioning operations; and (3) that there are two distinguishable learning processes.

Mediation

There has been a positive search for the classically conditioned response mediating instrumental behavior. Primarily, the experiments have attempted to show that indices of the classically conditioned mediating association (a CR) and of the instrumental response association (performance) co-varied in ways appropriate to the theory. According to the theory, considerable concommitance between the two response indices is required. Indeed, mediation theory would seem to minimally require the observations that: (a) the classical CR should precede the instrumental response in time, (b) the instrumental response should be more probable after the CR than after the absence of CR, (c) greater magnitude of instrumental response should be correlated with greater magnitudes of CR. We shall see if the data are consistent with these requirements.

Reward Training. Salivation has typically been selected as the index for observation in instrumental appetitive experiments, becasue it would seem to be the best indicator of appetitive classical conditioning. And, several experiments (Shapiro, 1960; Kintsch & Witte, 1962) have found a reasonable degree of parallelism between salivation and responding. Figure 1 shows the instrumental performances and salivation of two dogs trained by Kintsch and Witte (1962) to press a bar on an FI schedule of reward. Note that both the rate of instrumental responding and the rate of salivation increase as the time for the next reinforced response approaches (the so-called FI-scallop pattern).

Using a schedule that required spaced responding, a DRL schedule, Shapiro found that not only did salivation parallel instrumental responding but indeed always preceded each press (Shapiro, 1962). Shapiro also gave some test presentations of a tone previously used in appetitive classical conditioning. He observed that when the tone elicited salivation, instrumental responses also occurred even though the responses were not trained to the tone and were not functional in earning food during the tests. When such tone CSs failed to elicit salivation, no bar-presses occurred!

The preceding results plus those from several other experiments (e.g., Miller & DeBold, 1965) were very promising and clearly congruent with our expectations based upon two-process theory. Unfortunately for the theory, such correlations are not universal. For example, Williams (1965) has reported, using concurrent measurement techniques, that after extended training on an FR 33 schedule, pressing and salivation are not well correlated. Indeed, essentially no salivation was observed to occur during the first few presses!

More importantly, Ellison and Konorski (1965) demonstrated that salivation and instrumental responding could be completely disassociated. Using two serial stimuli, one as an S^D for the appetitive instrumental response on an FR 9 and one as a cue for the impending availability of food which was scheduled for delivery 9 seconds after completion of the FR 9, they found that the instrumental response only occurred during the S^D and salivation only occurred during the second stimulus -- complete separation and independence. Sample data from their experiment are presented in Figure 2. While Konorski used the data as a basis for elaborating other theoretical notions, the data clearly contravene the requirements of any theory which argues that the instrumental pressing response is mediated by conditioned salivation.

  Avoidance Training. Solomon and his students have made a number of comparisons across experiments of latencies of avoidance and latencies of cardiac CRs in dogs under various CS-US intervals. They found a high degree of similarity of latencies of the cardiac CR and the avoidance responses for each given CS-US interval (Church &

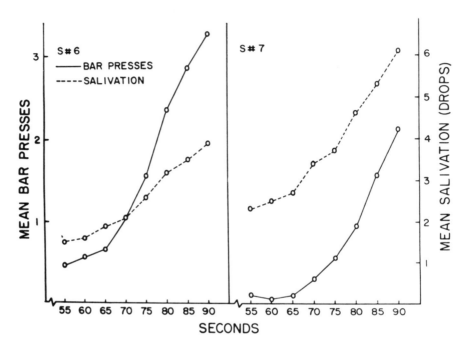

Figure 11.1   Concurrent rates of anticipatory salivation and of
               pressing a bar by two dogs to earn food on a FI 90-
               sec. schedule.   Note that increases in rates of bar-
               pressing are positively correlated with increases in
               the rates of salivation.   (After Kintsch & Witte,
               1962.)

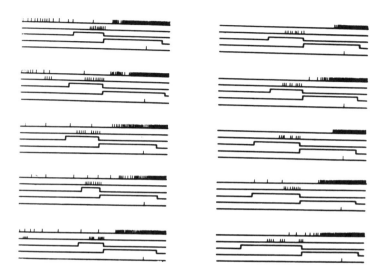

Figure 11.2 A representative series of five consecutive trials
(top to bottom) for each of two dogs (left: S-3;
right: S-4). The key to the five markers on each
trial are, from the top down: drops of salivation;
instrumental lever presses; the S^D for pressing; the
classical CS; food presentation. Note the negative
correlation between lever-pressing and salivation.
(After Ellison & Konorski, 1965.)

Black, 1958). In a more direct test, Black (1959) trained dogs to press a panel in order to avoid strong electric shocks. Skeletal-motor responses and changes in heart rate were recorded on each trial. Heart rate increases during the 5 sec following CS onset were observed to be greater than on those trials on which the dogs performed avoidances than on those trials when they did not make avoidance responses. And, Soltysik and Kowalska (1960) have reported similar relationships.

Of course, these data cannot be weighted very heavily in favor of two-process theory because of an inherent confounding. It is well known that physical activity places demands on the cardio-vascular system; any motor movements would themselves be expected to induce increases in heart rate, a point Obrist (1968) and others have made repeatedly. Hence, it is difficult if not impossible in these experiments to infer cause and effect from the observed cor-relation. A wide variety of experiments have tried to circumvent this confounding, often by use of neuro-muscular blockade with curare and monitoring EMGs in place of gross motor movements (e.g., Black, 1965), but none of these can be said to be successful in bringing into sharp focus the initiating question.

Despite inevitable confounding in active avoidance tasks that should always produce results favoring two-process theory, there have been some studies which nonetheless failed to observe the required concomittance. For example, while studying heart rate changes in dogs during extinction of instrumental avoidance responding, Black (1959) observed that the cardiac response tended to extinguish more quickly than the avoidance response.

More recently, Overmier (1966, 1968) carried out discriminative classical conditioning in which two CSs were paired with different durations or different qualities of USs. Later these CSs were used to evoke avoidance responses in a transfer-of-control paradigm. Al-though the dogs did not show differential degrees of cardiac condi-tioning, they nonetheless showed differential avoidance responding to the CSs, responding more rapidly to the one which had signaled the more severe US. Here again the required concomittance was not ob-tained when the motor test was separated from assessment of the CRs.

An alternative strategy to test the role of autonomic responses in the mediation of avoidance behavior has been to disrupt sympathe-tic autonomic functioning. A variety of techniques have been used including surgical (Wynne & Solomon, 1955), pharmacological (Auld, 1951) and immuno-sympathectomy (Wenzel, 1968). The logic here is that if avoidance responses are mediated by sympathetic activation, then blockade of such activation should block avoidance responding. In summary, researchers have found that sympathetic blockade does modestly disrupt or impair avoidance responding, but it does not prevent it. Data from a recent experiment by Lord, King, and Pfister (1976) are representative. Figure 3 shows that rats subjected to chemical sympathectomy (by 6-hydroxydopamine) were slower to learn a two-way avoidance task than normal rats, but they did learn it! Such results do not provide any strong support for the role of autonomic activity as the mediator of avoidance behavior, although as some have pointed out (Wenzel, 1968), the results are not clearly contradictory.

Our brief review of experiments exploring the role of selected conditioned autonomic responses as mediators of instrumental behavior in both appetitive and aversive domains makes it clear that no firm basis exists for, and indeed evidence against, assuming a tight causal relationship between the occurrence of peripheral autonomic CRs and instrumental behavior. One could always argue that we simply have not searched hard enough for the mediating response and that some such response will eventually be found. But this strategy has not been followed. Instead, the current trend has been to forsake the possibility of finding peripheral indices of the mediational state and assume that the antedating mediating response is some central representation of the reinforcer event. Further, the argument is that the observed CRs are only imprecisely related to the central mediational state.

Central Mediation Theory. Because no one has found a classically conditioned autonomic response that bears the necessary invariant relations to instrumental responding to qualify as a peripheral mediator of instrumental responding, theorists have recently tended, either explicitly or implicitly, toward a central (or at least unidentified) classically conditioned mediation theory. In general, such a theory assumes that classical conditioning of some central state takes place during the course of instrumental learning with the reinforcer for the instrumental learning acting as a UCS. The revised theory continues to assume that this classically conditioned state of affairs interacts with associative connections for specific responses, and hence partially determines what responses will occur and their vigor. Central mediational state theories, however, explicitly disavow that the classically conditioned mediational state is going to be indexed by muscles, or glands, or peripheral nervous system of the subject. Rather, these theories tend to assume that peripheral autonomic changes are at best only poor correlates of the central classically conditioned state, whatever it is (see Black, 1971; Obrist & Webb, 1967; Sheffield, 1965). Indeed, the conditioned central state may not be directly measurable at all.

Some investigators feel that, unless there is something measurable in the response systems of the subject which can be directly and

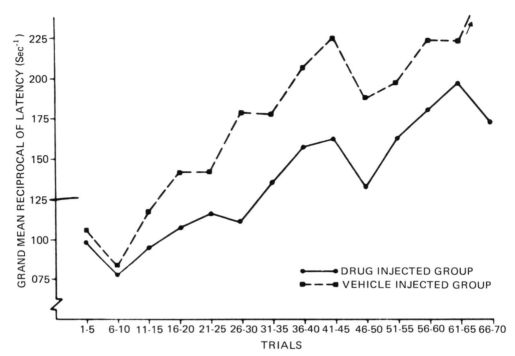

Figure 11.3   Avoidance learning by two groups of rats in a two-way
               shuttlebox.   The groups differed in that one (DRUG
               INJECTED) had been injected with 6-hydroxydopamine
               (2, 4, 5-trihydroxyphenylethylamine) four hours prior
               to training, a treatment which disrupts peripheral
               sympathetic nervous system function.   The horizontal
               line at 125 represents the CS-US interval; speeds
               above this line are avoidances.   (After Lord et al.,
               1976.)

positively identified as the classically conditioned mediating state, such a theory is without value. However, the status of the type of theory which positively disengages itself from the possibility of directly measuring its (central) explanatory entity is logically unassailable (see Logan, 1959; MacCorquodale & Meehl, 1948).

A mediation theory which does not contain a specification of the locus of physiological nature of the mediator merely reflects the unfortunate fact that we have not yet found the proper event to identify with the mediator -- and that we may never do so. But, the theory is not made less useful or less valid by this fact.

The shift in theoretical emphasis away from peripheral mediating responses to central mediational states has had important consequences. One is that the theory cannot be proved or disproved by experiments which attempt to concurrently measure some potential mediating response and the instrumental behavior. As a result, such concurrent measurement experiments have virtually disappeared from the contemporary scene. In their place have appeared experiments which seek to make such direct measurement of the embedded classical conditioning unnecessary. This is achieved by carrying out the classical conditioning to the desired level independently of the instrumental training. The question then is whether the CSs can modulate the instrumental behavior and whether the amount of control is a function of the parameters of the classical conditioning (e.g., Overmier, 1966). It is now well established that a CS stimulus which has never before been associated with a given instrumental response can come to control that response as a direct result of some response-independent classical conditioning operation (e.g., Overmier & Bull, 1969; Trapold & Overmier, 1972; Trapold & Winokur, 1967).

Such experiments imply that during the classical conditioning something gets learned and that whatever got learned during the classical conditioning has the capacity to influence instrumental behaviors. This is evidenced by the fact that control of the instrumental response is promptly transferred to the classical CS in the testing phase. An adequate theory must provide a consistent account for the facts of such transfer (from several experiments) that describes what is learned in the classical phase and the nature of the mechanism whereby this learning can later exert control over some instrumental responses. This process involves making assumptions about the nature of what is learned in the classical conditioning phase, assumptions about the conditions necessary for this kind of learning to occur, and assumptions about how the rate of this kind of learning is affected by various variables. *The ultimate arbiter of the adequacy of any particular set of assumptions is their ability to generate and integrate the facts about transfer as revealed by experimentation.*

Given such a theory, one might conduct research for the express purpose of finding the central nervous system locus of what was learned during the classical conditioning phase. Various changes that occur in the brain could be studied until one was found that corresponded to the learning <u>assumed</u> to occur during the classical conditioning phase of the <u>transfer</u> experiment, at which point the theory would be validated. *However, the only way the investigator can know when he has found an index corresponding to the classical conditioning phase learning is when he finds something that varies the way the theory claims the hypothetical mediator ought to vary.* If such an entity is not found, it may be because no such index exists. Nonetheless, we should not throw out the abstract theory, because it still provides our only predictive and integrative handle on the facts about transfer of training.

Modulation

Let us turn to the second basic question posed by two-process theory: Can classical conditioning operations carried out independently of the instrumental training modulate the instrumental behavior? As already suggested, such experiments seek to break apart the two procedures that are thought to be the basis of the two processes hypothesized, classical conditioning and response habit learning. The prototype experiment has three basic phases: (a) a classical conditioning phase in which the instrumental response is somehow precluded and a CS is paired with a reinforcer; (b) an instrumental training phase in which a specific response is trained by conventional techniques; and (c) a transfer-of-control test phase in which the CS is tested for its capacity to evoke and/ or modulate the separately trained instrumental response.

The typical findings from such experiments have been that the CS exercises strong control over the instrumental response in the test phase, despite the fact that the CS and response have never before been directly associated. The interpretation that two-process theory puts on such experiments is that the power of the CS to control the behavior is directly attributable to the classical conditioning of a central CS-mediator state link which can activate or modulate the mediator-instrumental response link established in the conventional training phase. Table III shows the hypothesized relationships between operations and theoretical mechanisms in the transfer of control experiment (homogeneous reinforcer case).

An illustrative experiment is one by Solomon and Turner (1962). They trained dogs to press a panel whenever a light S^D came on to avoid shocks to the feet. Then they totally immobilized the dogs

TABLE III

TRANSFER OF CONTROL: OPERATIONS AND THEORETICAL CHARACTERIZATION

PHASE	OPERATION	THEORY
(a)	CS —— shock	CS → fear
(b)	S^D —— R —— shock/avoid	S^D → fear → R_{avoid}
(c)	CS test presentations: observe if CS → R	CS → fear → R_{avoid} link 1 link 2

Note: The arrow (→) should be read as "evokes."

with curare and carried out discriminative classical conditioning
in which one tone was paired with shock while the other tone was
presented alone.   Later after full recovery from curare, all stimuli
were presented (without any further shocks) to test for the CS stimuli's control over the panel pressing behavior.   The result was that
the tone paired with shock now evoked fully the avoidance response
while the control tone did not.   A sample of these data are shown in
Figure 4.   In an extension of this study, Overmier and Bull (1969)
were able to show that such transfer of control is not dependent upon
the continued power of the original training stimulus to control the
avoidance response.   The last demonstration provides strong support
for the separability and independence of the mediational links
theorized to control behavior.

Rescorla and Solomon (1967), in an excellent review, indicated
that there are four major classes of transfer-of-control experiments.
These classes are derived from determination of whether the classical
conditioning is based upon appetitive or aversive USs and whether the
instrumental training is based upon appetitive or aversive reinforcers.   Table IV provides a schematic of the four resulting classes
of experiment.   The arrows show the direction of behavioral effect
expected on the basis of two-process theory and incorporating the
hypothesis that appetitive and aversive mediators are reciprocally
inhibitory.   The references in the table are to studies which have
in fact observed the results shown by the arrows.

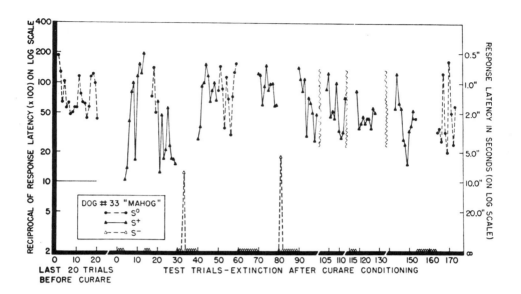

Figure 11.4   Transfer of control of avoidance responding to a
classical CS in a representative dog. The S⁰ is the
visual stimulus to which the avoidance response had
originally been trained; the last day's avoidance
training is shown at the left of the figure. Then
the dog was totally immobilized with curare and given
discriminative classical conditioning with the tone
S+ signalling shock USs while the S- was never followed by shocks. Finally, after recovery from the
drug, test presentations of the classical stimuli
were given to assess their degree of control of the
instrumental avoidance response. Note that from the
very first test presentation, the S+ exerted immediate and complete control by evoking the response
trained to S⁰; the S- by contrast exerted no control.
No shocks were ever administered during testing.
(After Solomon & Turner, 1962.)

TABLE IV

CLASSES AND OUTCOMES OF TRANSFER OF CONTROL EXPERIMENTS

| CLASSICAL CONDITIONING | INSTRUMENTAL TRAINING | |
	Appetitive Baseline	Aversive Baseline
Appetitive CS+	↑ (Estes, 1948)	↓ (Bull, 1970)
Aversive CS+	↓ (Estes & Skinner, 1941)	↑ (Solomon & Turner, 1962)

Note: The direction of the arrow indicates the effect of the classical CS when superimposed upon baseline behavior: the ↑ means increases while ↓ means decreases. The references in parentheses are to exemplary experiments.

To this point then, all seems to be congruent with two-process theory. Unfortunately not all transfer of control experiments have yielded the pattern presented in the figure, although clearly the bulk of them have. Still the exceptions are critical challenges to the viability of the theory. For example, Scobie (1972) presented CSs which had previously been paired with shock to rats while they were responding on a temporal avoidance schedule. The CSs had been paired with different intensities of shock. Two-process theory, emphasizing the motivational properties of the conditioned mediational states, would predict that the greater the US intensity used in classical conditioning the greater degree of facilitation of avoidance that we should see in the transfer test. This is not what he observed. He found that when the US in the classical conditioning was much stronger than the US used to train the avoidance response, then the CS produced suppression! Capaldi, Hovancik, and Friedman (1976) have observed a similar effect in the appetitive transfer class.

Certainly, such results argue against motivational interpretation of the mediational process. But they do not argue against a mediational interpretation! After all, they were derived from experiments which by their very nature require some mediational interpretation.

Recently, Trapold and Overmier (1972) argued that a mediational mechanism might just as well be based upon stimulus properties of the conditioned mediators. In so doing, they were harking back to the original intent of Hull (1930) in introducing a mediational mechanism. Based upon the assumption that the stimulus properties conditioned to the mediational state would be determined by all the parameters of the signaled US, they had predicted that one might get exactly the pattern of results later reported by Scobie and by Capaldi, et al., that is, less facilitation or even suppression by a CS paired with a reinforcer markedly different in magnitude from that used in the instrumental phase.

Is there any empirical basis for Overmier and Trapold's assumption that at least part of the mediation mechanism is based upon the distinctive stimulus properties of conditioned mediators? To provide one test of this idea, Overmier, Bull, and Trapold (1971) taught three groups of dogs a discriminative choice avoidance task. All dogs had to learn to perform R_1 in the presence of S_1 and R_2 in the presence of S_2. The groups differed with respect to the relationships between the signal and the US to be avoided. For two experimental groups, S_1 consistently warned of shock to one leg while S_2 warned of shock to the other; for the control group, S_1 and S_2 could be followed by shock to either leg on a random basis. This choice avoidance task was learned faster when S_1 and S_2 consistently signaled different USs than when there was no consistent relationship between them and the USs. The results are presented in Figure 5. They concluded that a stimulus consistently signaling a specific US comes, through classical conditioning, to evoke fear mediators specific to that US and that these specific fears themselves have stimulus-like properties that can function as cues just as any other stimuli can. Such additional reliable discriminative cue-stimuli would be expected to facilitate appropriate choice behavior in the experimental groups. By way of contrast, for the control group, both of the discriminative stimuli elicit the identical mediator based upon signaling both USs--each on half the trials; thus choice behavior for the control group should be impaired because of the ambiguous cues contributed by the mediator as to the correct response choice. Clearly, then, a view of the mediational mechanism as being based upon acquired stimulus properties has merit and can explain phenomena that elude a view of the mediational mechanism as being based upon motivational interactions.

Where does all of this leave us? The original characterization of two-process theory that we presented has not faired too well. The mediator does not seem to be directly assessable, though mediational

theory continues to be viable. In addition, the hypothesis that
a motivational mechanism underlies the mediation process, if not
wrong, clearly requires supplementation. There is one last general assumption that we have yet to explore. It is that two different learning processes underlie the hypothesized parallel associations.

  <u>Two Learning Processes</u>. Can we in fact distinguish two processes? There have been three bases proposed for distinguishing
between the classical and instrumental learning processes: (1) the
response classes subject to modification by a given procedure, (2)
the functional laws that obtain under a given procedure, and (3)
physiological manipulations that differentially modulate learning
under the two paradigms. We shall consider each briefly.

  Throughout the history of two-process theorizing, researchers
sought to distinguish between the classical and instrumental pro-

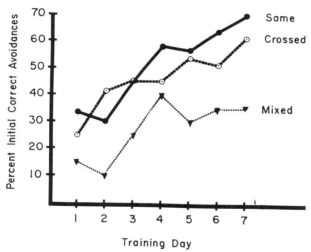

Figure 11.5 Course of learning in the discriminative choice avoid-
            ance task by three groups of dogs. The groups dif-
            fered in the consistency of the relationship between
            the two S-R sequences to be learned and the US that
            would follow that sequence if the animal did not make
            a correct choice. For the SAME and CROSSED groups, a
            specific, distinctive US was associated with each of
            the S-R choices, whereas for the MIXED group there
            was no specific US consistently associated with either
            of the S-R choices. (After Overmier, Bull, & Trapold,
            1971.)

cesses by reference to the responses modified. Miller and Konorski (1928) initially suggested the amount of sensory feedback accompanying the response as a basis; responses with little feedback (e.g., blood pressure) were thought to be only classically conditionable while those with substantial feedback (e.g., running) were thought to be instrumentally trainable. A variety of other suggestions have followed and these are presented in Table V. If we recall the examples considered in preceding chapters, it is clear that none of these are satisfactory. We have only to think of the classical conditioning of key-pecking (Williams & Williams, 1969) or the instrumental training of heart rate and salivation (Miller, 1969) to despair of the adequacy of the proposed dichotomizations.

Two caveats seem in order here, however. The first is that the response distinctions need not result in two non-overlapping sets of responses to have some value to a theorist. Nearly all arguments proceed on the basis of the non-overlapping case, with little attention given to other possibilities. The second is that because both classical and instrumental procedures may lead to the establishment of topographically similar responses does not insure the identity of the learned responses or that their acquisition was through the same process. Let us consider for a moment an experiment by Wahlsten and Cole (1965) which makes this clear. They subjected dogs to either classical conditioning (C) or instrumental avoidance training (I) procedures with the leg flexion

TABLE V

HYPOTHESIZED RESPONSE SYSTEM CHARACTERISTICS

THEORIST(S)	CLASSICAL	INSTRUMENTAL
Konorski	little sensory feedback	rich sensory feedback
Schlosberg	diffuse emotional	precise adaptive
Skinner	elicited	emitted
Mowrer	autonomic	skeletal motor
Kimble	involuntary	voluntary
Turner & Solomon	high reflexivity, short latency	low reflexivity

as the indicator response in both cases. The leg flexion response was learned under both paradigms. But, as Figure 6 makes clear, the latency characteristics differed dramatically as a function of procedure -- even when the dog had previously experienced the other procedure.

The functional laws of classical and instrumental conditioning have been offered as a basis for the process distinction (cf,

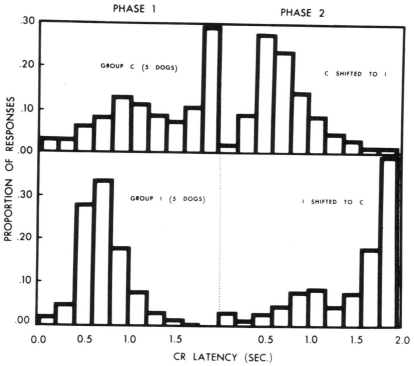

Figure 11.6   Distributions of leg flexion response latencies derived from groups given either classical conditioning (C) or instrumental avoidance training (I). Later each group was shifted to the other procedure. Not only did the avoidance trained dogs make more responses than the classically conditioned dogs (90% versus 50%) but, more important for our purposes, the latencies of those responses were distinctively different and specifically associated with the experimental procedure, as demonstrated by the shift in latency distribution concommitant with the shift in procedures. (After Wahlston & Cole, 1972.)

Rescorla & Solomon, 1967). In general, we have noted so far that
the same basic phenomena appear to result from both procedures.
One possible basis for detecting a difference has been with respect
to the effects of partial reinforcement. Partial reinforcement in
classical conditioning severely retards acquisition (Ost & Lauer,
1965) and has little effect on extinction. In contrast, we have
previously made much of the fact partial reinforcement may enhance
terminal acquisition in instrumental training and markedly increases
resistance to extinction.

Our enthusiasm here must be tempered, however, because careful comparisons of the same response under the two procedures have
not been provided. Of course, this comparison itself is complicated by the difficulty of insuring that you are working with the
same response -- a problem we have already illustrated by reference to the Wahlsten and Cole experiment above.

A final opportunity for salvation would seem to be in physiological manipulations which may allow dissociation of the two
learning processes. Given our frustrations on the other possibilities for distinguishing instrumental and classical processes, perhaps we should not be too hopeful. Yet here, I think, are some reasonably promising data. We will consider just two suggestive
examples.

The first example has to do with the interhemispheric transfer
of information in a natural split-brain preparation -- the goldfish. McCleary (1960) studied the interocular transfer of a classically conditioned heart-rate response and of an instrumental
avoidance response in goldfish whose eyes were covered with translucent contact lenses. He began by establishing the response to
a CS (beam of light) presented to one eye and then tested by presenting the CS to the other eye. The results differed strikingly
as a function of the paradigm used. Under the classical conditioning procedure, McCleary found prompt and essentially complete transfer of the heart rate reaction. In contrast, using the instrumental procedure only one fish showed any tendency to make the
avoidance response when the test stimulus was presented to the untrained eye. Here then, is one procedure that yields clear differences as a function of paradigm. Because such a finding is of
considerable theoretical importance, we need inquire as to its reliability and representativeness. In this regard we can note that
Meikle, Sechzer, and Stellar (1962) have reported a similar pattern
of results in a parallel experiment with split-brain cats. One
might reasonably argue, however, that the two response systems
(autonomic and motor) are quite different, and that this is the
basis for the differential transfer observed. Therefore, our second
example will be of special interest.

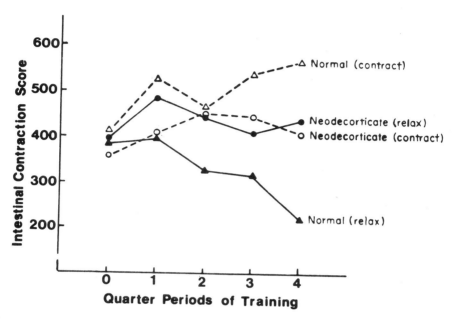

Figure 11.7 Demonstration of the difference between normal and neodecorticate rats in their ability to learn to either contract or to relax their visceral muscles as an instrumental avoidance response. While normal rats learned to do either of these, the neodecorticate rats learned neither. This differential instrumental learning is in sharp contrast to both groups' learning under classical conditioning relative to random control groups. (After DiCara, Braun, & Pappas, 1970.)

DiCara, Braun, and Pappas (1970) undertook to compare classical conditioning and instrumental training of the same response -- heart rate change -- in different groups of rats. Each group was subdivided and one half subjected to neocortical ablation prior to the learning experience. They found that while normal and neo-decorticate rats did not differ in the classical conditioning of heart rate, only the neo-decorticate rats failed to learn to alter their heart rates as an instrumental response to avoid shock. Their results are shown in Figure 7. To guard against the possibility that the results were unique to heart rate, they repeated the experiment using gastrointestinal contractions as the indicator response. Again, the same pattern was found. Both normal and neo-decorticate rats showed excellent classical conditioning but the neo-decorticates -- and only they -- failed to manifest instrumental learning. Because classical conditioning of the effectors can be achieved readily with the CS and US employed, the deficit must be an associative one related to the paradigmatic contingencies rather than a sensory-motor one. The results imply that an intact neocortex is necessary for instrumental learning but not for classical conditioning.

The preceding physiological experiments which dissociate learning under classical and instrumental paradigms are consistent with the hypothesis that classical and instrumental conditioning represent two different learning processes. Clearly, the viability of two-process theory is still an open question. I make this point vigorously because some commentators in the literature have recently argued that there is no viable basis for a two-process theory. The present physiological results belie this conclusion. Nonetheless, the totality of the tests of two-process theory has not been strongly confirmatory. Much more research and theory revision is required.

REFERENCES

Amsel, A. The role of frustrative nonreward in noncontinuous reward situations. PSYCHOLOGICAL BULLETIN, 1958, 55, 102-119.

Auld, F. The effects of tetraethylammonium on a habit motivated by fear. JOURNAL OF COMPARATIVE AND PHYSIOLOGICAL PSYCHOLOGY, 1951, 44, 565-574.

Bindra, D. A motivational view of learning, performance, and behavior modification. PSYCHOLOGICAL REVIEW, 1974, 81, 199-213.

Black, A. H. Heart rate changes during avoidance learning in dogs. CANADIAN JOURNAL OF PSYCHOLOGY, 1959, 13, 229-242.

Black, A. H. Cardiac conditioning in curarized dogs: The relation-
    ship between heart rate and skeletal behavior. In W. F. Pro-
    kasy (Ed.), CLASSICAL CONDITIONING: A SYMPOSIUM. New York:
    Appleton-Century-Crofts, 1965.

Black, A. H. Autonomic aversive conditioning in infrahuman subjects.
    In F. R. Brush (Ed.), AVERSIVE CONDITIONING AND LEARNING. New
    York: Academic Press, 1971.

Brogden, W. J. Sensory pre-conditioning. JOURNAL OF EXPERIMENTAL
    PSYCHOLOGY, 1939, 25, 323-332.

Bull, J. A. III. An interaction between appetitive Pavlovian
    CSs and instrumental avoidance responding. LEARNING AND MOTI-
    VATION, 1970, 1, 18-26.

Capaldi, E. D., Hovancik, J. R., & Friedman, F. Effects of ex-
    pectancies of different reward magnitudes in transfer from
    noncontingent pairings to instrumental performance. LEARNING
    AND MOTIVATION, 1976, 7, 197-210.

Church, R. M., & Black, A. H. Latency of the conditioned heart rate
    as a function of the CS-US interval. JOURNAL OF COMPARATIVE
    AND PHYSIOLOGICAL PSYCHOLOGY, 1958, 51, 478-482.

Crespi, L. P. Quantitative variation of incentive and performance
    in the white rat. AMERICAN JOURNAL OF PSYCHOLOGY, 1942, 55,
    467-517.

DiCara, L. V., Braun, J. J., & Pappas, B. A. Classical conditioning
    and instrumental learning of cardiac and gastrointestinal re-
    sponses following removal of neocortex in the rat. JOURNAL OF
    COMPARATIVE AND PHYSIOLOGICAL PSYCHOLOGY, 1970, 73, 208-216.

Dyal, J. A. Latent extinction as a function of the number and dura-
    tion of pre-extinction procedures. JOURNAL OF EXPERIMENTAL
    PSYCHOLOGY, 1962, 63, 98-104.

Ellison, G. D., & Konorski, J. An investigation of the relations
    between salivary and motor responses during instrumental per-
    formance. ACTA BIOLOGIAE EXPERIMENTALIS (Warsaw), 1965, 25,
    297-315.

Estes, W. K. Discriminative conditioning. II. Effects of a
    Pavlovian conditioned stimulus upon a subsequently established
    operant response. JOURNAL OF EXPERIMENTAL PSYCHOLOGY, 1948,
    38, 173-177.

Estes, W. K. The statistical approach to learning theory. In S.
    Koch (Ed.), PSYCHOLOGY: A STUDY OF A SCIENCE, Vol. 2. New York:
    McGraw-Hill, 1959.

Fowler, H., & Trapold, M. A. Escape performance as a function of delay of reinforcement. JOURNAL OF EXPERIMENTAL PSYCHOLOGY, 1962, 63, 464-467.

Gallon, R. L. Spatial location of a visual signal and shuttle box avoidance acquisition by goldfish (*Carassius auratus*). JOURNAL OF COMPARATIVE AND PHYSIOLOGICAL PSYCHOLOGY, 1974, 86, 316-321.

Gray, J. A. ELEMENTS OF A TWO-PROCESS THEORY OF LEARNING. London: Academic Press, 1975.

Guthrie, E. R. THE PSYCHOLOGY OF LEARNING. New York: Harper, 1935.

Hearst, E., & Jenkins, H. M. SIGN-TRACKING: THE STIMULUS-REINFORCER RELATION AND DIRECTED ACTION. Austin, Texas: Psychonomic Society, 1974.

Hull, C. L. Knowledge and purpose as a habit mechanism. PSYCHO-LOGICAL REVIEW, 1930, 37, 511-525.

Hull, C. L. Goal attraction and directing ideas conceived as habit phenomena. PSYCHOLOGICAL REVIEW, 1931, 38, 487-506.

Hull, C. L. A BEHAVIOR SYSTEM. New Haven: Yale University Press, 1952.

Kimble, G. A. HILGARD & MARQUIS' CONDITIONING AND LEARNING (2nd Edition). New York: Appleton-Century-Crofts, 1961.

Kintsch, W., & Witte, R. S. Concurrent conditioning of bar press and salivation responses. JOURNAL OF COMPARATIVE AND PHYSIO-LOGICAL PSYCHOLOGY, 1962, 55, 963-968.

Konorski, J. On the mechanism of instrumental conditioning. XVII International Congress of Psychology (Washington), 1963. [cf. ACTA PSYCHOLOGIA, 1964, 23, 45-59.]

Konorski, J. Some problems concerning the mechanism of instrumental conditioning. ACTA BIOLOGIAE EXPERIMENTALIS (Warsaw), 1964, 24, 59-72.

Konorski, J. INTEGRATIVE ACTIVITY OF THE BRAIN. Chicago: University of Chicago Press, 1967.

Konorski, J., & Szwejkowska, G. Reciprocal transformations of hetero-geneous conditioned reflexes. ACTA BIOLOGIAE EXPERIMEN-TALIS (Warsaw), 1956, 17, 141-165.

Logan, F. A. The Hull-Spence approach. In S. Koch (Ed.), PSYCHOL-OGY: A STUDY OF A SCIENCE, Vol. 2. New York: McGraw-Hill, 1959.

Lord, B. J., King, M. G., & Pfister, H. P. Chemical sympathectomy
    and two-way escape and avoidance learning in the rat. JOURNAL
    OF COMPARATIVE AND PHYSIOLOGICAL PSYCHOLOGY, 1976, 90, 303-316.

MacCorquodale, K., & Meehl, P. E. On the distinction between hypo-
    thetical constructs and intervening variables. PSYCHOLOGICAL
    REVIEW, 1948, 55, 95-107.

MacCorquodale, K., & Meehl, P. E. Edward C. Tolman. In W. K.
    Estes *et al.* (Eds.), MODERN LEARNING THEORY. New York:
    Appleton-Century-Crofts, 1954.

McCleary, R. A. Type of response as a factor in interocular trans-
    fer in the fish. JOURNAL OF COMPARATIVE AND PHYSIOLOGICAL
    PSYCHOLOGY, 1960, 55, 311-321.

Meikle, T. H., Sechzer, J. A., & Stellar, E. Interhemispheric
    transfer of tactile conditioned responses in corpus callosum-
    sectioned cats. JOURNAL OF NEUROPHYSIOLOGY, 1962, 25, 530-543.

Millenson, J. R., & deVilliers, P. A. Motivational properties of
    conditioned suppression. LEARNING AND MOTIVATION, 1972, 3,
    125-137.

Miller, N. E. Learning of visceral and glandular responses.
    SCIENCE, 1969, 163, 434-445.

Miller, N. E., & DeBold, R. C. Classically conditioned tongue-
    licking and operant bar-pressing recorded simultaneously in
    the rat. JOURNAL OF COMPARATIVE AND PHYSIOLOGICAL PSYCHOLOGY,
    1965, 59, 109-115.

Miller, S., & Konorski, J. Sur une forme particulière des reflexes
    conditionnels. LES COMPTES RENDUS SÉANCES DE LA SOCIÉTÉ
    BIOLOGIE, 1928, 99, 1155-1157.

Mowrer, O. H. On the dual nature of learning. HARVARD EDUCATIONAL
    REVIEW, 1947, 17, 102-148.

Mowrer, O. H. LEARNING THEORY AND BEHAVIOR. New York: John Wiley,
    1960.

Mowrer, O. H., & Solomon, L. N. Contiguity vs. drive-reduction in
    conditioned fear: The proximity and abruptness of drive reduc-
    tion. AMERICAN JOURNAL OF PSYCHOLOGY, 1954, 67, 15-25.

Obrist, P. A. Heart rate and somatic motor coupling during classical
    aversive conditioning in humans. JOURNAL OF EXPERIMENTAL PSY-
    CHOLOGY, 1968, 77, 180-193.

Obrist, P., & Webb, R. A. Heart rate during conditioning in dogs:
    Relationships to somatic-motor activity. PSYCHOPHYSIOLOGY,
    1967, 4, 7-34.

Ost, J. W. P., & Lauer, D. W. Some investigations of classical sali-
    vary conditioning in the dog. In W. F. Prokasy (Ed.), CLASSI-
    CAL CONDTIONING: A SYMPOSIUM. New York: Appleton-Century-
    Crofts, 1965.

Overmier, J. B. Instrumental and cardiac indices of Pavlovian fear
    conditioning as a function of UCS duration. JOURNAL OF COM-
    PARATIVE AND PHYSIOLOGICAL PSYCHOLOGY, 1966, 62, 15-20.

Overmier, J. B. Differential Pavlovian fear conditioning as a func-
    tion of the qualitative nature of the UCS: constant versus
    pulsating shock. CONDITIONAL REFLEX, 1968, 3, 175-180.

Overmier, J. B., & Bull, J. A. On the independence of the stimulus
    control of avoidance. JOURNAL OF EXPERIMENTAL PSYCHOLOGY,
    1969, 79, 464-467.

Overmier, J. B., & Bull, J. A. III. Influences of appetitive Pav-
    lovian conditioning upon avoidance behavior. In J. H. Rey-
    nierse (Ed.), CURRENT ISSUES IN ANIMAL LEARNING. Lincoln,
    Neb.: University of Nebraska, 1970.

Overmier, J. B., Bull, J. A., III, & Trapold, M. A. Discriminative
    cue properties of different fears and their role in response
    selection in dogs. JOURNAL OF COMPARATIVE AND PHYSIOLOGICAL
    PSYCHOLOGY, 1971, 76, 478-482.

Rescorla, R. A., & Solomon, R. L. Two-process learning theory:
    Relationships between Pavlovian conditioning and instrumental
    learning. PSYCHOLOGICAL REVIEW, 1967, 74, 151-182.

Schlosberg, H. The relationship between success and the laws of
    conditioning. PSYCHOLOGICAL REVIEW, 1937, 44, 379-394.

Scobie, S. R. Interaction of an aversive Pavlovian conditional
    stimulus with aversively and appetitively motivated operants
    in rats. JOURNAL OF COMPARATIVE AND PHYSIOLOGICAL PSYCHOLOGY,
    1972, 79, 171-188.

Seward, J. P. An experimental analysis of latent learning. JOURNAL
    OF EXPERIMENTAL PSYCHOLOGY, 1949, 39, 177-186.

Shapiro, M. M. Respondent salivary conditioning during operant
    lever pressing in dogs. SCIENCE, 1960, 132, 619-620.

Shapiro, M. M. Temporal relationship between salivation and lever
    pressing with differential reinforcement of low rates.
    JOURNAL OF COMPARATIVE AND PHYSIOLOGICAL PSYCHOLOGY, 1962, 55,
    567-571.

Sheffield, F. D. Relation between classical conditioning and instru-
    mental learning. In W. F. Prokasy (Ed.), CLASSICAL CONDITION-
    ING. New York: Appleton-Century-Crofts, 1965.

Solomon, R. L., & Turner, L. H. Discriminative classical condi-
     tioning under curare can later control discriminative avoid-
     ance responses in the normal state. PSYCHOLOGICAL REVIEW,
     1962, 69, 202-219.

Solomon, R. L., & Wynne, L. C. Traumatic avoidance learning: The
     principles of anxiety conservation and partial irreversibility.
     PSYCHOLOGICAL REVIEW, 1954, 61, 353-385.

Soltysik, S., & Kowalska, M. Studies on the avoidance conditioning:
     1. Relations between cardiac (type I) and motor (type II)
     effects in the avoidance reflex. ACTA BIOLOGIAE EXPERIMEN-
     TALIS (Warsaw), 1960, 20, 157-170.

Spence, K. W. BEHAVIOR THEORY AND CONDITIONING. New Haven: Yale
     University Press, 1956.

Spence, K. W., & Lippett, R. An experimental test of the sign-
     Gestalt theory of trial-and-error in learning. JOURNAL OF
     EXPERIMENTAL PSYCHOLOGY, 1946, 36, 491-502.

Taub, E., & Berman, A. J. Avoidance conditioning in the absence of
     relevant proprioceptive and exteroceptive feedback. JOURNAL
     OF COMPARATIVE AND PHYSIOLOGICAL PSYCHOLOGY, 1963, 56, 1012-
     1016.

Taub, E., & Berman, A. J. Movement and learning in the absence of
     sensory feedback. In S. J. Freedman (Ed.), THE NEUROPSYCHOLOGY
     OF SPATIALLY ORIENTED BEHAVIOR. Homewood, Ill.: Dorsey Press,
     1968.

Tolman, E. C., & Honzik, C. H. Introduction and removal of reward
     and maze performance in rats. UNIVERSITY OF CALIFORNIA PUB-
     LICATIONS IN PSYCHOLOGY, 1930, 4, 257-275.

Tolman, E. C., Ritchie, B. F., & Kalish, D. Studies in spatial
     learning. II. Place versus response learning. JOURNAL OF
     EXPERIMENTAL PSYCHOLOGY, 1946, 36, 221-229.

Trapold, M. A., & Overmier, J. B. The second learning process in
     instrumental learning. In A. H. Black & W. F. Prokasy (Eds.),
     CLASSICAL CONDITIONING II: CURRENT THEORY AND RESEARCH. New
     York: Appleton-Century-Crofts, 1972.

Trapold, M. A., & Winokur, S. W. Transfer from classical condition-
     ing and extinction to acquisition, extinction, and stimulus
     generalization of a positively reinforced instrumental re-
     sponse. JOURNAL OF EXPERIMENTAL PSYCHOLOGY, 1967, 73, 517-525.

Wahlston, D., & Cole, M. Classical avoidance training of leg flexion
     in the dog. In A. H. Black & W. F. Prokasy (Eds.), CLASSICAL
     CONDITIONING II: CURRENT RESEARCH AND THEORY. New York:
     Appleton-Century-Crofts, 1965.

Watson, J. B.   The place of the conditioned reflex in psychology.
     PSYCHOLOGICAL REVIEW, 1916, 23, 89-116.

Wenzel, B. M.   Behavioral studies of immunosympathetectomized mice.
     JOURNAL OF COMPARATIVE AND PHYSIOLOGICAL PSYCHOLOGY, 1968, 66,
     354-362.

Williams, D. R.   Classical conditioning and incentive motivation.
     In W. F. Prokasy (Ed.), CLASSICAL CONDITIONING.   New York:
     Appleton-Century-Crofts, 1965.

Williams, D. R., & Williams, H.   Auto-maintenance in the pigeon:
     Sustained pecking despite contingent non-reinforcement.
     JOURNAL OF THE EXPERIMENTAL ANLYSIS OF BEHAVIOR, 1969, 12,
     511-520.

Wynne, L. C., & Solomon, R. L.   Traumatic avoidance learning:
     Acquisition and extinction in dogs deprived of normal peripheral
     autonomic function.   GENETIC PSYCHOLOGY MONOGRAPHS, 1955, 52,
     241-284.

12. GENERALIZATION

M. E. Bitterman

University of Hawaii
Honolulu, Hawaii, U.S.A.

As both Thorndike and Pavlov discovered early, the effects of training are not entirely specific to the training conditions, but generalize to other conditions. For example, a pigeon reinforced for pecking a red key may as a result of the training show a strong tendency to peck a yellow key which it never has encountered before even though it is perfectly capable of distinguishing the two colors. Generalization, which is not an all-or-none phenomenon, does vary with the distinguishability of the training and testing situations, but distinguishability is not the only determinant of generalization, nor the most interesting one. Our principal concern in these experiments is to try to understand why it is that the amount of generalization may vary widely even when distinguishability remains constant.

In the prototypical generalization experiment, animals are trained with a given stimulus, S, in a specified context, X, and then tested in the same context with some property of S varied over a wide range (SX, $S'X$, $S''X$, and so forth). The main outcome of such an experiment is a generalization gradient, which is a plot of response to each of the stimuli used in the tests as a function of its physical similarity to the training stimulus. The gradient shown in Figure 1 comes from an experiment by Guttman and Kalish (1956), who trained pigeons on a VI 1-min schedule for pecking a key on which light of a given wavelength was projected and then gave them unreinforced tests with a variety of wavelengths, including the training wavelength, presented for fixed periods of time in quasi-random order. In these tests, the animals responded

maximally to the training wavelength and progressively less as
the difference between the training and testing wavelengths increased.

Generalization experiments differ markedly in design. Guttman
and Kalish used the <u>multiple-stimulus</u> method, in which an entire
gradient is obtained from a single group of animals, each animal
being tested with all the stimuli. In the <u>single-stimulus</u> method,
each point on a gradient is provided by a separate group, each
animal being tested with only a single stimulus. Since the effects
of experience during testing also may be expected to generalize,
the single-stimulus method provides a purer measure of the generalized effects of training, although it is less efficient than the
multiple-stimulus method in that it requires many more subjects.
The purest gradients (and the most costly in time and subjects)
are provided by experiments in which each animal has only a single
test trial with only a single stimulus (see, for example, Newman
& Grice, 1965). Where repeated trials are given, a decision must
be made as to whether or not they should be reinforced. In most
cases, test trials are unreinforced, but in some experiments they
are reinforced. On occasion, too, differential reinforcement is
used, with stimuli and treatments balanced over groups. In this
<u>discrimination</u> method, two groups of animals experience SX and then
are trained to discriminate between SX and $S'X$, with SX positive
for one group and $S'X$ positive for the second, the difference in
the rate at which the two groups master the discrimination affording
a sensitive inverse measure of generalization (Lashley & Wade, 1946).
Other variations in the design of generalization experiments will
be considered later.

SOURCES OF GENERALIZATION

Generalization may be due in large measure to <u>common stimuli</u>.
Since the animals are not trained with S alone, but with S in the
context of X, some of the response to $S'X$ after training with SX
may be elicited by X. Jenkins and Harrison (1960) found that pigeons
reinforced for pecking a white key in the presence of a 1000-Hz tone,
would respond about as readily in the presence of a 1500-Hz tone, or
even in the absence of any tone at all. Clearly the animals were responding to stimuli other than tone--such as the lighted key itself--
which had been present along with the tone in training. Generalization from SX to $S'X$ may be due also to the equivalence (or partial
equivalence) of S' and S. Jenkins and Harrison reinforced a second
group of pigeons for pecking the white key in the presence of a
1000-Hz tone but not in its absence. In subsequent tests, the
animals responded more to the 1500-Hz tone than to the white key
alone, although not as much as to the 1000-Hz tone. These results

require us to assume that the two tones were at least partially equivalent. Generalization due to equivalence or substitutability of stimuli, which is called <u>stimulus generalization</u>, has itself two distinguishable sources. Suppose that a pigeon is reinforced for pecking at a red circle. Tested with an orange circle, it may respond at least in part because of the common form, or, tested with a red square, it may respond to the common color. Here we have generalization based on <u>common attributes</u>. The second source of stimulus generalization is <u>attributive similarity</u>. The pigeon reinforced for pecking at a red circle may respond to the orange circle in part because of the similarity in color. Generalization based on attributive similarity is demonstrated by the orderly gradients of performance that often are obtained when some dimension of the training stimulus (such as angular orientation, size, or wavelength) is varied systematically in the test trials. An example of such a gradient has already been given in Figure 1.

The explanation of generalization based on attributive similarity which was proposed by Pavlov (1927) and which is widely accepted in principle even today (see, for example, Blough, 1975; Mackintosh, 1974; Rescorla, 1976) is that similar stimuli excite common neural elements. The reinforcement of Color 1, which excites, say, neural elements, a, b, and c, will result in the conditioning of each of those elements. Color 2, which excites elements b, c, and d, now also will evoke the conditioned response, although not to the same extent as does Color 1 because only two-thirds of its elements are conditioned. Color 3, which excites elements c, d, and e, will evoke the response to a still lesser extent. In general, if S now represents <u>only the attribute of the training stimulus in which we are interested</u>, such as its color, and if V_S represents the associative strength which that color has acquired in training, then the generalized associative strength of a different color, S', is given by an equation of the form

$$V_{S'} = G \cdot V_S \qquad\qquad (1)$$

where G is the proportion of neural elements excited by S that are excited also by S'. What is assumed here is that generalization is inversely related to resolving power. When $G = 1$, which is to say that the animal is entirely unable to distinguish between S and S', $V_{S'} = V_S$; with V_S constant, $V_{S'}$ approaches zero as G approaches zero.

If X, the common sensory context of S and S', now is defined broadly to include <u>common attributes as well as common stimuli</u>, and if we assume that

$$V_{SX} = V_S + V_X \qquad (2)$$

then, by equations 1 and 2,

$$V_{S'X} = G \cdot V_S + V_X \qquad (3)$$

where V_{SX} and $V_{S'X}$ represent the associative strengths of SX and $S'X$. The next step in the analysis of generalization is to make provision for the translation of V into performance. The common assumption (e.g. Newman & Grice, 1965) that performance is a simple multiplicative function of associative strength and the strength of a set of activating factors is expressed in the equation

$$P = A \cdot V \qquad (4)$$

where P is the measure of performance and A the level of activation. If $P_{SX} = A \cdot V_{SX}$ and $P_{S'X} = A \cdot V_{S'X}$, and if the distance from

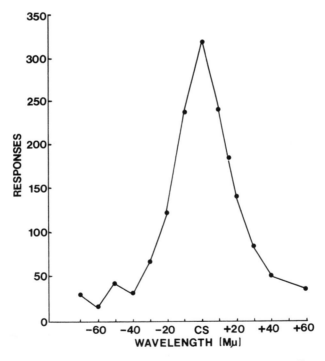

Figure 12.1   Wavelength generalization in pigeons after training at the point indicated as CS.   After Guttman and Kalish (1956).

S to S' is taken to be 1, the slope of the generalization gradient between the two points may be written

$$P_{SX} - P_{S'X} = A \cdot V_S + A \cdot V_X - A \cdot G \cdot V_S - A \cdot V_X$$
$$= A \cdot V_S \ (1 - G) \qquad (5)$$

which means that the gradient steepens with increase in activation, in the associative strength of S, or in resolving power.

Evidence that the slope of a generalization gradient is determined in part by the distinguishability of the stimuli used in the generalization tests comes from experiments with several different animals--by Ganz (1962) with monkeys, by Haber and Kalish (1963) with pigeons, and by Moore (1972) with rabbits. In some work with goldfish by Yarczower and Bitterman (1965), the wavelength-distinguishability function was found to be sharply asymmetrical in the region of 580 nm, decreases in wavelength being much less readily discriminated than increases in wavelength when the animals were trained with differential reinforcement. This asymmetry was mirrored in the generalization gradients shown in Figure 2; after reinforced training with 580 nm alone, there was much more rapid response to shorter wavelengths than to longer ones. What happens when A is varied is shown in Figure 3. Coate (1964) trained rats at several different levels of thirst (all animals at all levels in balanced order) and then divided them into groups, each of which was tested at a different level. The frequency of responding increased with drive in testing and so also did the steepness of the generalization gradient.

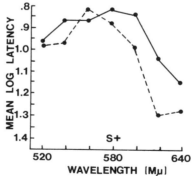

Figure 12.2 Asymmetrical wavelength generalization gradients for goldfish trained at 580 nm. After Yarczower and Bitterman (1965). The two curves were obtained in different experiments. Note that the latency scale is inverted.

Generalization experiments are not particularly interesting for what they tell us about G, which is much better estimated from the results of training with differential reinforcement. Nor is there any reason, if we are interested only in A, for looking at its effects on $P_{S'X}$ as well as on P_{SX}. The primary interest of generalization experiments lies in what they tell us about V_S-- about the development of what may be called attributive control. The problem of stimulus control has, of course, been encountered before in the analysis of compound conditioning, where the associative strength of a stimulus is estimated by removing it from the compound altogether. After training with TLC, where T and L are two stimuli, such as a tone and a light, presented in the context of a set of background stimuli, C, it follows from equations 2 and 4 that $P_{TLC} - P_{LC} = A \cdot V_T$. While we cannot simply remove an attribute of a training stimulus--every visual form must, for example, have some size--we can vary the attribute and estimate its associative strength from the resulting change in behavior. Where we do remove a stimulus from a compound, we measure the combined strengths of all the attributes of that stimulus.

As is evident from equation 5, the slope of a raw or absolute generalization gradient does not provide a very good measure of V_S. Suppose that $P_{SX} - P_{S'X}$ is greater after training under one condition than another and in both cases > 0. Even if G has been kept constant by using the same stimuli under both conditions, the difference may be due either to V_S or to A or to both. Drive is relatively easy to manipulate, of course, but it should be emphasized that drive is only one component of A. There are other components of A about which we know less, and some of them may vary with the conditions whose effects on attributive control are being studied. A common solution to this problem is to transform the absolute measures of response into relative measures and then to compare the slopes of the relative gradients. Response to each stimulus used in the generalization tests is expressed either as a ratio of response to the training stimulus (the training-stimulus transformation) or as a ratio of response to all stimuli combined (which may be called the areal transformation, since the denominator of the ratio is essentially a measure of area under the absolute generalization curve). What the slope of the relative gradient promises to provide is an activation-free index, not of V_S, but of V_S/V_{SX}-- an index of relative control by S--which is sufficient for many purposes and sometimes precisely what is required.

Justification for the training-stimulus transformation may be found in equations 1, 2, and 4. Since

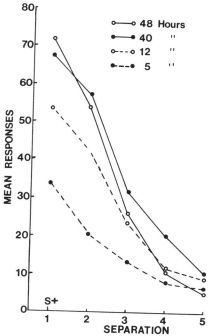

Figure 12.3 Generalization in rats tested after four different periods of water-deprivation. After Coate (1964).

$$\frac{P_{S'X}}{P_{SX}} = \frac{A(G \cdot V_S + V_X)}{A \cdot V_{SX}}$$

$$= 1 - \frac{V_S}{V_{SX}} \; (1 - G) \qquad (6)$$

and, since the transformed value of $P_{SX} = 1$, the slope of the relative gradient between S and S' is given by the equation

$$1 - \frac{P_{S'X}}{P_{SX}} = \frac{V_S}{V_{SX}} \; (1 - G) \qquad (7)$$

in which A does not appear. Equation 7 tells us that, with G constant, differences in the slopes of relative gradients measured

under different conditions reflect differences in the associative
strength of S relative to that of SX--the steeper the slope, the
greater the relative control by S. Analysis of the areal transformation, which is not quite so simple, leads to an equation of
comparable form. Of the two, the areal transformation is perhaps
to be preferred on statistical grounds, since the measure of response to all stimuli combined is more reliable than the measure
of response to the training stimulus alone.

Evidence that the slope of a relative gradient is independent
of A comes from experiments on drive, the best of which from the
methodological viewpoint is that of Coate (1964). The relative
gradients shown in Figure 4, which are areal transformations of
the absolute gradients of Figure 3, are essentially identical in
slope. Much the same results were obtained by Newman and Grice
(1965) in an experiment on hunger in rats, although Thomas and
King (1959), working with hunger in pigeons, found some tendency
for relative slope first to steepen and then to flatten again as
drive increased. Where relative slope varies with drive, we may
conclude either that drive does affect attributive control or that

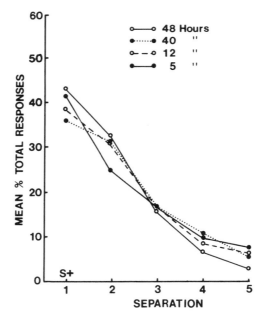

Figure 12.4   Relative generalization gradients (areal transformations) based on the same data as the absolute gradients shown in Fig. 3. After Coate (1964).

equation 7 is incorrect--that A does not in fact cancel out in the
relative transformation. It should be emphasized that the assumptions embodied in equations 1-4, upon which equation 7 is based,
are undoubtedly much too simple. A stimulus is treated as the sum
of its attributes, and a compound as the sum of its components,
although there is a good deal of evidence (to be considered later)
that parts interact to generate new properties of the whole.
Another assumption which may prove too simple is that associative
strength is reflected directly in performance, without the inter-
vention of attentional mechanisms. Evidence on this point also
will be considered later. An oversimplified assumption which it
may be well to deal with at once is that performance measures are
uniformly <u>capable</u> of reflecting all degrees of associative strength.
In fact, however, distorting effects of performance <u>floors</u> and
<u>ceilings</u> are always to be guarded against.

A good illustration of a performance ceiling--the insensitivity of a performance measure to relatively high values of V--
has been provided by Hoffman and Fleshler (1961). Using the conditioned suppression technique, they trained pigeons with a 1000-
Hz tone as the CS and then tested them in a series of extinction
sessions with tones of varying frequency. The bird whose data are
shown in Figure 5 generalized widely at first, which might suggest
no control at all by frequency, but the steepening of the gradient
with continued testing contradicts this assumption. A reasonable
interpretation is that the training frequency was more disturbing
than other frequencies even at the outset, but that the generalization gradient was flat because the animal could do no more than suppress completely. Only as the emotional response extinguished in
the course of repeated testing, and performance fell below the
ceiling, could differences in associative strength be manifested.
Hoffman and Fleshler also provide a good illustration of a performance floor--the insensitivity of a performance measure to
relatively small values of V. One pigeon showed no suppression
to a 450-Hz tone when very hungry, from which it might be inferred
that the tone had no generalized strength at all, except for the
fact that the same tone produced clear suppression when the animal
was less hungry. (The role of hunger in such an experiment, of
course, is to activate the behavior that the CS tends to suppress.)
It is interesting to see what happens when a threshold constant,
T is introduced into the performance equation. If

$$P = A \cdot V - T \qquad (8)$$

then

$$1 - \frac{P_{S'X}}{P_{SX}} = \frac{A \cdot V_S}{A \cdot V_{SX} - T} \qquad (1 - G) \qquad (9)$$

which means that the slope of the relative gradient is no longer
independent of A. The slope will change also, of course, with any
condition that affects T.

SOME DETERMINANTS OF ATTRIBUTIVE CONTROL

Even on the common assumption that the slopes of relative
gradients do reflect differences in attributive control, little
can be said with any confidence about the determinants of attributive control because available results are fragmentary and often
contradictory. To the question of how relative gradients change
with amount of training, for example, all possible answers have
been obtained. Thompson (1958) and Hoffeld (1962), working with
cats, found steepening with increased training; Margolius (1955),
working with rats, found flattening (more with certain performance
measures than with others); while Hearst and Koresko (1968), working with pigeons, found no consistent effect. The various experiments on this question differ in almost all possible ways, and
there has been no systematic study of how those differences may be
related to the differences in outcome. It seems reasonable to
anticipate steepening where there are salient common stimuli that
are differentially nonreinforced in training, such as the white
key in the experiment of Jenkins and Harrison (1960) on auditory

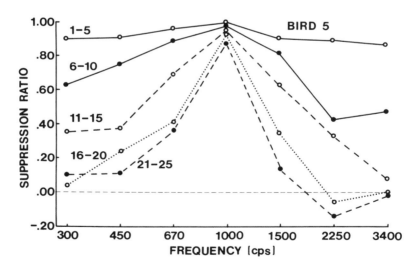

Figure 12.5 Generalization of conditioned suppression in succes-
             sive unreinforced tests after training at 1000 Hz.
             Each curve shows the pooled results of five tests.
             From Hoffman and Fleshler (1961).

frequency. A variable related to amount of training, at least as it affects frequency of reinforcement, is schedule of reinforcement, which has been studied in a variety of experiments. Hearst, Koresko, and Poppen (1964) reported higher levels of responding and steeper relative gradients for angular orientation in pigeons trained with denser VI schedules (30-sec or 1-min) as compared with sparser schedules (3- or 4-min). The same authors also reported that a DRL 6-sec schedule (requiring the animal to wait at least 6 sec between pecks) produced a much flatter relative gradient than did a VI 1-min schedule although there were no fewer reinforcements per session. This result follows from the fact that stimuli correlated with the passage of time (such as fading proprioceptive feedback from response) play an important role in DRL training.

Another question in which there has been considerable interest but to which there is no clear answer is about what happens to attributive control in the course of repeated unreinforced generalization tests (as the absolute level of responding declines). Some experiments with pigeons suggest that relative gradients steepen (Jenkins & Harrison, 1960) while others suggest that they remain essentially the same (Kalish & Guttman, 1957). Where steepening occurs, it is essential to consider the possibility that a ceiling effect is responsible. (If we had only a set of relative gradients based on the data of Figure 5, we might be tempted to conclude that relative control by frequency increased in the course of testing, but that conclusion would be incorrect.) Since what is varied in these experiments is amount of unreinforced experience with the entire set of test stimuli (rather than with SX alone), relative control might be expected to diminish. Absolute levels of responding vary also from subject to subject, and the absolute gradients of high and low responders differ in much the same way as the absolute gradients for all animals combined early and late in extinction. The fact that differences between the gradients of high and low responders disappear with relative transformation (Kalish & Guttman, 1957) suggests that the animals differ primarily in level of activation. Plotted in Figure 6 are the almost identical relative gradients for two groups of pigeons, one of which responded about five times as frequently as the other in the tests (Thomas & King, 1959).

A finding with interesting implications for the problem of forgetting is that relative gradients may flatten appreciably in the interval between training and testing. Perkins and Weyant (1958) trained four groups of rats in a runway of a given brightness and then tested them either 60 sec or one week later in a runway either of the same brightness or of a brightness different from that used in training. The crossover of the absolute gradients reproduced in Figure 7 makes it apparent that the relative gradients would show a corresponding difference in slope and that the difference in

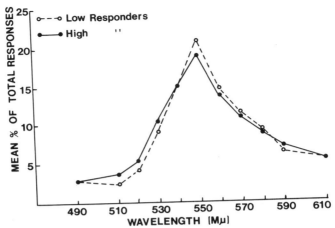

Figure 12.6   Relative wavelength generalization gradients (areal
              transformations) of pigeons with markedly different
              absolute rates of responding. From Thomas and King
              (1959).

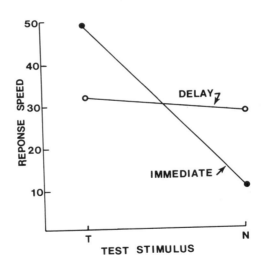

Figure 12.7   Running speeds of four groups of rats tested immedi-
              ately or after a delay of 1 week either in a runway
              of the same brightness as that used in training (T)
              or in a runway of different brightness (N). From
              Perkins and Weyant (1958).

slope is not an artifact either of a floor or of a ceiling effect. Flattening of relative gradients in the interval between training and testing has been found also for brightness generalization in rats trained to avoid shock and tested after 3 min or 24 hr (McAllister & McAllister, 1963) and for wavelength generalization in pigeons given VI training and then tested after 1 min or 24 hr (Thomas & Lopez, 1962). Here, at least, there is some consistency in the experimental results, although their meaning is far from clear. If the absolute associative strengths of all components of a complex stimulus decay in time at the same rate, there should be no change in their relative strengths.

The principal concern of contemporary research on generalization is with the effects of discriminative training on attributive control. We turn first to experiments like that of Jenkins and Harrison (1960), which have come to be called _interdimensional_ experiments for reasons that are not now worth considering. Their purpose is to study the effects of training to discriminate the presence from the absence of a stimulus on control by some attribute of that stimulus. Jenkins and Harrison trained an interdimensional group of pigeons with TKC positive and KC negative--where T was the 1000 Hz tone, K, the lighted key, and C, all other stimuli--and then compared its gradient for an attribute of T (frequency) with that of a group which had no KC-negative experience. Both groups also had some unreinforced experience with C, which means that both had some discriminative training, although C was not very salient and exposures to it were relatively brief. Only in the interdimensional group, in any case, was the presence or absence of T uniquely correlated with reinforcement. An interdimensional experiment of the same design by Farthing (1972) is more informative because generalization was studied in the absence as well as in the presence of K, and because absolute as well as relative gradients were presented. Here T was a vertical white line projected on K, a colored background. Shown in Figure 8 are the results of generalization tests with lines varying in angular orientation presented either on the colored ground (line-color compounds) or on the darkened key (line-only). A finding directly analogous to that of Jenkins and Harrison is that the relative orientation gradient was steeper for the interdimensional group when tests were made with line-color compounds.

Results of this sort do not seem very difficult to understand. Unreinforced experience with KC should certainly be expected to reduce its associative strength and so to increase relative control by T and by attributes of T. That control by K was substantial in the interdimensional group can be seen by comparing its absolute line-color and line-only gradients, which show higher overall performance in the presence of color (K), but the perform-

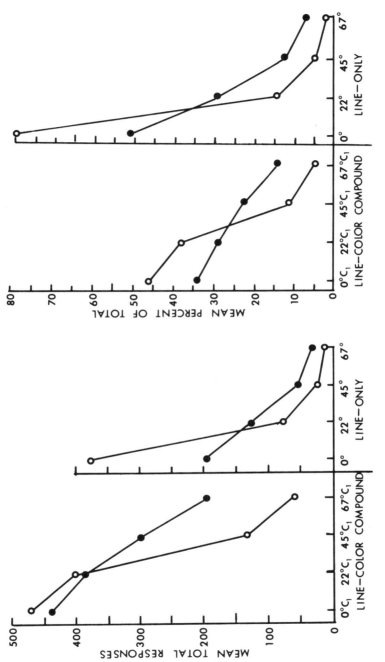

Figure 12.8 The effect of interdimensional discriminative training on generalization. Absolute gradients are shown at left and relative gradients (areal transformations) at right. The open circles are for a group of pigeons trained with a vertical line on a colored background as $S+$ and the colored background alone as $S-$. The closed circles are for a group with no $S-$ experience. After Farthing (1972).

ance of the group without KC-negative training seemed to be depressed even more by the removal of color. The Rescorla-Wagner principle of shared associative strength suggests, furthermore, that any strength lost by KC as a result of nonreinforcement should accrue to T, and on that account also relative control by T should be greater. A finding which suggests that V_T was higher in the interdimensional group is that its relative gradient was steeper even in the line-only condition. Another explanation is based on the old principle of selective attention, according to which attention to differentially reinforced stimuli is enhanced at the expense of attention to nondifferentially reinforced stimuli (Sutherland & Mackintosh, 1971). Attention is assumed to influence both the growth of V and its translation into performance. The interdimensional gradient should be steeper in this view because V_T should be greater and, even apart from its greater value, should have greater weight in the performance equation. The concept of selective attention, which does not seem to add very much at this point, will be taken up later in some detail.

The effects of discriminative training on attributive control have been studied also in so-called extradimensional experiments. What we measure in interdimensional experiments is control by an attribute of a stimulus which is relevant to the discriminative problem in the sense that reinforcement is related to its presence or absence. In extradimensional experiments, by contrast, we measure control by an attribute of an irrelevant stimulus, which may be present during the discriminative training but nondifferentially reinforced, or which may even not be present at all. Where the stimulus is present during the discriminative training, we describe the procedure as concurrent. Where the stimulus is encountered separately, either before or after the discriminative training, we describe the procedure as sequential. The results of these experiments have led to a new interpretation of the effects of discriminative training on generalization, which is that discriminative training heightens general attentiveness; discovering that "stimuli are significant," the animal becomes more attentive to and learns more about "all" of them (Tomie, Davitt, & Thomas, 1975).

In the first stage of a sequential experiment by Thomas, Freeman, Svinicki, Burr, and Lyons (1970), two groups of pigeons were trained with red and green keys. For the extradimensional group, green was reinforced on a VI 1-min schedule and red was unreinforced. For a second group--a pseudodiscrimination or PD group--half the presentations of each color were reinforced on the same schedule and half were unreinforced. It should be noted that, where the reference group in such an experiment is given PD training, the extradimensional group is commonly referred

to as a <u>true</u> <u>discrimination</u> or TD group. Then both groups were
given reinforced experience with a white vertical line on a green
ground, after which there were unreinforced generalization tests
with lines of different angular orientations on a dark ground.
The gradients, which are plotted in Figure 9, show better control
by angle in the TD than in the PD group. Results of this kind,
obtained also in goldfish (Tennant & Bitterman, 1975), conform
nicely to the view that discriminative training alters general
attentiveness which in turn determines how much is learned about
the attributes of subsequently encountered stimuli. Some work by
Honig (1974) points to effects of extradimensional training, not
on <u>learning</u> about the attributes of other stimuli, but on <u>perform-</u>
<u>ance</u> in generalization tests. The critical finding is that TD
and PD training can influence control by attributes of a stimulus
encountered even before that training has been given. The notion
of general attentiveness is, of course, broad enough to encompass
these effects on performance, which actually are not very large.
When a familiar stimulus is changed, according to Thomas, Freeman,
Svinicki, Burr, and Lyons (1970), the "more discriminating" TD
animals should be expected to respond less than the PD animals,
which have been taught to "ignore" differences.

An alternative account of the results of sequential experiments
is suggested by the principle of shared associative strength.

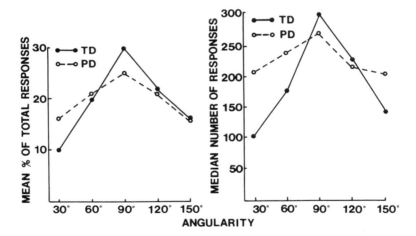

Figure 12.9   Generalization in pigeons given extradimensional *TD*
or *PD* training in a sequential experiment. Absolute
gradients are shown at right and relative gradients
(areal transformations) at left. From Thomas, Freeman, Svinicki, Burr, and Lyons (1970).

Where K_1 and K_2 are two stimuli presented in some common sensory context, C--that is, K_1C and K_2C--it may be assumed that VC will be less after TD training than after PD training. If the animals subsequently are given reinforced experience with some new stimulus, T, in the context of C, then--again by the principle of shared strength--V_T will be larger, both absolutely and relative to V_C, in the TD animals, and relative control by attributes of T also will be greater. Even where TC has been encountered before the extradimensional training, reduction in V_C is expected to increase V_T/V_{TC}. The same sort of explanation is suggested by the principle of selective attention, according to which attention to C should be less after TD than after PD training. Unfortunately, however, neither principle makes it possible to account for the results of analogous concurrent experiments--that is, experiments in which the TD or PD training is given with TK_1C and TK_2C. The concurrent experiments typically show steeper relative gradients for attributes of T despite the fact that T (along with C) is nondifferentially reinforced in both groups.

In a concurrent analogue of their sequential experiment, Thomas, Freeman, Svinicki, Burr, and Lyons (1970) trained TD and PD groups of pigeons with a white vertical line (T) on a green ground (K_1) and the same white vertical line on a red ground (K_2)--that is, with TK_1C and TK_2C. The gradients for both groups, which were tested with lines of different angular orientation on a dark ground, are shown in Figure 10. The finding of principal interest is that the relative gradient of the TD group was steeper than that of the PD group. Another interesting finding, which seems to contradict the principle of shared associative strength as well as the principle of selective attention, is that the absolute level of responding to the nondifferentially reinforced vertical line (TC) was greater if anything in the TD than in the PD group. Bresnahan (1970) reinforced a TD group of pigeons for responding to a tilted white line (K_1) on a colored ground (T) but not for responding to a line of opposite tilt (K_2) on the same ground. A PD group was equally often reinforced and nonreinforced for responding to the two stimuli, while a third group had reinforced experience with TK_1C but no unreinforced experience with TK_2C. In subsequent generalization tests with the positive line (K_1) on grounds of varying color, the relative gradient for the third group was reliably flatter than that of the TD group but reliably steeper than that of the PD group. These results are consistent with the interpretation that general attentiveness is increased by TD training and decreased by PD training. They suggest also that the results obtained in interdimensional experiments may be due at least in part to general attentiveness.

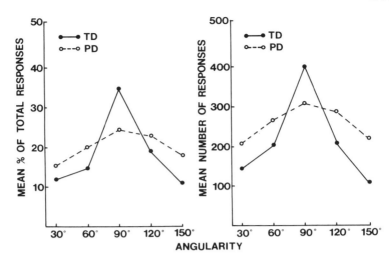

Figure 12.10   Generalization in pigeons given extradimensional
                training in a concurrent experiment. Absolute
                gradients are shown at right and relative gradients
                (areal transformations) at left. From Thomas,
                Freeman, Svinicki, Burr, and Lyons (1970).

An interesting implication of the general attentiveness view is
that relative gradients do not measure relative control--if they did,
discriminative training could not steepen relative gradients for
"all" features of the training situation--although just what they do
measure we are not told. If, however, we are prepared to give up
the assumption that differences in the slopes of relative gradients
necessarily indicate differences in relative control, there may be no
need for the notion of general attentiveness. The concurrent extradimensional results are difficult to understand in associative terms
only because they suggest that discriminative training increases
relative control by attributes of nondifferentially reinforced common
stimuli, but the effects of the training on relative control are inferred from the slopes of relative gradients. The admission that we
may have no unambiguous measure either of absolute or of relative
control leaves us, of course, in the unenviable position of being
unable to test the associative or any other interpretation of
generalization, but we know at least that our efforts should be concentrated on the development of such measures.

THE GENERALIZATION OF INHIBITION AND ITS INTERACTION
WITH EXCITATION

The only examples of generalization dealt with thus far are of
the excitatory or response-producing effects of reinforcement, but it

is easy to show that the response-suppressing effects of non-reinforcement also generalize. Farthing and Hearst (1968) reinforced
pigeons on a VI 1-min schedule for responding to a blank white key
but not to one on which a dark vertical line was projected, and then
gave extinction tests with lines of various angular orientations projected on the key as well as with the blank key. Absolute generalization gradients for three groups of animals, one tested after only
a single session of training to discriminate between the lined and
the blank key, a second after two sessions, and a third after eight
sessions, are plotted in Figure 11. Here change in stimulus produced,
not a decrement, but an increment in performance. It seems reasonable to conclude from these results that the vertical line acquired
some inhibitory or response-suppressing property which generalized to
lines of other orientations.

Inhibitory effects can be measured only in relation to excitatory
effects, because performance cannot be less than zero. In the experiment of Farthing and Hearst, reinforcement of response to the blank
key made it possible to see the inhibitory effect of the line. Another way to study the generalization of the effects of nonreinforcement is to reinforce all of the stimuli to be used in the generalization tests before giving unreinforced experience with one of them.
For example, Kling (1952) rewarded rats with food for responding to
circles of two sizes (the difference in size being varied from group

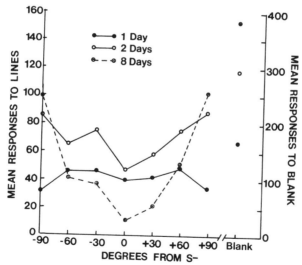

Figure 12.11   Generalization of inhibition in pigeons after dif-
ferent amounts of training with a blank white key as
S+ and a dark vertical line projected on the key as
S-. After Farthing and Hearst (1968).

to group), then extinguished response to one of the circles and
tested with the other. The more similar the two circles, the less
rapid was the response of the animal on the test trial. Sometimes
it is practical to use reinforced rather than unreinforced test
trials. An example of this technique, which measures what may be
called <u>resistance to reinforcement,</u> is provided by Karpicke and
Hearst (1975). After unreinforced experience with light of one wavelength, their pigeons responded much less rapidly to that wavelength
than to longer or shorter ones in a series of reinforced test trials.
It should not be thought, however, that excitatory gradients are in
any sense purer than inhibitory gradients; all but the simplest excitatory gradients are shaped by the interacting effects of reinforcement and nonreinforcement.

The traditional Pavlovian assumption is that a stimulus--and
presumably any attribute of a stimulus--may have both excitatory
(E) and inhibitory (I) properties at the same time, although what is
evidenced in behavior is only the net effect--the algebraic resul-
tant--of the two opposed processes. The net effect may be represented by V, which now is defined as E-I, and which becomes
negative when inhibition dominates. It may be well to look again
at equation 7 in the light of this change in the definition of V.
If V can be negative, so also can the control ratio (V_S/V_{SX}),
which seems reasonable enough when V_S is negative and V_X positive,
but not so reasonable when V_S is positive and V_X negative. For
practical purposes, however, we are restricted to working under
conditions in which P_{SX} and $P_{S'X}$ are positive and sufficiently
large to produce some measurable performance, since lesser P
values have no empirical consequences; where $P_{SX} > 0$, the control
ratio has the same sign as V_S. Another difficulty to be considered
is that, with V = E - I, equation 4 becomes $V = A \cdot E - A \cdot I$, which
implies that inhibitory as well as excitatory effects are intensified by factors such as drive. While there is no compelling
reason to reject this idea, the usual assumption (Hull, 1943)
is that $V = A \cdot E - I$. If E alone is multiplied by A, equation 7
must be replaced by the equation

$$1 - \frac{P_{S'X}}{P_{SX}} = \frac{A \cdot E_S - I_S}{A \cdot E_{SX} - I_{SX}} (1 - G) \qquad (10)$$

which means that the slope of the relative gradient is no longer
independent of A except when $I_S = I_X = 0$. Clearly, a good deal of
uncertainty remains as to the meaning of relative gradients.

In all experiments considered thus far, generalization was measured along a given dimension after experience with only a single point on that dimension. Suppose, however, that generalization is measured after differential reinforcement at two points on a dimension, say, with *SX* reinforced with *S'X* unreinforced. Such experiments are described as <u>intradimensional</u>, and the gradients they yield--which are called <u>postdiscrimination</u> gradients--reflect the generalization both of excitation and inhibition. The gradients of two groups of pigeons trained by Hanson (1959) are shown in Figure 12. For one group, response to 550 nm was reinforced on a VI schedule; for a second group, response to 550 nm (S+) was reinforced and response to 560 nm (S-) was unreinforced. The gradients for the two groups differ first of all in the location of their maxima, which in the postdiscrimination case is not at S+, but shifted away from it in the direction opposite S-. This phenomenon is called the <u>peak shift</u>. Another striking difference is in the overall level of responding, which is higher after intradimensional training.

The peak shift is reminiscent of a phenomenon which was discovered in early choice experiments. Consider a set of stimuli, *1, 2, 3, 4, ...,* varying in some dimension, say, size, and suppose that an animal has been trained to choose *2* rather than *3* when they are presented simultaneously--that is, choice of *2* is reinforced and choice of *3* is unreinforced. In subsequent tests with *1* and *2*, the animal may choose *1* despite the fact that *1* has never been reinforced, and in tests with *3* and *4* it may choose *3* despite the extensive unreinforced experience with *3*. Such results may suggest that the animal has learned a size-relation-- the apparent response to relationships is called <u>transposition</u>-- but Spence (1937) proposed that the results might be accounted for in terms of generalization, without the need for any assumption about the learning of relations. Figure 13 shows some hypothetical generalization gradients, an excitatory gradient peaking at S+ and an inhibitory gradient peaking at S-. The two gradients sum to give a gradient labeled (E-I) that illustrates the peak shift, although its overall level is not higher, but lower, than that of the E gradient. There is good reason, however, to believe that the peak shift and the high level of responding which characterize postdiscrimination gradients are separately determined, peak by the interaction of E and I gradients, and level by some other factor. The rate of response to S+ in the discriminatively trained animals is higher even during acquisition--a well-known if little understood phenomenon called <u>behavioral contrast</u> (Reynolds, 1961). As may be seen in Figure 13, the addition of a constant to the E-I curve is sufficient to produce a reasonable representation of the main features of the postdiscrimination gradient.

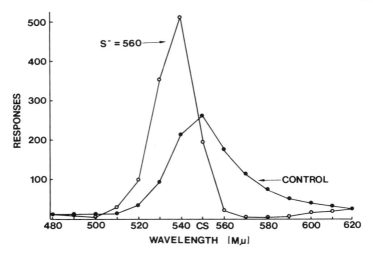

Figure 12.12 Wavelength generalization gradients for two groups
of pigeons reinforced for response to 555 nm, an
intradimensional group with unreinforced experience
at 560 nm and a control group without that experience. After Hanson (1959).

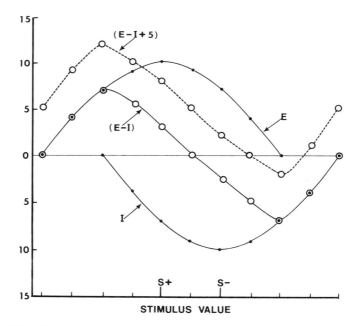

Figure 12.13 Hypothetical gradients of excitation (E) and inhibition (I) generated by discriminative training.

The analysis in terms of overlapping excitatory and inhibitory gradients suggests also that there should be a negative peak shift (a shift in the point of minimal responding away from S- in the direction opposite to S+). As already noted, choice experiments do show transposition in both directions, although experiments like that of Hanson do not show negative shifts because differences in inhibition are evident only where there is some substantial excitation to be opposed. Where a background of excitatory strength is provided, clear indication of a negative peak shift does, in fact, appear. Guttman (1965) reinforced pigeons on a VI 40-sec schedule for responding to each of 19 wavelengths ranging from 510-600 nm in steps of 5 nm, and then trained them to discriminate between 550 and 560 nm, with the shorter wavelength positive (VI 40-sec) for some animals and negative for others. After good discrimination was achieved, unreinforced generalization tests were made at 15 of the 19 wavelengths used in the first stage of the experiment, with the results shown in Figure 14. At the end of the first stage, the animals were responding at substantially the same rate to all of the wavelengths. In the curve for the first minute of generalization testing after discriminative training, both positive and negative peak shifts are evident.

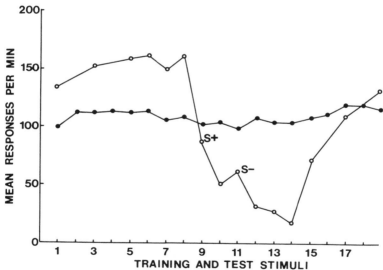

Figure 12.14   Response of pigeons in the course of reinforced training with 19 different wavelengths (flat curve) and in unreinforced generalization tests after discriminative training with *S*+ and *S*-.   From Guttman (1965).

Peak shifts can be derived also from a model like that of
Rescorla and Wagner (1972), which is applicable, not only to stimuli and to the attributes of stimuli, but to the hypothetical
neural consequences of stimulation (Rescorla, 1976). If Stimulus
1 activates neural elements a, b, c, Stimulus 2 activates b, c, d,
and so forth, and if the V of each stimulus is arrived at by summation, as in $V_1 = V_a + V_b + V_c$, we may ask what happens when an
animal is trained with, say, Stimulus 2 positive and Stimulus 3 ($c,d,$
e) negative. By the principle of shared associative strength, element b should gain a good deal of strength, the common elements
c and d considerably less, while V_e should become negative. That
would not be enough to produce a peak shift, since the change from
Stimulus 2 to Stimulus 1, for example, would substitute $V_a = 0$ for
$V_d > 0$. Blough (1975) has shown, however, that peak shifts can be
derived on the reasonable assumption that the gradient of arousal
over an array of neural elements which is produced by a given
stimulus is Guassian (with a maximum at the mid-element of the
array) rather than rectilinear. The degree to which any given
element is activated, which may be thought of as the salience
of the element, will vary, then, with the stimulus which activates
it. In the simple example already given, element c is activated
more strongly by Stimulus 2 than by Stimulus 3. Since salience
is assumed to determine both the rate at which associative strength
changes (as in the Rescorla-Wagner model) and the contribution of
the element to performance, the loss of element d in the shift
from Stimulus 2 to Stimulus 1 is more than compensated for by the
fact that b, the element which has acquired the greatest V,
is the one most strongly activated by Stimulus 1.

Blough has made some other reasonable changes in the Rescorla-
Wagner model, such as in the rule for applying up or down β--in
Blough's model, the choice depends on whether ($\lambda - V_S$) is positive
or negative--but what seems to be critical in the derivation of
peak shifts is the differential excitation of common elements. An
intriguing feature of Blough's derivation is that $V > 0$, which
is to say that no inhibitory process is assumed. It is possible,
therefore, to account for peak shifts found after discriminative
training in which both stimuli are reinforced, but one more frequently than the other. In an experiment by Guttman (1959), pigeons
were trained with two wavelengths, 550 nm reinforced on a VI 1-min
schedule and 570 nm on a VI 5-min schedule. Although the rate of
response to 570 nm was substantial (about 20/min), the postdiscrimination gradient was very much like one obtained by Hanson (1959)
after training with the same wavelengths but with 570 nm unreinforced. To account for these results in terms of the summation of
E and I gradients, it would be necessary to assume (with Guttman)
that a stimulus associated with the weaker of two reinforcement
schedules acquires inhibitory properties. If so, response in the

presence of the VI 5-min wavelength would have to be due to the excitatory properties of common attributes and common stimuli.

From the postdiscrimination gradient, it is but one small further step to the _sustained_ or _steady_ state gradient, which is measured under conditions in which <u>all</u> the stimuli used are-- and continue to be--differentially reinforced. The steady-state procedure is not a two-stage but a one-stage procedure, in which generalization is studied while discriminative training is in progress. Suppose that an animal is trained with a series of lines tilted from the vertical, with all lesser tilts reinforced and all greater tilts unreinforced (Malone & Staddon, 1973). Differential response to two stimuli which are treated identically in such an experiment--as, for example, two lines of lesser tilt-- must be due to differences in the generalization of excitation and inhibition developed at other points on the continuum. Shown in Figure 15 are some results reported by Blough (1975) for pigeons trained with a relatively low probability of reinforcement for response to a series of wavelengths shorter than 597 nm and with a higher probability of reinforcement for response to a series of longer wavelengths. While the reason for the shape of the gradients may not be intuitively apparent, both the trough and the peak can be derived readily form Blough's model or from the simple summation of excitatory and inhibitory gradients such as those of Figure 13. Because of their stability, sustained gradients offer a particularly valuable opportunity for the quantitative analysis of generalization.

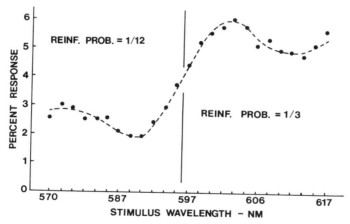

Figure 12.15 Sustained generalization gradient of a group of pigeons reinforced with a relatively low probability for response to wavelengths shorter than 597 nm and with a higher probability for response to longer wavelengths. From Blough (1975).

REFERENCES

Blough, D. S.   Steady-state data and a quantitative model of operant
    generalization and discrimination.   JOURNAL OF EXPERIMENTAL
    PSYCHOLOGY: ANIMAL BEHAVIOR PROCESSES, 1975, 104, 3-21.

Bresnahan, E. L.   Effects of extradimensional pseudodiscrimination
    and discrimination training upon stimulus control.   JOURNAL
    OF EXPERIMENTAL PSYCHOLOGY, 1970, 85, 155-156.

Coate, W. B.   Effect of deprivation on postdiscrimination stimulus
    generalization in the rat.   JOURNAL OF COMPARATIVE AND PHYSIO-
    LOGICAL PSYCHOLOGY, 1964, 57, 134-138.

Farthing, G. W.   Overshadowing in the discrimination of successive
    compound stimuli.   PSYCHONOMIC SCIENCE, 1972, 28, 29-32.

Farthing G. W., & Hearst, E.   Generalization gradients of inhibition
    after different amounts of training.   JOURNAL OF THE EXPERI-
    MENTAL ANALYSIS OF BEHAVIOR, 1968, 11, 743-752.

Ganz, L.   Hue generalization and hue discriminability in *Macaca*
    mulatta.   JOURNAL OF EXPERIMENTAL PSYCHOLOGY, 1962, 64, 142-150.

Guttman, N.   Generalization gradients around stimuli associated with
    different reinforcement schedules.   JOURNAL OF EXPERIMENTAL
    PSYCHOLOGY, 1959, 58, 335-340.

Guttman, N.   Effects of discrimination formation on generalization
    measured from a positive-rate baseline.   In D. I. Mostofsky
    (Ed.), STIMULUS GENERALIZATION.   Stanford: Stanford University
    Press, 1965.

Guttman, N., & Kalish, H. I.   Discriminability and stimulus general-
    zation.   JOURNAL OF EXPERIMENTAL PSYCHOLOGY, 1956, 51, 79-88.

Haber, A., & Kalish, H. I.   Prediction of discrimination from
    generalization after variations in schedule of reinforcement.
    SCIENCE, 1963, 142, 412-413.

Hanson, H. M.   Effects of discrimination training on stimulus
    generalization.   JOURNAL OF EXPERIMENTAL PSYCHOLOGY, 1959, 58,
    321-334.

Hearst, E., & Koresko, M. B.   Stimulus generalization and amount of
    prior training on variable-interval reinforcement.   JOURNAL OF
    COMPARATIVE AND PHYSIOLOGICAL PSYCHOLOGY, 1968, 66, 133-138.

Hearst, E., Koresko, M. B., & Poppen, R.   Stimulus generalization
    and the response-reinforcement contingency.   JOURNAL OF THE
    EXPERIMENTAL ANALYSIS OF BEHAVIOR, 1964, 7, 369-380.

Hoffeld, D. R.   Primary stimulus generalization and secondary extinc-
    tion as a function of strength of conditioning.   JOURNAL OF
    COMPARATIVE AND PHYSIOLOGICAL PSYCHOLOGY, 1962, 55, 27-31.

Hoffman, H. S., & Fleshler, M.   Stimulus factors in aversive con-
     trols: The generalization of conditioned suppression.   JOURNAL
     OF THE EXPERIMENTAL ANALYSIS OF BEHAVIOR, 1961, 4, 371-378.

Honig, W. K.   Effects of extradimensional discrimination training
     upon previously acquired stimulus control.   LEARNING AND
     MOTIVATION, 1974, 5, 1-15.

Hull, C. L.   PRINCIPLES OF BEHAVIOR.   New York: Appleton-Century-
     Crofts, 1943.

Jenkins, H. M., & Harrison, R. H.   Effect of discrimination training
     on auditory generalization.   JOURNAL OF EXPERIMENTAL PSYCHOL-
     OGY, 1960, 59, 246-253.

Kalish, H. I., & Guttman, N.   Stimulus generalization after equal
     training on two stimuli.   JOURNAL OF EXPERIMENTAL PSYCHOLOGY,
     1957, 53, 139-144.

Karpicke, J., & Hearst, E.   Inhibitory control and errorless dis-
     crimination learning.   JOURNAL OF THE EXPERIMENTAL ANALYSIS
     OF BEHAVIOR, 1975, 23, 159-166.

Kling, J. W.   Generalization of extinction of an instrumental response
     to stimuli varying in the size dimension.   JOURNAL OF EXPERI-
     MENTAL PSYCHOLOGY, 1952, 44, 339-346.

Lashley, K. S., & Wade, M.   The Pavlovian theory of generalization.
     PSYCHOLOGICAL REVIEW, 1946, 53, 72-87.

Mackintosh, N. J.   THE PSYCHOLOGY OF ANIMAL LEARNING.   New York:
     Academic Press, 1974.

Malone, J. C., Jr., & Staddon, J. E. R.   Contrast effects in main-
     tained generalization gradients.   JOURNAL OF THE EXPERIMENTAL
     ANALYSIS OF BEHAVIOR, 1973, 19, 167-179.

Margolius, G.   Stimulus generalization of an instrumental response
     as a function of the number of reinforced trials.   JOURNAL OF
     EXPERIMENTAL PSYCHOLOGY, 1955, 49, 105-111.

McAllister, W. R., & McAllister, D. E.   Increase over time in the
     stimulus generalization of acquired fear.   JOURNAL OF EXPERI-
     MENTAL PSYCHOLOGY, 1963, 65, 576-582.

Moore, J. W.   Stimulus control: Studies of auditory generalization
     in rabbits.   In A. H. Black & W. F. Prokasy (Eds.), CLASSICAL
     CONDITIONING II: CURRENT THEORY AND RESEARCH.   New York:
     Appleton-Century-Crofts, 1972.

Newman, J. R., & Grice, G. R.   Stimulus generalization as a function
     of drive level, and the relation between two measures of
     response strength.   JOURNAL OF EXPERIMENTAL PSYCHOLOGY, 1965,
     69, 357-362.

Pavlov, I. P.   CONDITIONED REFLEXES.   Oxford: Oxford University
     Press, 1927.

Perkins, C. C., Jr., & Weyant, R. G. The interval between training and test trials as a determiner of the slope of generalization gradients. JOURNAL OF COMPARATIVE AND PHYSIOLOGICAL PSYCHOLOGY, 1958, 51, 596-600.

Rescorla, R. A. Stimulus generalization: Some predictions from a model of Pavlovian conditioning. JOURNAL OF EXPERIMENTAL PSYCHOLOGY: ANIMAL BEHAVIOR PROCESSES, 1976, 102, 88-96.

Rescorla, R. A., & Wagner, A. R. A theory of Pavlovian conditioning. In A. H. Black & W. F. Prokasy (Eds.), CLASSICAL CONDITIONING II: CURRENT THEORY AND RESEARCH. New York: Appleton-Century-Crofts, 1972.

Reynolds, G. S. Behavioral contrast. JOURNAL OF THE EXPERIMENTAL ANALYSIS OF BEHAVIOR, 1961, 4, 57-71.

Spence, K. W. The differential response in animals to stimuli varying within a single dimension. PSYCHOLOGICAL REVIEW, 1937, 44, 430-444.

Sutherland, N. S., & Mackintosh, N. J. MECHANISMS OF ANIMAL DIS-CRIMINATION LEARNING. New York: Academic Press, 1971.

Tennant, W. A., & Bitterman, M. E. Extradimensional transfer in the discriminative learning of goldfish. ANIMAL LEARNING & BEHAVIOR, 1975, 3, 201-204.

Thomas, D. R., Freeman, F., Svinicki, J. G., Burr, D. E. S., & Lyons, J. Effects of extradimensional training on stimulus generalization. JOURNAL OF EXPERIMENTAL PSYCHOLOGY MONOGRAPHS, 1970, 83, No. 1, Part 2.

Thomas, D. R., & King, R. A. Stimulus generalization as a function of level of motivation. JOURNAL OF EXPERIMENTAL PSYCHOLOGY, 1959, 57, 323-328.

Thomas, D. R., & Lopez, L. J. The effects of delayed testing on generalization slope. JOURNAL OF COMPARATIVE AND PHYSIO-LOGICAL PSYCHOLOGY, 1962, 55, 541-544.

Thompson, R. F. Primary stimulus generalization as a function of acquisition level in the cat. JOURNAL OF COMPARATIVE AND PHYSIOLOGICAL PSYCHOLOGY, 1958, 51, 601-606.

Tomie, A., Davitt, G. A., & Thomas, D. R. Effects of stimulus similarity in discrimination training upon wavelength generalization in pigeons. JOURNAL OF COMPARATIVE AND PHYSIOLOGICAL PSYCHOLOGY, 1975, 88, 945-954.

Yarczower, M., & Bitterman, M. E. Stimulus generalization in the goldfish. In D. I. Mostofsky (Ed.), STIMULUS GENERALIZATION. Stanford: Stanford University Press, 1965.

13. DISCRIMINATION

M. E. Bitterman

University of Hawaii
Honolulu, Hawaii   USA

Discriminative training is designed to produce differential response to stimuli by associating them with different treatments or contingencies. Consider, for example, an experiment by Woodard, Ballinger, and Bitterman (1974), who trained pigeons to peck a key which was lighted for a period of 8 sec on each trial with one of three colors (red, green, or blue) presented in quasi-random orders. When the animals were pecking readily at all colors, the discriminative training was begun. For one of the colors, at least one peck during the 8-sec period was required to produce food at the end of the period (reward); for a second color, a single peck prevented food, which came at the end of the period only if the animal failed to peck (omission); while a third color never was followed by food (extinction). There were 10 trials with each color and 10 blank trials in each session. In Figure 1, performance in the discriminative training is plotted in terms of two measures--the probability of at least one response on any trial, and the total number of responses per trial. It should be noted that the two measures of performance were not entirely equivalent. One was more sensitive to the difference between omission and extinction, while the other was more sensitive to the difference between omission and reward.

Discriminative training has three main uses, of which the first is to measure distinguishability, not only of the simplest attributes of stimuli, but of higher-order attributes of compounds and arrays of stimuli. We may ask, for example, whether an animal can distinguish such properties as oddity or number. Where distinguishability is at

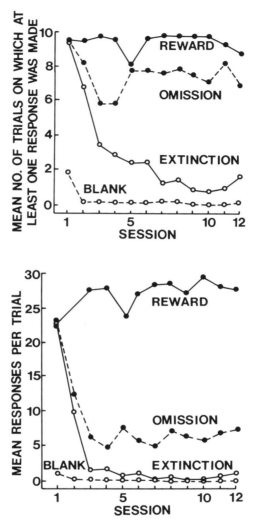

Figure 13.1   Response to three colors, each associated with a different contingency, and response on blank trials.
From Woodard, Ballinger, and Bitterman (1974).

issue, care must be taken in the choice of contingencies (Stebbins, 1970). The second main use of discriminative training is in the analysis of contingencies. The purpose of the Woodard experiment was to study omission, a contingency which provides stimulus-reinforcement contiguity without response-reinforcement contiguity. The effectiveness of stimulus-reinforcement contiguity is shown by greater response to the omission color than to the extinction color, which was used to control for the effects of reinforcement independently of contiguity. While contingencies often are compared in between-groups experiments, within-groups comparisons may be more convenient, and they provide evidence of interactions among contingencies (such as behavioral contrast) which between-groups comparisons, of course, do not. The third main use of discriminative training is in the analysis of stimulus control. Along with the differentially reinforced ("relevant") stimuli that are responsible for differential response, there always are common ("irrelevant") stimuli or attributes that tend to produce a common response, and a great deal of research has been directed to the question of how control by the relevant stimuli develops. This question, which we began to consider in the chapter on generalization, will be taken up again in the chapter on attention.

STIMULUS-DISPLAYS, RESPONSES, CONTINGENCIES

The most widely used discriminative training procedures are the unitary or go/no-go procedures. The stimuli to be discriminated (the discriminanda) are presented successively (now one, now another), a single indicator-response (such as salivating or pecking a key) is defined, and discrimination is evidenced by differences in the latency, magnitude, or rate of that response. In the simplest case, two stimuli are used, one (S+) reinforced on some schedule--with or without a response-contingency--and the second (S-) unreinforced, but a variety of other treatments may be programmed. The two stimuli may be correlated with different magnitudes, probabilities, or delays of reinforcement; for example, a rat may find one pellet of food in the goal box at the end of a black runway but eight pellets in the goal box at the end of a white runway (Bower, 1961). There is no reason why classical and instrumental contingencies should not be intermixed in such an experiment; for example, one stimulus may be paired with food independently of the animal's behavior while response to the other prevents the delivery of food (Schwartz & Williams, 1972). Nor is there any reason why there should not be more than two stimuli and two contingencies involved in the training. An example of a three-contingency problem is provided by the Woodard experiment already cited.

To be sure that the animals are responding differentially to the discriminanda (rather than to other features of the training situation), and that the differential response is, in fact, attributable to the different treatments, it is necessary to take a few simple precautions. Clearly, stimuli and treatments must be balanced. In the simplest case of two stimuli and two treatments, each of the two possible combinations of stimuli and treatments should be used for half the animals. The order in which the two stimuli are presented also must be balanced. It is not appropriate merely to alternate them, as often is done in free-operant experiments, because the magnitude of response to a stimulus may be influenced by whether the preceding stimulus is positive or negative (Wilton & Clements, 1971; Mackintosh, Little, & Lord, 1972). The preferred procedure is to schedule at least a brief interpresentation interval and to balance the order of stimuli so that each is preceded by itself as often as by the other. Some suitable quasi-random orders for two stimuli have been developed by Gellermann (1933) and by Fellows (1967). The familiar free-operant discrimination procedure is open also to the criticism that the first reinforcement during any presentation of S+ may itself become a stimulus for further responding. In a target-striking experiment with goldfish, Tennant and Bitterman (1973) presented two colors in quasi-random orders after brief interpresentation intervals. When half the presentations of each color were reinforced on a VI 30-sec schedule and the rest were unreinforced--note that color and schedule were uncorrelated in this pseudodiscrimination procedure--the animals responded more during the reinforced presentations than during the unreinforced presentations of each color. The solution to the problem is to terminate each presentation of S+ after the first reinforcement and each presentation of S- after a comparable period without reinforcement, the two stimuli succeeding each other in quasi-random order after an intertrial interval (Woodard & Bitterman, 1974). The adoption of these precautions converts the free-operant procedure to a discrete-trials procedure in which rate is measured. The measurement of rate is not, it should be noted, a defining characteristic of free-operant training (Platt, 1971).

A useful and interesting finding is that discrimination may be sharpened by _requiring_ repeated responding to the stimuli, which, of course, permits the measurement of rate, but it should not be assumed that rate is necessarily a more sensitive or reliable measure than latency. In a discrete-trials experiment with pigeons in which two readily discriminable colors were associated with different probabilities of reinforcement, there was no difference in latency of responding when the first peck at a color produced reinforcement or ended the trial; but when the required number of pecks was increased to 10 or 20, a difference in response time which could be traced entirely to a difference in the latency of the first peck was found (Graf, Bullock, & Bitterman, 1964). Despite a substantial difference in the associative strengths of S+ and S-, unitary training may yield no evidence of discrimination because the associative strength of common stimuli is sufficient to bring performance to the ceiling. A reasonable interpretation of the Graf results is that the ratio schedule reduced the associative strength of common stimuli enough to bring response to S- below the performance ceiling. It is clear, in any event, that nondifferential reinforcement alone does not neutralize common stimuli in the sense of reducing their associative strength to zero--recall, for example, the interdimensional experiment of Farthing (1972), whose results are shown in Figure 8 of the chapter on generalization-- nor does the principle of shared associative strength require neutralization. A computer simulation by Rescorla and Wagner (1972) of discriminative training with AX positive and BX negative (where X represents the set of common stimuli) shows V_X to be substantially greater than 0 at asymptote.

The effects of common stimuli tend to be balanced out by the simultaneous-choice procedures--another widely used set of discriminative training procedures--in which the stimuli to be discriminated are presented simultaneously and the animal is required to choose between them. Suppose, for example, that a rat in the Lashley apparatus is confronted with a pair of cards (one black the other white) and rewarded for jumping to the black. To be sure that the animal actually is discriminating between the cards in such an experiment, it is necessary to unconfound cards and positions-- the positive card must be at right on half the trials and at left on the remaining trials in quasi-random (e.g. Gellermann) orders. Many animals show marked position preferences in the early stages of simultaneous-choice training--for example, a rat may go always to the left-hand card irrespective of its brightness--but the preference for position is supplanted by a preference for S+ as training continues. In principle, of course, simultaneous-choice problems can be mastered on the basis of attraction to the positive stimulus alone, but tests with the negative stimulus presented either separately or together with a novel stimulus that has been substituted for the positive usually give evidence of aversion to the negative stimulus as well (see, for example, Derdzinski & Warren, 1969; Mandler, 1968).

While unitary experiments are as likely to be classical as instrumental, simultaneous-choice experiments are instrumental. In the simplest case, reinforcement of some sort is contrasted with extinction, but two different probabilities, magnitudes, or delays of reinforcement may be used where the primary interest is

not in the distinguishability of the stimuli. For example, goldfish may be trained with red and green targets, choice of red being rewarded on a random 70% of trials and choice of green on the
remaining trials--a 70:30 problem (Woodard & Bitterman, 1973).
Choice of S- may simply go unrewarded, or it may be punished, as
in the original form of the Lashley apparatus where a jump to the
incorrect card, which is locked in place, precipitates the animal
into a net below. Choice experiments are discrete-trials experiments for the most part, but free-operant procedures may be used
as well. Consider, for example, a pigeon trained in a two-key
apparatus with response to red reinforced on a VI 1-min schedule
and response to green on a VI 3-min schedule (Nevin, 1969). A
complicating feature of experiments of this sort is that the probability of reinforcement for response to one stimulus increases with
the time spent in responding to the other.

    An option in choice experiments of the most common kind,
where one of the alternatives presented on a given trial is
"correct" (reinforced) and the other is "incorrect" (unreinforced),
is whether to use a <u>correction</u> or <u>noncorrection</u> procedure. In non-
correction, the trial ends with the animal's first choice, whether
or not it is correct. In correction, each trial ends with a reinforced response whether or not there has been an error, since
the animal has the opportunity to correct any erroneous choice.
Where repetitive errors (errors after the first one on each trial)
are possible, it is essential to distinguish them from initial
errors, because the two measures may have quite different meanings.
Repetitive errors may decline substantially in the course of training before the probability of initial error falls below the chance
level, because the animal may learn to shift responses after an
incorrect choice before learning very much about the discriminanda
where they are not strikingly different. Correction serves a
variety of purposes. One is to obviate the possibility that a
strong position preference which an animal may have at the start
of training will be maintained by reinforcement on a random half
of the trials. Correction is useful also where exact control
over the distribution of reinforcements is required. A difficulty encountered in the use of correction is that, where the
response is not very effortful and the interval between error
and opportunity to make the correct response is brief, the animal
may fail to discriminate stimuli that under other conditions are
readily discriminated. Bullock and Bitterman (1962), training
pigeons in a two-key situation, found poor performance in a series
of problems when there was immediate opportunity for correction
after error, but performance improved markedly when opportunity
to make the correct response was postponed for 6 sec by darkening
and inactivating the keys. Another way to sharpen performance
in choice problems (as in unitary problems) is to introduce a

ratio requirement; the trial is ended, not with the first peck at one of the colors but with, say, the fifteenth peck (Williams, 1971a).

Performance in discriminative tasks depends also on the structure of the training situation. It is obvious to begin with that the difficulty of a task must be increased when the relevant stimuli are not immediately evident to the animal. In some experiments, the stimuli are deliberately withheld until the animal makes some _observing response_, as when rats trained in an automated simultaneous-choice situation are required to step on a treadle at one end of the apparatus to turn on the stimuli at the opposite end (Coate & Gardner, 1965). One reason for introducing such a contingency is to have the animal in a certain place or at a certain distance from the stimuli when they are presented in order to standardize the conditions of presentation. In the treadle example just given, the contingency gets the animal away from the stimulus-panels and gives it an opportunity to see both of them before making a choice on any trial. In other cases, the interest of the experimenter may be to make the observing response (or some feature of it) explicit in an effort to understand how the response is developed and maintained; the assumption is that some such response, whether measured or not, may play an important role in all discriminative training with localized stimuli (Wyckoff, 1952).

Experiments on visual discrimination in simultaneous-choice situations show that performance is best when spatial contiguity of stimulus, response, and reward is maximal. The performance of monkeys in the Wisconsin apparatus is markedly impaired when the locus of the stimulus-object is separated from that of response and reward, as when an animal trained in a black-white discrimination must displace, not the black or the white block, but one or the other of two gray blocks adjacent to them, with reward under the gray block near S+ (McClearn & Harlow, 1954). Performance is disrupted also when the locus of reward is separated from that of stimulus and response, and when the locus of response is separated from that of stimulus and reward (Miller & Murphy, 1964; Cowey & Weiskrantz, 1968). Working with monkeys in an automated version of the Wisconsin apparatus, Polidora and Fletcher (1964) found spatial contiguity of stimulus and response to be much more important than contiguity of response and reward. Analogous results have been obtained with other animals in other situations. For rats in the Lashley apparatus, the difficulty of a simultaneous-choice problem with two quite distinguishable stimuli is enormously increased if the animals are required to jump, not to the positive card, but to a gray card adjacent to it (Wodinsky, Varley & Bitterman, 1954). In the conventional key-pecking apparatus for

pigeons, the locus of reward (the feeder) is different from that
of stimulus and response (the keys on which the discriminanda are
projected). Performance in color discrimination improves when,
instead of being required to choose between two differentially
lighted keys, the animal is required to choose between the differentially lighted apertures of two feeders, insertion of the head
into the correct aperture producing elevation of the grain tray
(Gonzalez, Berger, & Bitterman, 1966a).

Simultaneous-choice situations differ from unitary situations
in the opportunity they afford for direct comparison of the stimuli
to be discriminated. Where those stimuli are rather different,
opportunity for comparison seems to add little (Grice, 1949), but
it may help considerably where the stimuli are more similar.
Saldanha and Bitterman (1951) trained two groups of rats in the
Lashley jumping apparatus with four stimuli--two sets of stripes
differing in thickness and two gray cards differing in brightness--
which were paired in the manner illustrated in Figure 2. For the
comparison group, either the two striped cards or the two grays
were presented on each trial, while, for the noncomparison group,
the positive thickness was paired with the negative gray and the
positive gray was paired with the negative thickness. For both
groups, the four pairs of stimuli comprising the problem were presented equally often in quasi-random order. The results showed
more rapid learning by the comparison animals than by the noncomparison animals, most of which failed to learn at all. The comparison animals performed well when, after reaching criterion,
they were shifted to the noncomparison problem, which shows that
they were perfectly able to tell one gray from the other, and one
stripe-thickness from the other, in separate presentations. What
the opportunity for comparison may have done is to call the relatively subtle differences between the stimuli to the attention of
the animals. Experiments of like design demonstrate that a striking irrelevant difference between simultaneously presented stimuli
may impair performance. For example, rats trained in the jumping
apparatus to discriminate between two forms learn more rapidly when
both are black on some trials and both white on others than when
one form is white and the other black on each trial (Wortz &
Bitterman, 1953). Results of this sort suggest that selection of
a training method is not simply a matter of convenience. Some
interesting features of discriminative learning may be missed
entirely unless simultaneous problems are studied.

Although most experiments on discriminative learning have
been done either in unitary or in simultaneous-choice situations,
some interesting use has been made also of successive-choice situations, in which the discriminanda are presented successively but

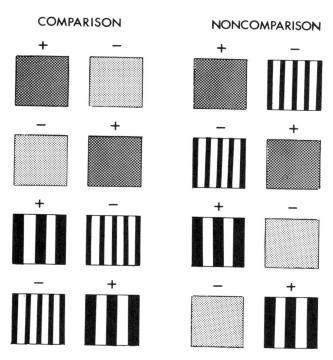

Figure 13.2 Simultaneous-choice problems requiring the discrimination of brightness and stripe-thickness with the stimuli paired either to permit or to prevent direct comparison. After Saldanha and Bitterman (1951).

choice is measured. Consider, for example, a T-maze, with gray
arms and a stem that is sometimes black and sometimes white, in
which rats are rewarded for turning, say, right when the stem is
black and left when the stem is white; or a Lashley apparatus with
two cards, both black on some trials and both white on others, in
which rats are rewarded for jumping, say, to the right card when
both are black and to the left card when both are white. Successive-choice procedures resemble simultaneous-choice procedures in
the use of a choice measure but (like unitary procedures) provide
no opportunity for comparison of the stimuli to be discriminated.
The importance of opportunity for comparison is indicated again by
a study of the relative difficulty of simultaneous and successive
problems. Working with rats in the jumping apparatus, MacCaslin
(1954) found successive problems to be somewhat more difficult than
simultaneous problems when readily distinguishable stimuli were
used. When less distinguishable stimuli (like those of the Saldanha
experiment) were used, the difference in difficulty increased substantially with most of the successive animals (like the Saldanha
noncomparison animals) failing to learn at all. A feature of the
successive-choice method which sharply distinguishes it from both
of the other methods is that the discriminanda themselves are not
differentially associated with reinforcement. The question of how
successive-choice problems are learned will be considered later.

DISCRIMINATION OF TRACES, COMPOUNDS, AND RELATIONS

Experiments on discriminative learning have been concerned
primarily with present stimuli, but there have been some interesting experiments also on delayed response, in which the discriminanda
are no longer present when the indicator-response is made. Where
the purpose of such work is to study traces or memories of the
discriminanda, care must be taken to ensure that the traces afford
the only basis for differential response. Suppose, for example,
that pigeons are trained in a successive-choice problem with three
keys; the center key is illuminated with one of two different
stimuli (say, two forms) and then turned off for a period of time
(the delay interval), after which the white side keys are turned on
and the animal rewarded for pecking one or the other of them depending on which form was previously displayed on the center key (Smith,
1967). It is possible here for the animal to orient to the correct
side key when a form is presented, to maintain that orientation
during the delay interval, and so to respond correctly when the
opportunity arises even if no trace of the form remains. Evidence
of postural mediating responses appeared in some of the earliest
experiments on delayed response (Hunter, 1913).

The use of postural orientation can be ruled out by nonspatial procedures of several different kinds. In <u>delayed</u> <u>matching</u>, the discriminanda are presented on the side keys after the delay interval, with their positions varying from trial to trial in quasi-random fashion, and the animal is rewarded for pecking the one that was earlier presented on the center key. Figure 3 shows the accuracy of delayed matching by pigeons in a red-green problem as a joint function of the time of exposure of the sample and the delay interval (Roberts & Grant, 1976). In <u>delayed</u> <u>alternation</u>, two stimuli are presented on each trial (say, two key colors) with their positions varied from trial to trial, and the animal is rewarded for choosing the one <u>not</u> chosen on the immediately preceding trial; here the intertrial interval functions as the delay interval (Williams, 1971b). While nonspatial procedures do not permit mediation by postural orientation, they do not, of course, eliminate the possibility that other interval-bridging responses will develop. An interesting example of the use of such responses by pigeons has been provided by Blough (1959).

While experiments on discriminative learning have dealt for the most part with simple attributes of stimuli, such as color or form, the animals studied give evidence also of capacity to discriminate more complex properties of the stimulus-arrays with which they are confronted--properties arising from conjunctions of stimuli or of attributes. The simplest evidence of conjunctive properties comes from work on <u>stimulus-compounding</u>. Pavlov (1927) found, for example, that by reinforcing a compound stimulus (AB) but not its components (A or B) when they were presented separately, he could train a dog to respond to the compound but not to the components. Reinforcement of the components but not the compound would produce response only to the components. A reasonable explanation of these findings--which as Pavlov recognized, contradicts the simple idea that the properties of a complex stimulus can be derived by summing the properties of its components--is that the two components together excite neural elements not excited by either alone. In this view, discrimination is based on the presence or absence of neural events generated only by the compound. While it may be suspected that discrimination in an experiment of this sort is based merely on the total number of elements excited, the phenomenon of <u>attributive</u> <u>compounding</u> requires us to assume discrimination on the basis of elements specific to particular compounds rather than on the total number of elements. For example, in an experiment on eyelid conditioning by Saavedra (1975), rabbits were trained to discriminate A_1B_1 or A_2B_2 positive from A_1B_2 or A_2B_1 negative, where A_1 and A_2 were auditory stimuli while B_1 and B_2 were visual stimuli. This kind of training sometimes is said to produce <u>conditional</u> <u>discrimination</u>.

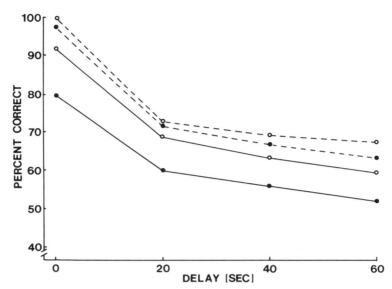

Figure 13.3   Delayed matching in pigeons as a function of the de-
              lay interval. The bottom curve is for 1-sec expo-
              sures of the sample, the second for 4-sec exposures,
              the third for 8-sec exposures, and the top curve for
              14-sec exposures. From Roberts and Grant (1976).

        Once elements unique to particular compounds have been postu-
lated, compounding is open to much the same common-elements treatment as is generalization, the only difference being that the
elements common to compounds and components are those uniquely
excited by the separate components. Ignoring generalization on the
assumption that we are using quite dissimilar components, we may
say that two components, A and B, excite sets of elements a or b
separately, but a, b, and x jointly (Rescorla, 1973). The idea of
compound-specific elements which are simply added to the elements
excited by the separate components of the compound is not the same
as Hull's idea that the afferent processes excited by the components
are themselves changed by interaction in the compound--the principle
of afferent neural interaction (Hull, 1943). The difference in
formulation is reflected in a difference in notation: instead of
abx, Hull wrote $\breve{a}\breve{b}$, where $\breve{a}$ and $\breve{b}$ are a and b as modified by
interaction. The two ideas have somewhat different implications.
For example, Hull predicted that, with A and B separately reinforced
to asymptotic levels, the first presentation of the compound might
evoke a lesser response than either A or B alone--an example of what
he called "spontaneous patterning." According to the abx analysis,
there should be no such decrement, since the "integrity" of the

elements is assumed to be retained in the compound (Rescorla, 1973). According to both interpretations, the difficulty of compound problems should vary considerably with the degree of interaction-- the salience of x, or the extent to which $\breve{a}$ and $\breve{b}$ differ from a and b--and where the interaction is substantial the product of the interaction should acquire significant associative value even with- out differential reinforcement of components and compounds. When- ever two stimuli are presented together, there are interactive possibilities which it is dangerous to ignore.

The solution of successive-choice problems has been explained in terms of the discrimination of compounds involving spatial loca- tion (Spence, 1952). In a simultaneous problem that requires direct response to the discriminanda (as in a conventional Lashley or Wisconsin apparatus), the animal learns to choose one of them inde- pendently of position--independently even of the specific positions occupied in training. Chimpanzees trained with black and white objects aligned horizontally continue to show the established preference when tested with the objects aligned vertically (Nissen, 1950). In a successive problem, by contrast, training with objects in horizontal alignment affords no basis for dealing with the same objects in vertical alignment, since response neither to brightness nor to position is differentially reinforced. There is, however, differential reinforcement of response to object- position compounds, and what the animal may be learning to do is to choose, say, white-right rather than white-left, and black-left rather than black-right. In this view, successive problems are more difficult than simultaneous problems because the compounds are less distinguishable than their components, but there is some contrary evidence. Bitterman and McConnell (1954) trained two groups of rats in the Lashley apparatus. One group had two consecu- tive simultaneous problems and the other had two consecutive successive problems with readily distinguishable stimuli which were balanced over problems. As the learning curves plotted in Figure 4 show, the simultaneous group was superior to the succes- sive in the first problem, but the successive group improved so much more from the first problem to the second that the difference be- tween the two groups disappeared in the second problem. These results suggest that a successive problem is more difficult to begin with, not because of poorer distinguishability, but because it requires an alternative strategy (or way of looking at the situation) which naive animals are less likely to adopt but which, once it has been adopted, transfers readily to successive problems with new stimuli. A good summary of the rather complex literature dealing with this question is provided by Mackintosh (1974).

Other experiments show discrimination, not of absolute proper- ties--either of separate stimuli or of compounds--but discrimination

of relations. We have already considered the phenomenon of transposition, which early suggested that performance in simultaneous-
choice situations may be based on relational properties, and the
subsequent proposal of Spence (1937) to account for transposition
in terms of the generalization of excitation and inhibition. Not
only is the relational assumption unnecessary, Spence argued, but
it is contradicted by the fact that the probability of transposition
declines as the absolute properties of the stimuli used in the test
become more and more different from those of the training pair.
(This phenomenon, known as the "distance effect." can be derived
from the summed gradients.) Although it is still reasonable to
think that generalization works as Spence proposed, whether the
discriminanda are presented simultaneously or successively, there
is evidence also of response to relations in simultaneous presentations. Using the Saldanha technique illustrated in Figure 2,
Thompson (1955) trained two groups of rats with pairs of stimuli so
chosen that the comparison problem was no easier than the noncomparison problem, and then he tested for transposition. As Spence's
theory predicts, there was some transposition even after noncomparison training, but--contrary to the theory--there was more transposition after comparison training. Riley, Ring, and Thomas (1960)
trained rats in a simultaneous-choice situation either with or
without a barrier between the stimuli that made it impossible for
the animal to see them at the same time. Here learning was more
rapid under conditions that permitted comparison, especially when

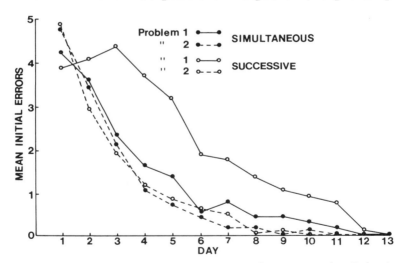

Figure 13.4   The performance of a group of rats trained in two
              consecutive simultaneous-choice problems and of
              another group trained in two consecutive successive-
              choice problems. From Bitterman and McConnell (1954).

the difference between the stimuli to be discriminated was small, but again subsequent tests (without the barrier) showed more transposition after comparison than after noncomparison training. The performance of the two groups in the transposition tests is shown in Figure 5.

As Spence (1942) noted, a critical test of his theory of transposition is provided by the so-called intermediate-size problem, or any problem involving three points on some sensory dimension in which the animal is reinforced only for choosing the intermediate point. Consider, for example, an animal trained to criterion in a problem involving points 2, 3, and 4 on some continuum, with 3 reinforced and the other two unreinforced. If the animal comes to choose 3 only because the net associative strength (E-I) of 3 is greater than that of 2 or 4, it cannot choose 2 when tested with 1, 2, and 3, or 4 when tested with 3, 4, and 5--that is, it cannot transpose. In fact, however, transposition of the intermediate-size problem has been reported both in chimpanzees (Gonzalez, Gentry, & Bitterman, 1954) and in rhesus monkeys (Gentry, Overall, & Brown, 1959). Despite the inability of Spence's theory to account for these data, aspects of the results that support the notion of interacting generalization gradients should not be overlooked. As the gradient analysis predicts, the intermediate-size problem is quite difficult--much more so, as Lashley (1938) discovered in early work with rats--than a problem which requires the animal to go to the largest of three stimuli or to the smallest of the three. The fact of transposition in the intermediate-size problem does not show that Spence's assumptions are wrong, but only that they are insufficient--that it may be necessary also to postulate a relational process. The conventional transposition experiment is not, however, the best way to study the capacity of an animal to discriminate relations, since the training permits the development of absolute preferences that may interfere with the expression of relational preferences in the subsequent tests. A better method is to train the animals in problems that permit only relational solutions.

Discrimination of relations somewhat more complex than intra-dimensional relations such as in size or brightness also has been demonstrated. Suppose, for example, that rats are trained in a three-window Lashley apparatus with two white cards and one black card on some trials, and with one white card and two black cards on other trials, response to the odd brightness in each of the six possible arrays (*WBB, BWB, BBW, BWW, WBW,* and *WWB*) being rewarded. This so-called oddity problem cannot be mastered either in terms of brightness alone, position alone, or brightness and position combined, and the possibility that the animal is discriminating between

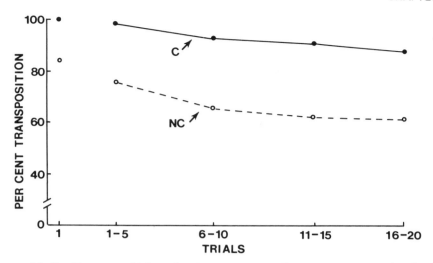

Figure 13.5   Transposition in two groups of rats, one trained with
              (C) and the other without (NC) opportunity for direct
              comparison of the stimuli. After Riley, Ring, and
              Thomas (1960).

the six separate arrangements of stimuli, intuitively unlikely, is
ruled out by transfer tests. Rats trained to criterion in two or
three such problems, each with a different set of stimuli, have
been reported to show better than chance performance in the earliest
trials with a new set (Wodinsky & Bitterman, 1953). Moon and Harlow
(1955) used the Wisconsin apparatus to train monkeys in a series of
two-position oddity problems--problems like those already described
except that the odd stimulus never appears in the middle position--
with six trials per problem and a new set of stimulus-objects in
each problem. As shown in Figure 6, the percentage of correct
choice on the very first trial of each problem rose gradually to
about 88% as compared, of course, with a chance level of 50%.

     If, in a problem involving the same four arrangements of
stimuli as the two-position oddity problem (say, *WBB*, *BBW*, *BWW*, and
WWB), the animal is reinforced for the "non-odd" choice, we have
matching from sample, and the solution of such problems also may
generalize to new sets of stimuli. For example, Nissen, Blum, and
Blum (1948) trained chimpanzees with a row of three stimulus-objects
and food under that one of the two side-objects which matched the
central object. Tested with new sets of objects, the animals showed
highly significant preferences for the matching alternatives.
Robinson (1955) trained chimpanzees in a two-choice Wisconsin
apparatus with each stimulus-object containing two colored figures
that were either the same or different (for example, the block over

one food well might contain two blue circles and the block over the
other might contain one blue circle and one red triangle) and
rewarded the animals for choosing the block that contained two
identical figures. The relational character of the solution of
this <u>same-difference</u> problem was demonstrated by the fact that it
generalized to new figures. In a recent experiment by Malott and
Malott (1970), pigeons trained in a same-difference color discrimination showed transfer of the solution to new colors.

The critical question posed by such results is whether they
require some new learning principles or whether the entire burden
of explanation can be borne by assumptions about stimulus-processing.
The most parsimonious view is that sensory events have a variety of
complex properties as well as simple properties, and that the animal
learns about complex properties in the same way as about simple ones.
Consider, for example, the work of Herrnstein and Loveland (1964),
who trained pigeons in a free-operant situation to discriminate
photographs containing human beings from photographs which did not,
or the work of Lubow (1974), who trained pigeons to discriminate
photographs of man-made objects from photographs of natural objects.
Although the pigeons were credited with conceptual ability in both
cases, the hypothesis may be entertained that the animals were

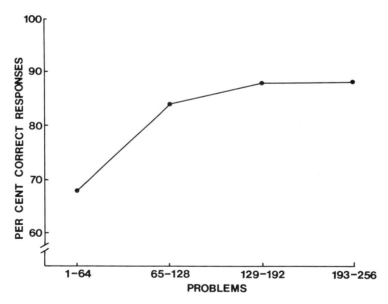

Figure 13.6   Trial-1 performance of monkeys trained in a long
              series of six-trial two-position oddity problems.
              From Moon and Harlow (1955).

responding to certain higher-order features of the stimulus arrays
--higher-order properties generated by afferent interaction. Lubow
himself suggested that much of the performance of his animals could
be accounted for on the assumption they were discriminating the
presence or absence of straight lines and right-angles. This
modern work is highly reminiscent of earlier work in which the
ability of rats and other animals to develop "concepts" of form,
such as triangularity, was studied in tests with triangles in a
variety of different sizes, orientations, and contexts (Fields,
1932; Lashley, 1938). Just how far the perceptual interpretation
can be carried, it is difficult to say. Consider, for example,
the work of Hicks (1956), who trained rhesus monkeys to discriminate
between cards containing three assorted geometric figures differing
in size, proportion, and arrangement from cards containing one, two,
four, or five such figures. If comparable results were obtained
with numbers other than three, the assumption of "counting" would
be difficult to avoid, although counting itself may be viewed as a
source of sensory feedback about which the animal learns in the
same way as about the features of simple component problems. It
might be well in any case to restrict the use of terms like "concept
formation" to instances of relational learning that cannot be attributed to the discrimination of complex sensory properties.

Some promising results have been obtained by Premack (1971) in
his efforts to teach an artificial language to a young chimpanzee
named Sarah. The words of the language were plastic chips which
were manipulated (placed on a magnetized "language board") both
by the experimenter and by the subject. After learning to use chips
meaning red and yellow, Sarah could be trained to use chips meaning
color of and not color of, being required, for example, to replace
the interrogative chip (?) with the chip meaning color of when asked
red/?/apple, but with the chip meaning not color of when asked red/
?/banana. In subsequent tests with chips representing new objects
(or with the objects themselves) and with questions of different
form, Sarah showed considerable transfer, being able, for example,
to respond with the chip meaning no to the question ?/red/color of/
feather. Considerable caution must, of course, be exercised in the
interpretation of these results, based as they are on relatively informal work with a single animal under conditions of intimate sub-
ject-experimenter interaction. Rumbaugh, Gill, and Glaserfeld (1973)
have reported that it is possible to do such experiments with illuminated pushbuttons instead of chips and with a computer replacing
the experimenter; it will be interesting to see whether results
like those reported for Sarah can be obtained under more objective
conditions. Although new optimism about the intellectual capabilities
of chimpanzees has been generated by recent work on "language," it is
only fair to note that the results obtained are not always impressive
(Fouts, Chown, & Goodin, 1976).

INTERPROBLEM TRANSFER

Improvement in performance over a series of discriminative problems such as that illustrated in Figure 6 often is described as
the formation of a <u>learning set</u>--in this case, an <u>oddity learning
set</u>. The meaning is that the animals learn about certain common
features of the series of problems. In the oddity case, presumably,
the animals learn about a higher-order property common to the different sets of stimuli, although the problems have other common
features as well. Animals exposed in the Wisconsin apparatus to
a series of simultaneous-choice problems each involving two objects
randomly chosen from a large assortment of objects also show striking improvement in performance from one block of problems to the
next (<u>object-discrimination learning set</u>). Here, of course, there
is no basis for better-than-chance performance on Trial 1, but the
improvement takes the form of an increase in the rate of learning
in each problem which may be evidenced as early as Trial 2. Shown
in the left-hand portion of Figure 7 are within-problems object-
discrimination learning curves for a group of blue jays based on
performance in consecutive blocks of 32 unrelated simultaneous-
choice problems (Kamil, Jones, Pietrewicz, & Mauldin, 1977) and in
the right-hand portion some comparable curves for rhesus monkeys
based on performance in consecutive blocks of 50 such problems
(Harlow & Warren, 1952). The birds were trained with three-dimen-
sional objects to a criterion of eight successive correct responses

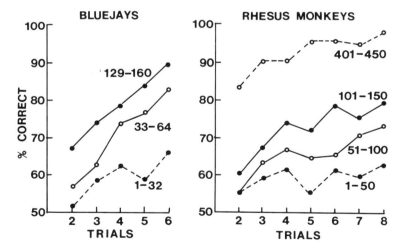

Figure 13.7   Interproblem improvement in blue jays and rhesus mon-
              keys trained in simultaneous-choice problems with un-
              related objects. After Kamil, Jones, Pietrewicz, and
              Mauldin (1977); Harlow and Warren (1952).

in each problem; the monkeys were trained with two-dimensional patterns and a fixed number of trials (eight) per problem.

Harlow (1950) traced improvement in performance over problems involving unrelated objects to the elimination of a set of common error-producing tendencies. In early problems, for example, monkeys may show strong preferences for one or the other of a pair of objects which, when the preferred object happens to be incorrect, produce stimulus perseveration errors. These errors tend to disappear in later problems. So also do differential cue errors, which stem from a tendency to choose the negative object when it is in the position occupied by the positive object on the immediately preceding trial. The reduction in this tendency is perhaps to be understood in terms of the nondifferential reinforcement of spatial location over the series of problems. Inexperienced monkeys also show a strikingly persistent tendency to "try out" both stimulus-objects, as a result of which Trial-2 performance is better on the average after an error on Trial 1 than after the animal has happened by chance to choose the correct object (Harlow, 1959). It also has been proposed that the animals develop what is called a win-stay or win-stay/lose-shift strategy, which means that they learn to respond on each trial in terms of the events of the immediately preceding trial carried over as stimulus-traces or in short-term memory (Restle, 1958). This interpretation is supported by the finding that the superior performance of experienced animals (birds as well as primates) depends to a considerable extent on the use of relatively brief intertrial intervals (Bessemer & Stollnitz, 1971; Kamil & Mauldin, 1975).

Another interesting example of interproblem transfer is provided by experiments on serial reversal learning. Here the same two stimuli are used throughout, one of them reinforced in odd-numbered problems and the other reinforced in even-numbered problems, with training in each problem continued either for a fixed number of trials or until some criterion of mastery has been achieved. Some representative results for rats trained by Gatling (1952) in a Lashley apparatus modified to eliminate punishment for error are shown in Figure 8. The discriminanda were light gray and dark gray cards simultaneously presented, with light gray positive in the original problem (R_0), dark gray positive on the first reversal (R_1), and so forth. Ten trials per day were given, and training in each problem was carried to a criterion of two successive days on each of which there were at least nine correct choices. The curve shows that the animals experienced considerable difficulty in the early reversals but that, as reversal training continued, their performance improved markedly. A curve of this sort, which is plotted in terms of performance over reversals, is called a between-reversals curve.

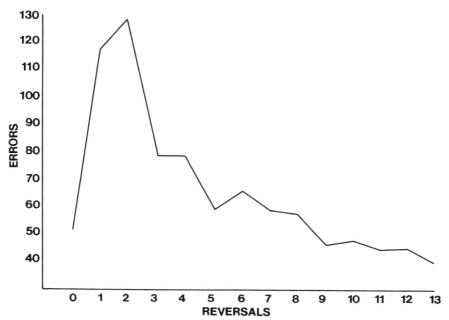

Figure 13.8   Discrimination reversal in rats trained in a simultaneous-choice problem.   The curve shows improvement
over reversals in terms of mean errors to criterion.
From Gatling (1952).

   Reproduced in Figure 9 are the early segments of some of the
within-reversals learning curves of a group of pigeons trained for
40 trials per day in a series of two-day red-green reversals
(Bitterman, 1968).   Since the positive and negative stimuli were
reversed every two days, each odd-numbered day except the first was
a reversal (R) day and each even-numbered day was a nonreversal (NR)
day.   The curves show the probability of error on the very first
10 trials (of Day 1, R_0) and on the first 10 trials of each of four
other days.   The most striking feature of these curves is a decline
in the extent to which the preference established on each day carries
over to the next day.   To begin with, there was a great deal of
carryover, which produced good performance on NR days and poor performance on R days.   After 120 problems, however, there was little
carryover, with both the R and NR curves beginning at about the
chance level and declining rapidly over the next few trials in
essentially identical fashion.   Note that the improvement in performance on R days was accompanied by a deterioration of performance
on NR days, although there was improvement in performance over
reversals for each problem as a whole.   It has been suggested by

Gonzalez, Berger, and Bitterman (1966b) that the decline in carryover from day to day is due to interference-produced forgetting--
to what students of human learning have called proactive interference; as a result of contradictory experiences, the animals can
no longer remember at the start of each day which stimulus was
rewarded on the previous day.

A finding which requires a different explanation is that performance in later reversals may become even better than performance in R_0. As in learning set experiments with unrelated
stimuli, there is an increase in the rate of learning within problems, and it is possible that the improvement may have the same
basis as in learning set experiments. This interpretation is supported by the fact that serial reversal training may improve performance in a subsequent series of problems with unrelated stimuli.
Schusterman (1964) obtained such results in work with chimpanzees,

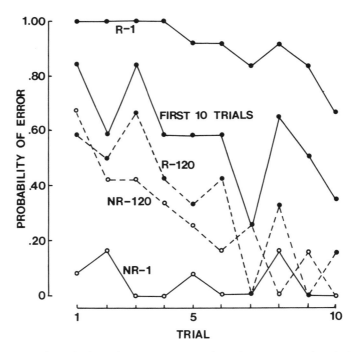

Figure 13.9 Discrimination reversal in pigeons trained in a simul-
            taneous-choice problem. The curves show performance
            on the first 10 trials of the original problem, two
            reversal sessions (R_1 and R_{120}), and two nonreversal
            sessions (NR_1 and NR_{120}). From Bitterman (1968).

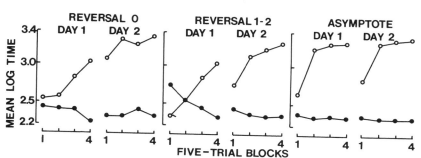

Figure 13.10 Discrimination reversal in pigeons trained in a unitary problem. Latency of response to each stimulus is shown. After Woodard, Schoel, and Bitterman (1971).

Kamil, Jones, Pietrewicz, and Mauldin (1977) with jays, and Warren (1966) with rhesus monkeys, although not with cats. Whatever the improvement is due to, it must have some basis other than the development of a win-stay/lose-shift strategy if that strategy requires relatively short intertrial intervals. Working with rats, North (1959) found progressive improvement in the reversal of a simultaneous brightness discrimination even when the intertrial interval was as long as one day. Decreased retention between days would work against improvement here, of course, since it would slow progress to the criterion of learning in each reversal (which was 11 correct choices in 12 trials, with no error in the last eight).

While most experiments on serial reversal learning have been done under simultaneous-choice conditions, unitary procedures permit a more refined analysis of improvement. Woodard, Schoel, and Bitterman (1971) trained pigeons in a series of two-day red-green reversals by a discrete-trials procedure with response defined as 20 pecks and time to complete the ratio as the measure. Selected within-reversals learning curves plotted in terms of five-trial blocks are shown in Figure 10. In R_0, the discrimination appeared rapidly on Day 1 and developed further on Day 2, with little intersession decrement in retention. Performance in the early reversals (R_{1-2}) continued to reflect considerable intersession retention, the reversal day showing a characteristic crossover. At asymptote, however, the picture is radically different. Trial-by-trial plots show that the animals begin each day by responding rapidly to both colors, continue to respond rapidly to whichever color happens to be positive, and slow down sharply to the negative. The only indication of intersession retention at asymptote (which is reflected in the S- curve for blocks of trials in Figure 10) is that the slowing of response to S- is somewhat more precipitous on Day 2 than on Day 1. Here, then, is the between-sessions decrement in retention which is found

in choice experiments. The increased speed of learning within
reversals which is found in choice experiments is evident here also
in the more rapid separation of the S+ and S- curves on Day 1 at
asymptote than on Day 1 of R_0. What this unitary experiment shows
that choice experiments do not prepare us for is that both of the
changes which occur over the series of reversals are changes primarily
in the effect of nonreinforcement. The loss in retention from one
session to the next is a loss in the reluctance to respond to the
negative stimulus of the preceding session. The increase in the
speed of learning is an increase in the rate at which reluctance
to respond to the negative stimulus develops in each session. None
of the theories of progressive improvement prepares us for these
findings. It is conceivable that an understanding of improvement
must wait on an understanding of the asymptotic performance, various
models of which have been quantitatively evaluated by Woodard and
Bitterman (1976).

Comparative experiments once seemed to indicate meaningful relations between taxonomic status and measures of interproblem transfer. Harlow (1959) concluded, for example, that the performance of
rhesus monkeys in learning set experiments is superior to that
of squirrel monkeys, and that primates perform better than mammals
of other orders. It is now evident, however, that the performance
of any given species in a learning set experiment may vary enormously
with training conditions which there seems to be no way of equating
across species; one of the most important factors is the distinguishability of the stimuli employed (Warren, 1973). Sutherland and
Mackintosh (1971) have concluded that there are characteristic differences in the amount of improvement shown by various animals
studied in experiments on serial reversal learning--that rats
improve more than pigeons, and pigeons more than certain fishes.
For reasons not well understood, the fishes studied do fail to
show any improvement at all under a wide variety of circumstances
(Behrend, Domesick, & Bitterman, 1965), but there are circumstances
under which they show substantial improvement (Engelhardt, Woodard,
& Bitterman, 1973). In reversal learning as in learning set experiments, the performance of any given species may vary markedly with
training conditions, and again the differentiation required is an
important variable. For example, the reversal performance of rats
is vastly better in olfactory than in visual or auditory tasks
(Nigrosh, Slotnick, & Nevin, 1975). Hodos (1970) has suggested
that a phylogeny of plasticity might be constructed on the basis
of the best performance of each species in learning set experiments,
but the suggestion is unacceptable. If performance is a function
of variables other than plasticity, the fact that the "best" performance of one species exceeds the "best" performance of another
cannot safely be attributed to a difference in plasticity until
the other variables are equated.

As might be expected from the results of experiments on the effects of extradimensional discriminative training on generalization which were described in the chapter on generalization, experience in only a single discriminative problem may markedly influence performance in a subsequent problem with unrelated stimuli. Training goldfish in a unitary situation to discriminate two auditory frequencies, Tennant and Bitterman (1975) found much better performance after TD than after PD training with two colors. Another instance in which training in a single problem can influence performance in a subsequent problem with different stimuli is provided by the McConnell experiment on change in the relative difficulty of simultaneous- and successive-choice problems which has already been considered (Figure 4). An even more extreme change in the relative difficulty of these problems was found by Gonzalez and Shepp (1961), who gave serial reversal training to two groups of rats in a modified Lashley apparatus. The successive group made more errors in the early reversals but improved more rapidly than the simultaneous group, making fewer errors in the later reversals. Interproblem transfer clearly is a multifaceted phenomenon of which we now have only the most rudimentary grasp. It will be considered further in the chapter on attention, evidence for which rests heavily on the outcome of transfer experiments.

REFERENCES

Behrend, E. R., Domesick, V. B., & Bitterman, M. E. Habit reversal in the fish. JOURNAL OF COMPARATIVE AND PHYSIOLOGICAL PSYCHOLOGY, 1965, 60, 407-411.

Bessemer, D. W., & Stollnitz, F. Retention of discriminations and an analysis of learning set. In A. M. Schrier & F. Stollnitz (Eds.), BEHAVIOR OF NONHUMAN PRIMATES. Vol. 4. New York: Academic Press, 1971.

Bitterman, M. E. Reversal learning and forgetting. SCIENCE, 1968, 160, 99-100.

Bitterman, M. E., & McConnell, J. V. The role of set in successive discrimination. AMERICAN JOURNAL OF PSYCHOLOGY, 1954, 67, 129-132.

Blough, D. S. Delayed matching in the pigeon. JOURNAL OF THE EXPERIMENTAL ANALYSIS OF BEHAVIOR, 1959, 2, 151-160.

Bower, G. H. A contrast effect in differential conditioning. JOURNAL OF EXPERIMENTAL PSYCHOLOGY, 1961, 62, 196-199.

Bullock, D. H., & Bitterman, M. E. Habit reversal in the pigeon. JOURNAL OF COMPARATIVE AND PHYSIOLOGICAL PSYCHOLOGY, 1962, 55, 958-962.

Coate, W. B., & Gardner, R. A. Sources of transfer from original training to discrimination reversal. JOURNAL OF EXPERIMENTAL PSYCHOLOGY, 1965, 70, 94-97.

Cowey, A, & Weiskrantz, L. Varying spatial separation of cues, response, and reward in visual discrimination learning in monkeys. JOURNAL OF COMPARATIVE AND PHYSIOLOGICAL PSYCHOLOGY, 1968, 66, 220-224.

Derdzinski, D., & Warren, J. M. Perimeter, complexity, and form discrimination learning by cats. JOURNAL OF COMPARATIVE AND PHYSIOLOGICAL PSYCHOLOGY, 1969, 68, 407-411.

Engelhardt, F., Woodard, W. T., & Bitterman, M. E. Discrimination reversal in the goldfish as a function of training conditions. JOURNAL OF COMPARATIVE AND PHYSIOLOGICAL PSYCHOLOGY, 1973, 85, 144-150.

Farthing, G. W. Overshadowing in the discrimination of successive compound stimuli. PSYCHONOMIC SCIENCE, 1972, 28, 29-32.

Fellows, B. J. Chance stimulus sequences for discrimination tasks. PSYCHOLOGICAL BULLETIN, 1967, 67, 87-92.

Fields, P. E. Concerning the discrimination of geometrical figures by white rats. JOURNAL OF COMPARATIVE PSYCHOLOGY, 1932, 14, 63-77.

Fouts, R. S., Chown, B., & Goodin, L. Transfer of signed responses in American sign language from vocal English stimuli to physical object stimuli by a chimpanzee. LEARNING AND MOTIVATION, 1976, 7, 458-475.

Gatling, F. The effect of repeated stimulus reversals on learning in the rat. JOURNAL OF COMPARATIVE AND PHYSIOLOGICAL PSYCHOLOGY, 1952, 45, 347-351.

Gellerman, L. W. Chance orders of alternating stimuli in visual discrimination experiments. JOURNAL OF GENETIC PSYCHOLOGY, 1933, 42, 206-208.

Gentry, G. V., Overall, J. H., & Brown, W. L. Transposition responses of rhesus monkeys to stimulus-objects of intermediate size. AMERICAL JOURNAL OF PSYCHOLOGY, 1959, 72, 453-455.

Gonzalez, R. C., Berger, B. D., & Bitterman, M. E. A further comparison of key-pecking with an ingestive technique for the study of discriminative learning in pigeons. AMERICAN JOURNAL OF PSYCHOLOGY, 1966, 79, 217-225. (a)

Gonzalez, R. C., Berger, B. D., & Bitterman, M. E. Improvement in habit-reversal as a function of amount of training per reversal and other variables. AMERICAN JOURNAL OF PSYCHOLOGY, 1966, 79, 517-530. (b)

Gonzalez, R. C., Gentry, G. V., & Bitterman, M. E. Relational discrimination of intermediate size in the chimpanzee. JOURNAL OF COMPARATIVE AND PHYSIOLOGICAL PSYCHOLOGY, 1954, 47, 385-388.

Gonzalez, R. C., & Shepp, B. E. Simultaneous and successive discrimination-reversal in the rat. AMERICAN JOURNAL OF PSYCHOLOGY, 1961, 74, 584-589.

Graf, V., Bullock, D. H., & Bitterman, M. E. Further experiments on probability matching in the pigeon. JOURNAL OF THE EXPERIMENTAL ANALYSIS OF BEHAVIOR, 1964, 7, 151-157.

Grice, G. R. Visual discrimination learning with simultaneous and successive presentation of stimuli. JOURNAL OF COMPARATIVE AND PHYSIOLOGICAL PSYCHOLOGY, 1949, 42, 365-373.

Harlow, H. F. Analysis of discrimination learning by monkeys. JOURNAL OF EXPERIMENTAL PSYCHOLOGY, 1950, 40, 26-39.

Harlow, H. F. Learning set and error factor theory. In S. Koch (Ed.), PSYCHOLOGY: A STUDY OF A SCIENCE, Vol. 2. New York: McGraw-Hill, 1959.

Harlow, H. F., & Warren, J. M. Formation and transfer of discrimination learning sets. JOURNAL OF COMPARATIVE AND PHYSIOLOGICAL PSYCHOLOGY, 1952, 45, 484-489.

Herrnstein, R. J., & Loveland, D. H. Complex visual concept in the pigeon. SCIENCE, 1964, 146, 549-551.

Hicks, L. An analysis of number-concept formation in the rhesus monkey. JOURNAL OF COMPARATIVE AND PHYSIOLOGICAL PSYCHOLOGY, 1956, 49, 212-218.

Hodos, W. Evolutionary interpretation of neural and behavioral studies of living vertebrates. In F. O. Schmidt (Ed.), THE NEUROSCIENCES: SECOND STUDY PROGRAM. New York: Rockefeller University Press, 1970.

Hull, C. L. PRINCIPLES OF BEHAVIOR. New York: Appleton-Century, 1943.

Hunter, W. S. The delayed reaction in animals and children. BEHAVIORAL MONOGRAPHS, 1913, 2, 21-30.

Kamil, A. C., Jones, T. B., Pietrewicz, A., & Mauldin, J. E. Positive transfer from successive reversal training to learning set in bluejays (*Cyanocitta cristata*). JOURNAL OF COMPARATIVE AND PHYSIOLOGICAL PSYCHOLOGY, 1977, 91, 79-86.

Kamil, A. C., & Mauldin, J. E. Intraproblem retention during learning-set acquisition in bluejays (*Cyanocitta cristata*). ANIMAL LEARNING & BEHAVIOR, 1975, 3, 125-130.

Lashley, K. S. The mechanism of vision. XV. Preliminary studies on the rat's capacity for detail vision. JOURNAL OF GENERAL PSYCHOLOGY, 1938, 18, 123-193.

Lubow, R. E. High-order concept formation in the pigeon. JOURNAL OF THE EXPERIMENTAL ANALYSIS OF BEHAVIOR, 1974, 21, 475-483.

MacCaslin, E. F. Successive and simultaneous discrimination as a function of stimulus-similarity. AMERICAN JOURNAL OF PSY-CHOLOGY, 1954, 67, 308-314.

Mackintosh, N. J. THE PSYCHOLOGY OF ANIMAL LEARNING. New York: Academic Press, 1974.

Mackintosh, N. J., Little, L., & Lord, J. Some determinants of behavioral contrast in pigeons and rats. LEARNING AND MOTIVATION, 1972, 3, 148-161.

Malott, R. W., & Malott, M. K. Perception and stimulus generaliza-tion. In W. C. Stebbins (Ed.), ANIMAL PSYCHOPHYSICS. New York: Appleton-Century-Crofts, 1970.

Mandler, J. M. The effect of overtraining on the use of positive and negative stimuli in reversal and transfer. JOURNAL OF COMPARATIVE AND PHYSIOLOGICAL PSYCHOLOGY, 1968, 66, 110-115.

McClearn, G. E., & Harlow, H. F. The effect of spatial contiguity on discrimination learning by rhesus monkeys. JOURNAL OF COMPARATIVE AND PHYSIOLOGICAL PSYCHOLOGY, 1954, 47, 391-394.

Miller, R. E., & Murphy, J. V. Influence of the spatial relation-ships between the cue, reward, and response in discrimination learning. JOURNAL OF EXPERIMENTAL PSYCHOLOGY, 1964, 67, 120-123.

Moon, L. E., & Harlow, H. F. Analysis of oddity learning by rhesus monkeys. JOURNAL OF COMPARATIVE AND PHYSIOLOGICAL PSYCHOLOGY, 1955, 48, 188-194.

Nevin, J. A. Interval reinforcement of choice behavior in discrete trials. JOURNAL OF THE EXPERIMENTAL ANALYSIS OF BEHAVIOR, 1969, 12, 875-885.

Nigrosh, B. J., Slotnick, N. M., & Nevin, J. A. Olfactory discri-mination, reversal learning, and stimulus control in rats. JOURNAL OF COMPARATIVE AND PHYSIOLOGICAL PSYCHOLOGY, 1975, 89, 285-294.

Nissen, H. W. Description of the learned response in discrimination behavior. PSYCHOLOGICAL REVIEW, 1950, 57, 121-131.

Nissen, H. W., Blum, J. S., & Blum, R. A. Analysis of matching beha-vior in chimpanzees. JOURNAL OF COMPARATIVE AND PHYSIOLOGICAL PSYCHOLOGY, 1948, 41, 62-74.

North, A. J. Discrimination reversal with spaced trials and distinctive cues. JOURNAL OF COMPARATIVE AND PHYSIOLOGICAL PSYCHOLOGY, 1959, 52, 426-429.

Pavlov, I. P. CONDITIONED REFLEXES. Oxford: Oxford University Press, 1927.

Platt, J. Discrete trials and their relation to free-behavior situations. In H. H. Kendler & J. T. Spence (Eds.), ESSAYS IN NEOBEHAVIORISM. New York: Appleton-Century-Crofts, 1971.

Polidora, V. J., & Fletcher, H. J. An analysis of the importance of S-R spatial contiguity for proficient primate discrimination performance. JOURNAL OF COMPARATIVE AND PHYSIOLOGICAL PSYCHOLOGY, 1964, 57, 224-230.

Premack, D. Language in a chimpanzee? SCIENCE, 1971, 172, 808-822.

Rescorla, R. A. Evidence for "unique stimulus" account of configural conditioning. JOURNAL OF COMPARATIVE AND PHYSIOLOGICAL PSYCHOLOGY, 1973, 85, 331-338.

Rescorla, R. A., & Wagner, A. R. A theory of Pavlovian conditioning. In A. H. Black & W. F. Prokasy (Eds.), CLASSICAL CONDITIONING II: CURRENT THEORY AND RESEARCH. New York: Appleton-Century-Crofts, 1972.

Restle, F. Toward a quantitative description of learning set data. PSYCHOLOGICAL REVIEW, 1958, 65, 77-91.

Riley, D. A., Ring, K., & Thomas, J. The effect of stimulus comparison on discrimination learning and transposition. JOURNAL OF COMPARATIVE AND PHYSIOLOGICAL PSYCHOLOGY, 1960, 53, 415-421.

Roberts, W. A., & Grant, D. S. Studies of short-term memory in the pigeon using the delayed matching to sample procedure. In D. L. Medin, W. A. Roberts, & R. T. Davis (Eds.), PROCESSES OF ANIMAL MEMORY. Hillsdale: Lawrence Erlbaum, 1976.

Robinson, J. S. The sameness-difference discrimination problem in chimpanzee. JOURNAL OF COMPARATIVE AND PHYSIOLOGICAL PSYCHOLOGY, 1955, 48, 195-197.

Rumbaugh, D. M., Gill, T. V., & von Glasersfeld, E. C. Reading and sentence completion by a chimpanzee (Pan). SCIENCE, 1975, 182, 731-733.

Saavedra, M. A. Pavlovian compound conditioning in the rabbit. LEARNING AND MOTIVATION, 1975, 6, 314-326.

Saldanha, E. L., & Bitterman, M. E. Relational learning in the rat. AMERICAN JOURNAL OF PSYCHOLOGY, 1951, 64, 37-53.

Schusterman, R. J. Successive discrimination-reversal training and
    multiple discrimination training in one-trial learning by
    chimpanzees. JOURNAL OF COMPARATIVE AND PHYSIOLOGICAL PSY-
    CHOLOGY, 1964, 58, 153-156.

Schwartz, B., & Williams, D. R. The role of response-reinforcer
    contingency in negative automaintenance. JOURNAL OF THE
    EXPERIMENTAL ANALYSIS OF BEHAVIOR, 1972, 17, 351-357.

Smith, L. Delayed discrimination and delayed matching in pigeons.
    JOURNAL OF THE EXPERIMENTAL ANALYSIS OF BEHAVIOR, 1967, 10,
    529-533.

Spence, K. W. The differential response in animals to stimuli
    varying within a single dimension. PSYCHOLOGICAL REVIEW,
    1937, 44, 430-444.

Spence, K. W. The basis of solution by chimpanzees of the inter-
    mediate size problem. JOURNAL OF EXPERIMENTAL PSYCHOLOGY,
    1942, 31, 257-271.

Spence, K. W. The nature of the response in discrimination learning.
    PSYCHOLOGICAL REVIEW, 1952, 59, 89-93.

Stebbins, W. C. (Ed.) ANIMAL PSYCHOPHYSICS. New York: Appleton-
    Century-Crofts, 1970.

Sutherland, N. S. & Mackintosh, N. J. MECHANISMS OF ANIMAL DIS-
    CRIMINATION LEARNING. New York: Academic Press, 1971.

Tennant, W. A., & Bitterman, M. E. Asymptotic free operant discri-
    mination reversal in goldfish. JOURNAL OF COMPARATIVE AND
    PHYSIOLOGICAL PSYCHOLOGY, 1973, 82, 130-136.

Tennant, W. A., & Bitterman, M. E. Extradimensional transfer in
    the discriminative learning of goldfish. ANIMAL LEARNING
    & BEHAVIOR, 1975, 3, 201-204.

Thompson, R. Transposition in the white rat as a function of stimu-
    lus comparison. JOURNAL OF EXPERIMENTAL PSYCHOLOGY, 1955,
    50, 185-190.

Warren, J. M. Reversal learning and the formation of learning sets
    by cats and rhesus monkeys. JOURNAL OF COMPARATIVE AND PHY-
    SIOLOGICAL PSYCHOLOGY, 1966, 61, 421-428.

Warren, J. M. Learning in vertebrates. In D. A. Dewsbury & D. A.
    Rethlingshafer (Eds.), COMPARATIVE PSYCHOLOGY. New York:
    McGraw-Hill, 1973.

Williams, B. A. Color alternation learning in the pigeon under
    fixed-ratio schedules of reinforcement. JOURNAL OF THE
    EXPERIMENTAL ANALYSIS OF BEHAVIOR, 1971, 15, 129-140. (a)

Williams, B. A. Non-spatial delayed alternation by the pigeon.
JOURNAL OF THE EXPERIMENTAL ANALYSIS OF BEHAVIOR, 1971, 16,
15-21.   (b)

Wilton, R. N., & Clements, R. O. Behavioral contrast as a function
of the duration of an immediately preceding period of extinction. JOURNAL OF THE EXPERIMENTAL ANALYSIS OF BEHAVIOR, 1971,
16, 425-428.

Wodinsky, J., & Bitterman, M. E. The solution of oddity-problems
by the rat. AMERICAN JOURNAL OF PSYCHOLOGY, 1953, 66, 137-
140.

Wodinsky, J., Varley, M. A., & Bitterman, M. E. Situational determinants of the relative difficulty of simultaneous and successive discrimination. JOURNAL OF COMPARATIVE AND PHYSIOLOGICAL
PSYCHOLOGY, 1954, 47, 337-340.

Woodard, W. T., Ballinger, J. C., & Bitterman, M. E. Autoshaping:
Further study of automaintenance. JOURNAL OF THE EXPERI-
MENTAL ANALYSIS OF BEHAVIOR, 1974, 22, 47-51.

Woodard, W. T., & Bitterman, M. E. Further studies of probability
learning in goldfish. ANIMAL LEARNING & BEHAVIOR, 1973, 1,
25-28.

Woodard, W. T., & Bitterman, M. E. A discrete-trials/fixed-interval
method of discrimination training. BEHAVIOR RESEARCH METHODS
AND INSTRUMENTATION, 1974, 6, 389-392.

Woodard, W. T., & Bitterman, M. E. Asymptotic reversal learning in
pigeons: Mechanisms for reducing inhibition. JOURNAL OF
EXPERIMENTAL PSYCHOLOGY: ANIMAL BEHAVIOR PROCESSES, 1976,
2, 57-66.

Woodard, W. T., Schoel, W. M., & Bitterman, M. E. Reversal learning
with singly presented stimuli in pigeons and goldfish.
JOURNAL OF COMPARATIVE AND PHYSIOLOGICAL PSYCHOLOGY, 1971,
76, 460-467.

Wortz, E. C., & Bitterman, M. E. On the effect of an irrelevant
relation. AMERICAN JOURNAL OF PSYCHOLOGY, 1953, 66, 491-493.

Wyckoff, L. B., Jr. The role of observing responses in discrimination learning. PSYCHOLOGICAL REVIEW, 1952, 59, 431-442.

14. ATTENTION

M. E. Bitterman

University of Hawaii
Honolulu, Hawaii U.S.A.

The simplest theory of discriminative learning is the conti-nuity or conditioning-extinction theory, whose development usually is attributed to Hull (1929) and Spence (1936) although it owes much in the first instance to Pavlov (1927). The main assumptions of the theory have already been encountered in the associative analysis of generalization gradients. Response to a complex stim-ulus is a function of its associative strength, which is in turn an algebraic function of the associative strengths of its compo-nents, interaction products being treated as a special set of components. The associative strengths of all components are in-cremented on a reinforced trial and decremented on an unreinforced trial. An important feature of the early theory is that these changes in the associative strengths of the components were assumed to occur independently; the principle of shared associative strength is, of course, of much more recent origin (Rescorla & Wagner, 1972). Differential response to stimuli, say, SX and $S'X$, produced by discriminative training is attributed to a difference between V_S and $V_{S'}$ resulting from differential reinforcement, although V_X must be considered as well if we wish to understand the absolute level of response to each stimulus. Generalization also plays an important role; reinforcement of SX increments $V_{S'}$ as well as V_S and V_X while nonreinforcement of $S'X$ decrements V_S as well as $V_{S'}$, and V_X. With more similar stimuli, differential response develops less rapidly because the difference between V_S and $V_{S'}$ develops less rapidly.

Some early observations of Lashley (1929) and Krechevsky (1932) led them to another view. The learning curve of a rat trained to

445

discriminate brightness in Lashley's jumping apparatus did not
typically show gradual improvement over its entire course, but a
long period of chance performance followed by a relatively abrupt
transition to errorless performance. During the first period--
the so-called "presolution period"--rats usually jumped to a
preferred position, which suggested to Lashley that they might be
attending to position and learning nothing about brightness. The
proper direction of attention, he proposed, is the key to the
solution of a discriminative problem. Krechevsky trained rats in
an insoluble simultaneous-choice problem with black and white
stimuli varying in position from trial to trial. Although the
choices of the animals were reinforced and nonreinforced purely
at random (for which reason the problem could be termed insoluble),
some rather systematic behavior was observed. For example, an
animal might show a strong preference for the left position (independently of brightness) over a long series of trials, then shift
to a preference for white (independently of position), then shift
back to the earlier preference or to a new one (say, for black),
and so forth. Krechevsky's idea was that such fluctuations in
performance reflect fluctuations in attention; the animal attends
first to one, then to another feature of the situation until, in
soluble problems, the relevant features are discovered.

From these observations of Lashley and Krechevsky came the
noncontinuity or attention theory of discriminative learning,
which was taken up and elaborated by others, notably in more
recent years by Sutherland and Mackintosh (1971). The principal
assumption is that the animal deals selectively with the components
of a complex stimulus, attending to some and ignoring others.
Response is determined, not by the associative strengths of all
of the components combined, but only by the strengths of the
components to which the animal is attending, and only they are
assumed to change on any trial -- a rewarded response to BL
(black on the left) increments V_B if the animal is attending to
brightness and V_L if it is attending to position. Furthermore,
attention itself is assumed to be modifiable; in mastering a discriminative problem, the animal learns to which components it
must attend as well as what they signify.

The several versions of attention theory which have been
proposed differ substantially in detail. Attention in the Lashley-
Krechevsky version is at the level of specific attributes (the
animal attends to black, white, left or right), but attention for
Sutherland and Mackintosh (1971) is dimensional (the animal attends
to brightness or position). In the language of Sutherland and
Mackintosh, animals respond to the outputs of dimensional analyzers.
As any analyzer is "strengthened" (by differential reinforcement of
response to particular outputs) or "weakened" (by nondifferential

reinforcement)--that is, as attention to the dimension is increased or decreased--there is a corresponding increase or decrease in the control exerted by all of its outputs and in the speed with which their associative strengths are altered. Lashley and Krechevsky seemed to think rats capable of attending only to one attribute of a stimulus at a time (<u>either</u> black <u>or</u> left), but the idea was shown to be incorrect. For example, while a rat being trained to discriminate brightness still is responding systematically to a preferred position, it may begin to show different latencies of response to the positive and negative brightnesses as they appear in that position (Eninger, 1953). In the Sutherland-Mackintosh theory, animals are able to deal with several dimensions of stimulation (that is, to respond in terms of the outputs of several different analyzers) concurrently. Analyzer strength is treated as a continuous variable, and analyzers compete only in that their strengths are assumed to sum to a constant value, which means that any given analyzer can gain in strength only at the expense of others. This assumption--known as the <u>inverse hypothesis</u>--is, of course, incompatible with the assumption of general attentiveness which we encountered in considering the effects of extradimensional discriminative training on generalization gradients. Mackintosh himself has suggested recently (Mackintosh, 1975a) that the inverse hypothesis be discarded. The only assumption critical to attention theory and supported by the data, he argues, is that training may change attention to a stimulus apart from its associative properties.

REDUNDANT STIMULI

The first line of evidence to be reviewed comes from discrimination experiments with stimuli differing in more than one respect. Consider, for example, the discriminative analogue of overshadowing in compound conditioning. Lovejoy and Russell (1967) trained two groups of rats in a simultaneous-choice situation, an overshadowing group with rectangles differing both in brightness and in orientation, and a control group with rectangles differing only in orientation. In subsequent tests with rectangles of the same brightness, the overshadowing group showed no differential response to orientation. Tennant and Bitterman (1975a) trained two groups of carp in a unitary (free operant) situation, an overshadowing group with lines differing both in color and in orientation, and a control group with lines differing only in orientation. Here again the overshadowing group showed no differential response to orientation when tested with lines of the same color, although in both cases, of course, enough training was given to produce discrimination of orientation in the control groups. The performance of the carp on the last day of training and in the subsequent tests is shown in Figure 1. While results of this sort often are taken as evidence against the independence assumption of the early continuity

theorists, they are not entirely convincing. The poorer test performance of the overshadowing animals may be due to the fact that
during training they make many fewer responses to the negative
stimulus than do the controls because they soon begin to discriminate between the two stimuli on the basis of the readily-distin-
guishable overshadowing features. Wagner (1969) has suggested that
classical rather than instrumental training procedures be used in
these experiments on the ground that reinforcement and nonreinforcement are effective independently of response in classical
conditioning, but the instrumental findings are interesting in
any case as a basis for explicating opposing views.

The attentional explanation of, say, the Tennant results may
rest on the assumption that carp are naturally inclined to attend
to color rather than to orientation--that what Sutherland and
Mackintosh have called the "basal strength" of the color analyzer
is greater than that of the orientation analyzer. Since color is
relevant, the overshadowing group continues to attend to color and
learns little about orientation, while the control group must shift
its attention to orientation. Another possibility is that the overshadowing group attends to color rather than to orientation, not
because of a difference in the basal strengths of the two analyzers,

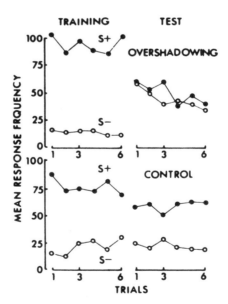

Figure 14.1   Responses of carp to S+ and S- on each trial of the
               last training day and the test day in the overshadow-
               ing experiment of Tennant and Bitterman (1975a).

but because the two colors (red and green) are much more readily distinguishable than the two orientations (30 and 60 degrees). A difference in distinguishability certainly is suggested by the training data shown in Figure 1, and there is some evidence from a unitary experiment with pigeons by Miles and Jenkins (1973) that overshadowing does indeed tend to increase with the distinguishability of the overshadowing stimuli. The reason, according to Sutherland and Mackintosh, is that the more readily the animal detects differences in the consequences (correlates) of different outputs of an analyzer, the more strongly will the analyzer be "switched in." The explanation is dangerously circular, however, since the readiness with which an animal detects differences in the consequence of different outputs of an analyzer is itself said to be determined by the strength of the analyzer. It may be reasonable in a simultaneous-choice experiment such as that of Lovejoy and Russell to suppose that attention is drawn more readily to a large difference in the overshadowing dimension than to a small one, but in unitary (successive) problems the differences are not perceived directly.

It is possible to account for overshadowing without reference to attention if the independence principle is abandoned in favor of the principle of shared associative strength (Rescorla & Wagner, 1972). For convenience, the Hull-Spence continuity theory will be called the Hull-Spence theory and the revised version the Rescorla-Wagner theory. The Rescorla-Wagner version still is a continuity theory in that it envisions only one learning process rather than two. Differences in salience which produce differences in associative strength are assumed, but salience itself is not said to be modified by learning. In the Rescorla-Wagner continuity theory, as in attention theory, there are two sources of overshadowing, one of which, of course, is salience, and the other is similarity. More salient overshadowing stimuli acquire a proportionately greater share of the available associative strength. The overshadowing of more similar stimuli by less similar stimuli can be derived from the Rescorla-Wagner model by applying it at the level of neural elements--as Blough (1975) and Rescorla (1976) have done-- on the assumption that more similar stimuli activate a greater number of common elements. It might be feasible, in fact, to dispense with the concept of salience altogether and to account for the differential conditionability of stimuli or attributes in terms of the number of equally-salient elements contributed by each.

Animals trained as in overshadowing experiments to discriminate stimuli differing in two respects often are found to learn more rapidly than animals trained with stimuli differing in either one alone. Consider, for example, some work with rats by Eninger (1952), who trained an auditory group to turn in one direction at

the choice-point of a T-maze when an auditory stimulus was presented and in the opposite direction when it was not; a visual
group to turn in one direction when the stem was black and in the
opposite direction when it was white; and an audiovisual group
with both sets of cues combined. As Figure 2 shows, the audio-
visual group learned more rapidly than either of the other two,
an outcome which can be understood readily in terms of continuity
theory. In the Hull-Spence version, the components independently
acquire associative strength which may summate to produce a
greater asymptotic difference in response to the discriminanda
as well as more rapid improvement in the redundant group, although
the asymptotic effect may be obscured by a performance ceiling.
In the Rescorla-Wagner version, the asymptotic difference in
response must be the same (by the principle of shared associative
strength), but the approach to asymptote still will be more rapid
in the redundant group.

    To explain results such as Eninger's in terms of attention,
it may be assumed that some rats are more likely at the outset
to attend to the visual components, while others are more likely
to attend to the auditory components (Sutherland & Mackintosh,
1971). In the redundant problem, animals of both kinds will be
capable of rapid acquisition since the dominant analyzer is relevant, but in either of the nonredundant problems some of the animals

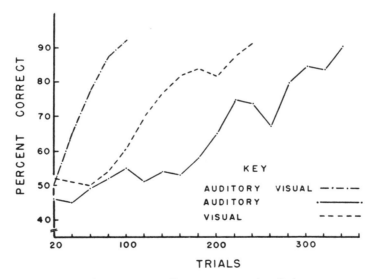

Figure 14.2   Learning curves for rats trained in a successive-
              choice problem with visual cues alone, auditory cues
              alone, or both combined. (From Eninger, 1952.)

will be at an initial disadvantage. Where one of the two analyzers
may be regarded as the stronger one for all animals, training in
the redundant problem should not, according to attention theory,
produce better performance than training with the dominant components alone, which is the result obtained by Warren (1953) in work
with monkeys. Stimuli differing in color were not discriminated
more readily than stimuli differing in color and form, color and
size, or color, form, and size, but much more readily than stimuli
differing in form, size, or form and size. To account for these
results in terms of continuity theory, it must be said that, given
the salience and distinguishability of the various attributes together with the performance ceiling, the associative strengths
acquired by form and size were too small to have a measurable
effect. None of the theories explains why form and size, separately
discriminable but much less readily so than color, did not when
combined in the absence of a difference in color produce more
rapid improvement than either alone.

   An implication both of Rescorla-Wagner continuity theory and
of attention theory, but not of Hull-Spence continuity theory, is
that when an animal is trained to discriminate between stimuli
differing in two relevant dimensions, the more it learns about
one difference, the less it will learn about the other. The
Rescorla-Wagner deduction is based, of course, on the principle
of shared associative strength. The deduction from attention
theory is based on the inverse hypothesis, which leads in fact to
the expectation that the analyzer which is the stronger at the
outset should continue with training to gain relative strength in
"runaway" fashion. Sutherland and Mackintosh were willing, however,
to consider the possibility that the competition for strength among
analyzers is not universal, but that in certain instances the
strengthening of one analyzer may even result in the strengthening
of related analyzers--a complication which does little for the
attractiveness of the theory. The available data are not at all
convincing. Some experiments with redundant two-dimensional problems do seem to show that the amount of control acquired by each
dimension varies inversely with the amount of control acquired by
the other. Sutherland and Holgate (1966) trained rats in a
simultaneous-choice situation to discriminate between rectangles
differing both in brightness and in orientation (such as black
vertical and white horizontal). In subsequent tests with rectangles
differing only in brightness or only in orientation, scores on tests
for brightness were found to be negatively correlated with scores
on tests for orientation, although the correlations were not very
large. Similar results were obtained by Warren and Warren (1969)
for monkeys, but not by Mumma and Warren (1968) for cats, nor by
Miles and Jenkins (1973) for pigeons.

Blocking is another phenomenon of compound conditioning that is found in discriminative experiments with redundant stimuli. Tennant and Bitterman (1975a) trained goldfish to discriminate successively presented 45° lines differing in color, one green and the other blue, and then to discriminate between horizontal and vertical lines of the same two colors--say, green horizontal and blue vertical. For the blocking group, the positive color of the first stage of training continued to be positive in the second and performance was very good from the outset. For the control group, the positive color of the first stage was negative in the second and the animals therefore made many errors, but the training of both groups was continued until the control animals mastered the discrimination. Then there were unreinforced test trials with white horizontal and vertical lines in which the control group performed well but the blocking group gave little evidence of horizontal-vertical discrimination--angle had been blocked by color. In a subsequent experiment of the same general design with the same animals, the blocking of color by angle also was demonstrated. The Sutherland-Mackintosh explanation of these results is that the analyzer strengthened in the first stage of training continues successfully in the second stage to control the performance of the blocking group and learning about the new features of the stimuli thus is prevented. For the control group, the analyzer strengthened in the first stage is weakened at the start of the second (when its outputs no longer successfully predict reinforcement); that provides an opportunity for other analyzers to come into play and for the animals to learn about other features of the stimuli. The results can be explained also, of course, in terms of Rescorla-Wagner continuity theory--that is, in terms of competition for associative strength rather than for attention. Nor is a Hull-Spence interpretation clearly ruled out, since blocking experiments of this kind are open to the same objection as overshadowing experiments: the blocking animals make very few unreinforced responses to the negative stimulus during training.

Mackintosh (1975a) has suggested that both overshadowing and blocking can be understood without either the inverse hypothesis or the principle of shared associative strength if the salience of a stimulus--for which he uses the Rescorla-Wagner symbol, α -- is assumed to increase when it predicts trial-outcomes better than any other stimulus and to decrease when it does not. If associative strength increases as a function of salience--as it is assumed to do in the Rescorla-Wagner model--the associative strength of the less salient element of a reinforced compound in an overshadowing experiment should increase less rapidly in the early trials, its salience should decrease correspondingly, and its associative strength should increase still less rapidly in subsequent trials.

Blocking, too, is accounted for in terms of the loss of salience, since the blocked stimulus is the poorer predictor. In discriminative training, the most excitatory component of the positive stimulus and the most inhibitory component of the negative stimulus are assumed to be the best predictors of trial-outcomes. It follows from the theory that overshadowing and blocking can never be complete because the overshadowed or blocked feature continues to change in associative value until its salience has been reduced to zero, and Mackintosh considers the possibility that to deal with cases in which overshadowed or blocked stimuli are found to exercise no control whatever it may be necessary to assume that salience is a determinant of performance (as is assumed in conventional attention theory). It also follows that there should be no overshadowing or blocking on a single compound trial--by contrast, the principle of shared associative strength permits one-trial blocking but not one-trial overshadowing, while the inverse hypothesis permits both. Mackintosh's own experiments on these questions are inconclusive. In one set of experiments (1971), he saw what seemed to be both one-trial overshadowing and one-trial blocking; in a later set (1975b), persistent effort failed to produce one-trial blocking.

The germ of yet another variant of attention theory is to be found in the work of Rescorla (1971) dealing with the effects of prior unreinforced presentation of a stimulus on subsequent conditionability. Finding it difficult to convert such a stimulus into a conditioned inhibitor as well as to give it excitatory properties by pairing it with a reinforcer, Rescorla suggested that the salience of the stimulus must have been reduced by the treatment. What we have here is an attention theory with the principle of shared associative strength (which Rescorla does not propose to abandon) substituted for the inverse hypothesis. The modification of salience contemplated both by Rescorla and by Mackintosh may seem to represent a return to the early idea of attention to attributes rather than dimensions, but Mackintosh (1975a) has suggested that change in salience may generalize over an attributive dimension in the same way as change in associative strength is assumed to do. The Mackintosh and the Rescorla versions of attention theory are to be distinguished from that of Sutherland and Mackintosh. The common implication of the three theories, but not of the continuity theories, is that experience with stimuli may change the readiness with which they can subsequently be discriminated quite apart from any changes in their response-eliciting properties that may be produced by the experience. This implication has been examined in a series of experiments to which we now turn.

ACQUIRED DISTINCTIVENESS

Pavlov (1927) found that his dogs were unable to discriminate between certain closely similar stimuli, such as tones of 2324 and 2600 Hz, if trained with them from the outset, but only if the dogs were trained first with more dissimilar stimuli and the difference was then gradually reduced. This phenomenon, which came to be known as <u>transfer along a continuum</u> and which is itself important for the theory of discriminative learning, will be taken up later. For the present, we need only consider the fact that after such training the same closely similar stimuli could be discriminated readily when the reinforcement was changed from food to acid, which meant, of course, a sharp change in the nature of the response. These results suggest that the distinctiveness of stimuli can be altered independently of the responses elicited by them.

Some formal work on the problem was done several decades later by Lawrence (1949, 1950). In the most convincing of his experiments, rats were trained in a simultaneous-choice situation to discriminate either between black and white compartments, wide and narrow compartments, or compartments with rough and smooth floors, and then trained in a successive-choice situation with two of the three pairs of stimuli, one pair relevant and the other irrelevant. For example, a rat might be trained in the successive problem to turn right when both alternative paths were white and rough or when both were white and smooth, but to turn left when both were black and rough or when both were black and smooth (here white and black were the relevant stimuli, rough and smooth irrelevant). On the assumption that the simultaneous pretraining would "emphasize" whichever stimuli were used, Lawrence predicted positive transfer when those stimuli were relevant in the second problem, but negative transfer when they were irrelevant. The reference group was one for which the stimuli used in the first problem were not used in the second. What Lawrence found was positive but not negative transfer--the group for which the stimuli of the first problem were relevant in the second did better than the other two, which did not differ. Even these results were not consistent from one set of stimuli to another; there was substantial positive transfer only when the relevant stimuli of the second problem were black and white, but none at all when they were wide and narrow. Since the wide and narrow stimuli were by far the most difficult to discriminate in the first problem, one might think that they would profit most from emphasis.

Despite substantial interest in Lawrence's findings, there has been little comparable work of any substance since. Siegel (1967) trained rats in a simultaneous-choice problem with brightness relevant and floor-texture irrelevant, or the reverse, and then trained them in a successive-choice problem with the relevant stimuli of the first problem either relevant or irrelevant in the second. Although performance in the second problem was better when the same stimuli were relevant in both, the behavior of the animals suggested that the transfer was not, in fact, response-independent (as Lawrence assumed), but related to specific orienting responses carried over from the first problem to the second. Similar observations have been made by others (Pullen & Turney, 1977). A rat trained in a simultaneous problem may not simply approach S+ (say, white), but orient first to a preferred position (say, right), proceding if white is on the right, but shifting if black is there. Such an animal is at a clear advantage in a subsequent successive problem if required to go right to white and left to black, although at a disadvantage if the opposite responses are required. It is not easy to understand, as Sutherland and Mackintosh have noted, why this sort of spatial anchoring should (in a properly balanced experiment) result in a net advantage for animals trained with the same stimuli relevant in the two problems, but it would seem better to pursue the question of acquired distinctiveness with situations much more different in their response-requirements and different even in the reinforcement employed.

If the changes in the distinctiveness of stimuli and the changes in their response-eliciting properties that are said to occur in the course of differential reinforcement are based on different processes, it is conceivable that the two changes may go on at different rates, which is precisely what Sutherland and Mackintosh have assumed in order to account for the overlearning reversal effect or ORE. Reid (1953) trained rats to discriminate black and white in a simultaneous-choice situation, one group to the criterion of nine out of 10 correct choices, a second group for 50 additional trials, and a third group for 150 additional trials. (Training beyond a criterion is called overtraining, and the learning which takes place during such training is called overlearning.) When the animals then were required to reverse the discrimination, Reid found a facilitating effect of overtraining-- the longer the animals had been trained to choose one brightness in the first problem, the more readily they learned to choose the opposite brightness in the second. The Sutherland-Mackintosh explanation of this paradoxical phenomenon is that analyzers grow less rapidly in strength than do their "response attachments"-- the connections of their outputs to responses. (In terms of Mackintosh's or Rescorla's attention theory, salience may be said to change more slowly than V.) If, in an experiment such as Reid's, the strength of the response attachments is asymptotic at criterion while that of the brightness analyzer is not, further

training will not increase the strength of the response attachments
to be reversed, but it will strengthen the analyzer and so facilitate reversal, since the rate of change in the strength of response
attachments is assumed to increase with the strength of the analyzer.

As Sutherland and Mackintosh have noted, the ORE is a rather
elusive phenomenon, failing to occur more often than it occurs in
the great variety of experiments designed to study it. Attention
theory does not, however, require the ORE under all conditions.
If, for example, the relevant analyzer is already strong at the
outset of training, it may reach asymptotic strength before the
attachments do. According to the theory, the ORE is more likely
to appear when there is a salient irrelevant dimension than when
there is not. The choice of criterion also is important; it must
not be so high that analyzer strength is already at asymptote when
reversal training begins. It is conceivable, furthermore, that
the difference in the rates at which analyzers and response attachments are strengthened may be greater under some conditions than
others. To account for the fact that the ORE is more likely to
appear when large rather than small rewards are used, Sutherland
and Mackintosh have suggested that increasing the size of the
reward facilitates the strengthening of attachments more than that
of analyzers. It cannot be said, however, that Sutherland and
Mackintosh have succeeded very well in accounting for the tremen-
dous inconsistency in the data. There simply are too many experiments, such as those of Lukaszewska (1968), which fail to show the
ORE under conditions that seem to meet all the requirements of the
theory.

Other explanations of the ORE have been proposed. One is
that the ORE is no more than the overlearning-extinction effect--
which, as we have seen earlier, also is more likely to occur with
large reward than with small--and has nothing to do with attention
(Birch, Ison & Sperling, 1960). Another is that the aversiveness
of S- declines during overtraining because the animals have little
unreinforced experience with it (D'Amato & Jagoda, 1961). There
is some evidence, however, that rats trained in simultaneous-choice
situations learn first about S+ and continue to learn about S- during overtraining (Hall, 1973; Mandler & Goldberg, 1975). In reject-
ing both alternatives, Sutherland and Mackintosh noted that overtraining often facilitates despite a more persistent tendency on
the part of the overtrained animals to choose the former S+ early
in reversal training. Shown in Figure 3 are the reversal curves
of two rats trained to discriminate black and white, one to criterion and the other beyond, in a simultaneous-choice situation.
Presumably typical, they were presented by Sutherland and Mackintosh

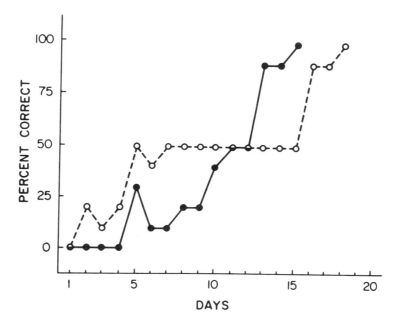

Figure 14.3 Reversal learning curves of a rat trained to criterion
            (open circles) and another which had been overtrained
            (filled circles).   (From Sutherland & Mackintosh,
            1971.)

to show that the benefits of overtraining may be evidenced only later in reversal training. After giving up its preference for the previously positive brightness, the criterion animal began to show a strong position preference (which produced a string of 50% points on its curve), while the overtrained animal did not. The Sutherland-Mackintosh explanation is that overtraining strengthens the relevant analyzer and so decreases the probability that control will be shifted to an irrelevant analyzer (such as the position analyzer) during reversal. It is to be regretted that there is not very much detailed information in the literature on the course of reversal learning, critical features of individual performance being easily obscured in averaged learning curves.

In some experiments, overtraining in one discriminative problem has been found to improve performance in an unrelated problem. Mandler (1968) reported, for example, that overtraining in a black-white discrimination improved the performance of rats in a subsequent horizontal-vertical discrimination. The implication of such results is that the ORE may be nothing more than a general practice effect. It should be noted, however, that Mackintosh (1962) found precisely the opposite results in another situation; overtraining in a black-white problem facilitated reversal but impaired performance in a subsequent horizontal-vertical problem. An interesting way in which to control for amount of experience in the training situation is to train animals in two problems concurrently, giving more trials with one than with the other, and then to reverse one or the other. In an experiment of this design with rats trained in an automated simultaneous-choice situation, Coate and Gardner (1965) found that speed of reversal increased as the amount of experience in the situation increased, but decreased as the amount of experience with the stimuli of the reversal problem increased. This work, unfortunately, is only of methodological interest. Since the same dimension of stimulation (angular orientation of lines) was used in both problems, the effect of training in both may have been to increase attention to orientation.

Sutherland and Mackintosh have proposed also that improvement in serial reversal learning is due in part to increased attention to the relevant stimuli. While the relevant analyzer (which, of course, remains relevant over the entire series) will be weakened to some extent in the early trials of each reversal as a result of incorrect "predictions" about the consequences of response, Sutherland and Mackintosh explain, it will soon begin to be strengthened again and may well be stronger at the end of the reversal than at the start. Irrelevant analyzers continue at the same time to be weakened. Evidence against the attentional interpretation comes from an experiment by Gonzalez and Bitterman (1968), who trained pigeons in a simultaneous-choice situation with red and green keys. Color was relevant and position irrelevant on odd-numbered days (half with red positive and half with green, in Gellermann order), while position was relevant and color irrelevant on even-numbered days (half with right positive and half with left). Although differential reinforcement in one dimension always involved non-differential reinforcement in the other, the animals improved in both dimensions concurrently, and their asymptotic performance in the color problem did not differ from that of a second group of pigeons which simply rested on even-numbered days. A third group, nondifferentially reinforced with respect both to color and position on even-numbered days, did show impairment of performance on color days, providing another example of the deleterious effect of pseudo-discrimination training which has been attributed to a decline in general attentiveness. To account for concurrent improvement in color and position reversals in attentional terms, it is not enough, of course, simply to discard the inverse hypothesis, since differential reinforcement in each dimension was accompanied by nondifferential reinforcement in the other.

The pigeons of the Gonzalez experiment continued in each color problem to show a preference for the positive color of the preceding color problem despite interpolated position training in which, of course, color was nondifferentially reinforced. The continuity explanation is that not enough nondifferential reinforcement was given to eliminate the preference entirely. The Sutherland-Mackintosh explanation is that the preference was protected by a shift in attention. A more troublesome finding is that the preference was undiminished by the interpolated position training, being no less in the group which was trained to position on even-numbered days than in the group which was rested. It was as though attention shifted so rapidly from color to position at the outset of each position problem that there was no time at all for change in the associative properties of the colors, and the problem for attention theory is to explain this precipitous shift. Continuity theory provides no explanation. The pigeon results are reminiscent of some earlier results for rats in a simultaneous-choice situation (Lawrence & Mason, 1955; Goodwin & Lawrence, 1955). In the first stage of training, brightness was relevant and another dimension irrelevant. In the second stage, there was a dimensional shift--the previously irrelevant dimension became relevant and brightness irrelevant. In the third stage, with brightness relevant once more, the animals showed a clear preference for the positive brightness of the first stage that was undiminished by overtraining in the second stage. These results are troublesome for continuity theory but not for attention theory, since they do not imply that the brightness preference was entirely unaffected by nondifferential reinforcement.

As Sutherland and Mackintosh have noted, there is very little evidence from discrimination experiments that nondifferentially reinforced experience with stimuli makes them difficult to discriminate--evidence, that is, of acquired <u>nondistinctiveness</u>. Lawrence's approach to the question has already been described. Another approach is illustrated by the work of Waller (1971), who trained rats in a simultaneous-choice situation to discriminate brightness (black and white) either with or without an irrelevant difference in texture (rough and smooth); for animals trained without the difference in texture, an intermediate value was used. Then the difference in brightness was eliminated (all stimuli were gray), and the animals were trained to discriminate rough and smooth. The irrelevant texture difference in the first stage was found to impair performance both in the brightness problem of the first stage and in the texture problem of the second. The results for the first stage suggest competition for attention, but the variation in texture may have impaired performance simply by increasing the number of different stimuli with which the animals were required to deal. The results for the second stage cannot be attributed to reduced attention to texture without a control group exposed in the brightness discrimination to irrelevant variation in some different dimension. Other experiments suggest that discriminability may be <u>heightened</u> by nondifferential reinforcement. Bitterman and Elam (1954) trained several groups of rats in the Lashley jumping apparatus either with a pair of gray cards or with a pair of horizontally and vertically striped cards, rewarding response to both members of the pair, and then trained all groups to discriminate the horizontal and vertical stripes. The difficulty of discrimination increases as the amount of previous nondifferential reinforcement increased, whether with the grays or with the stripes, which points to a nonspecific retarding effect of nondifferential reinforcement; but the performance of the groups pretrained with stripes was substantially better than that of groups pretrained with grays, which points to a specific facilitating effect of experience with the stripes. Some further (and better balanced) experiments of this sort might be instructive.

DIMENSIONAL EFFECTS

The experiments to be considered here differ from those on acquired distinctiveness in that they look for effects of differentially and nondifferentially reinforced experience with stimuli, not on the discriminability of those stimuli, but on the discriminability of novel stimuli varying in the same dimension. Work with novel stimuli may make it easier to distinguish changes in discriminability from changes in associative properties, although stimulus generalization is always, of course, to be reckoned with. In Mackintosh's (and presumably also in Rescorla's) attention theory, it is generalization that makes dimensional transfer of acquired distinctiveness possible, since changes in salience are assumed to generalize in the manner of changes in V.

A dimensional effect in which there has been a considerable amount of interest is Pavlov's transfer along a continuum. The name for the phenomenon comes from Lawrence (1952), who replicated it with rats in a simultaneous-choice situation. The discriminanda were a series of gray cards varying in brightness, A, B, C, ... G, selected pairs of which were used, with the lighter always correct for half the animals and the darker for the rest. A "hard discrimination" group (HDG) was trained for 80 trials with C and D. An "abrupt transition" group (ATG no. 1) had 30 trials with A and G followed by 50 trials with C and D. A "gradual transition" group (GTG) had 10 trials with A and G, 10 with B and F, and 10 with C and E, followed by 50 trials with C and D. Another "abrupt transition" group (ATG no. 2) had 50 trials with C and E followed by 30 trials with C and D. The learning curves reproduced in Figure 4 show that performance in the final trials was better in animals pretrained with less similar stimuli than in the animals trained from the outset with C and D. The attentional interpretation is that training with less similar stimuli helps to focus the animal's attention on the relevant features of the task. In the language of Sutherland and Mackintosh, the strength of the brightness analyzer grows more rapidly during training with less similar stimuli and output attachments therefore develop more rapidly in subsequent training with more similar stimuli. It may be possible, however, to account for the results without reference to attention.

One alternative explanation is in terms of stimulus generalization. Assume gradients of excitation and inhibition around the positive and negative points on the continuum like those shown in Figure 13 of the chapter on generalization. It follows then that training at more widely separated points, say, reinforcement at a point two units to the left of S+ in Figure 13 and nonreinforcement at a point two units to the right of S-, will result in a greater difference in the (E-I) values of S+ and S- than will training at S+ and S-. Specifically, (E-I) will be greater than 5 at S+ and less than -5 at S-; the derivation is easily made by eye or with the help of a piece of tracing paper. Some evidence in support of this interpretation has been reported by Sweller (1972), who trained rats in a simultaneous-choice situation, an "Easy" group with black and white stimuli, and a "Hard" group with light and dark grays; each animal in the Hard group was trained as long as was required for a paired Easy animal to master the problem. Then both groups were trained to criterion with the grays, light gray positive if

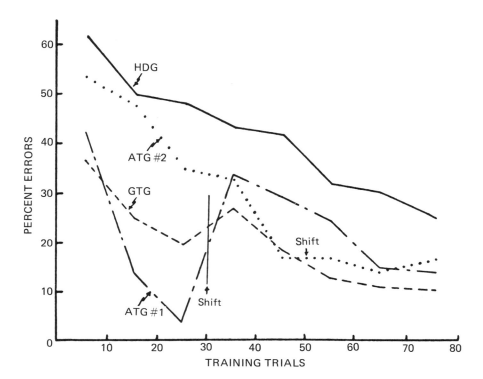

Figure 14.4   Learning curves of rats trained to discriminate a
               pair of closely similar grays either from the outset
               (HDG) or after experience with more dissimilar grays.
               (From Lawrence, 1952.)

<u>dark</u> gray or black had been positive in the previous problem, and
<u>dark</u> gray positive if <u>light</u> gray or white had been positive. At
the outset of the second problem, the Easy animals averaged only
1.67 correct choices in 10 trials as compared with 4.33 for the
Hard animals, which is just what the gradient analysis predicts--
the Easy animals make more errors because they have a stronger
(generalized) preference for the gray which is now negative. Even
so, however, the Easy group improved more rapidly than the Hard
and reached criterion sooner. These results suggest that stimulus
generalization plays an important role in transfer along a continuum
but does not provide a complete account of it.

Further support for the generalization interpretation has been
provided by Turney (1976), who trained rats with 0° and 90° lines
for six days (by which time they had made substantial progress) and
then shifted them to a discrimination between 30° and 60° lines.
If the new problem was "compatible" from the viewpoint of generalization (that is, if 30° was positive for animals previously trained
to 0°, and 60° was positive for animals previously trained to 90°),
the animals made uninterrupted progress to the criterion, performing
much better than control animals trained with the more similar stimuli from the outset. If the new problem was "incompatible," performance was markedly impaired, as in Sweller's experiment, but here
remained substantially inferior to that of the control group. In
another experiment, Turney trained rats with 0° and 90° lines, again
for six days, and then shifted them to -60° and +60° lines, a more
difficult problem in which there could be no differential effect of
generalization. With the shift, performance fell sharply to the
level of control animals trained from the outset on the more difficult problem, and both groups then progressed at the same rate. With
line-orientation as the relevant dimension in both problems, attention theory might have led us to expect better performance in the
easy-to-hard group; since there was considerable transfer to 30°
and 60° from training with vertical and horizontal, it cannot be
argued that there is a different analyzer for diagonal lines.

The possibility also has been suggested that whatever "transfer
along a continuum" cannot be accounted for in terms of stimulus
generalization is not transfer along a continuum at all--that is,
not dimensionally specific. Evidence that training in an easy
problem may facilitate performance in a more difficult problem
involving another dimension of stimulation has been provided by
Tennant and Bitterman (1975b). A group of goldfish was trained in
a free-operant situation with a red target to discriminate between
tones of 400 and 600 Hz. Another group, which was trained first to
discriminate between blue and green targets (a much easier problem),
outperformed the control group when shifted to the auditory problem.

An explanation suggested by the inverse hypothesis is that irrelevant analyzers are weakened more when the problem is easy than when
it is difficult because the relevant analyzer is stronger, and the
principle of shared associative strength also suggests that there
will be less control by common stimuli. Some relevant data come
from a classical eyelid conditioning experiment with rabbits by
Haberlandt (1971), who used compound stimuli made up of a common
vibratory component, a common visual component, and one of six tones
differing in frequency (T_1, T_2, ... T_6). A group trained from the
outset with T_3 and T_4 did not do as well as one trained first with
T_1 and T_6 and then with T_2 and T_5 before being shifted to T_3 and T_4.
The difference in performance could be traced to a lower probability
of response to the negative compound on the part of the easy-to-hard
group, which also responded less to the common vibratory and visual
components of the compounds when the intensity of the tones was
decreased to the point at which they were indistinguishable. It
should be clear, of course, from the effects of extradimensional
discriminative training on generalization that neither the inverse
hypothesis nor the principle of shared associative strength provides
a complete account of nonspecific transfer. It may be well to note
also that there are some other experiments on transfer along a continuum which, like Turney's, give no evidence of nonspecific facilitation. Working with pigeons, Marsh (1969) found that the discrimination between 520 and 530 nm was facilitated by pretraining with
less similar wavelengths, but not by pretraining with the same wavelengths when the problem was made easier by the addition of a difference in brightness.

The most convincing evidence of selective attention comes from
experiments on <u>intradimensional versus extradimensional transfer</u>
(or ID versus ED transfer). Suppose that animals are trained to
discriminate red and yellow horizontal and vertical lines, with
color relevant and orientation irrelevant for half the group (<u>e.g.</u>
red horizontal and red vertical positive, yellow horizontal and
yellow vertical negative), but with orientation relevant and color
irrelevant for the rest. Then the animals are trained with green
and blue diagonals, with color relevant for half and orientation
relevant for half (in a 2 X 2 design). By attention theory, performance in the second problem should be better in animals for which the
relevant dimension of both problems is the same (color-color or
orientation-orientation) than in animals for which the relevant
dimension of one problem is the irrelevant dimension of the other
(color-orientation or orientation-color)--that is, ID transfer
should be better than ED transfer. If, for example, training with
color relevant and orientation irrelevant in the first problem
strengthens attention to color and weakens attention to orientation,
the second problem should be mastered more quickly when color is

relevant again than when orientation is relevant. Care must be
taken, of course, in the choice of stimuli to rule out the possibility that ID transfer is based at least in part on stimulus
generalization. In this case, red and yellow must be so different
from green and blue for the animals that training with red or with
yellow positive does not produce a significant preference for green
or for blue. Even so, the choice of positive stimuli is balanced
over subjects so that any slight affinities will hinder as often as
help.

An experiment of this design by Shepp and Eimas (1964) was
done with rats in the Wisconsin apparatus, the stimulus objects
varying in form and in a superimposed pattern of stripes. The
curves reproduced in Figure 5 show that discrimination of form in
the second problem was better when form rather than stripes had
been relevant in the first, and that discrimination of stripes was
better when stripes rather than form had been relevant in the first.
Better ID than ED transfer has been found also with monkeys trained
in the Wisconsin apparatus with stimuli differing in color and form
(Shepp & Schrier, 1969). The picture is complicated somewhat, however, by the fact that experiments with pigeons have produced
ambiguous results (Mackintosh & Little, 1969; Couvillon, Tennant,
& Bitterman, 1976) and with goldfish results that are entirely
negative (Tennant & Bitterman, 1973). It should be noted that the
negative results for goldfish were obtained in a unitary (free
operant) situation, while the positive results for other animals
were obtained in simultaneous-choice situations, but attention
theory predicts better ID than ED transfer in unitary situations
as well. The picture is complicated further by the results of
Turrisi, Shepp, and Eimas (1969), who repeated the rat experiment
with no variation in the alternative dimension during the first
problem (the animals discriminated different forms with the same
stripes or different stripes on the same form). In the second
problem, with variation in both dimensions, it did not matter
whether or not the relevant dimension was the same as before.
These results must suggest to attention theorists that those of
the first experiment were due to reduction in attention to the
irrelevant dimension, and that nondifferentially reinforced experience with at least two points on a dimension is necessary to reduce
attention to the dimension. It should be clear in any case that
none of the experiments requires the assumption that attention to
a dimension is strengthened by the differential reinforcement of
stimuli varying in that dimension.

There are, in fact, some experiments on dimensional shifting
which seem directly to contradict the idea that animals deal with
multidimensional problems by abstracting relevant dimensions of

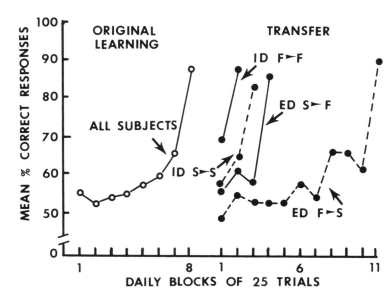

Figure 14.5   Intradimensional as compared with extradimensional
              transfer in rats trained with stimulus objects vary-
              ing in form (F) and pattern of stripes (S). These
              are "backward" learning curves, averaged not for each
              calendar day but for days relative to that on which
              any given animal reached the criterion; the technique
              is thought to provide a better picture of performance
              in the presolution period. (From Shepp & Eimas, 1964.)

stimulation. Consider, for example, a goldfish trained with red and yellow horizontal and vertical lines, red positive and orientation irrelevant, which then is shifted to vertical positive and color irrelevant. The shift does not disrupt responding to all four stimuli as attention theory might lead us to expect. Instead, the animal continues to respond at a high level to red vertical (which remains positive) and at a low level to yellow horizontal (which remains negative) while gradually increasing its level of response to yellow vertical and reducing it to red horizontal, which suggests that the animal is dealing with the colored lines, not dimensionally, but individually, as stimulus compounds. Results of this sort have been obtained with rats, turtles, and pigeons in simultaneous-choice situations (Graf & Tighe, 1971; Tighe & Frey, 1972; Tighe & Graf, 1972) as well as with goldfish in a unitary (free operant) situation (Tennant & Bitterman, 1973).

In all, discrimination experiments have yielded very little evidence that the salience of stimuli or the readiness with which their associative properties change may itself be modified by training. We can say only that the experiments have produced a few interesting phenomena which seem to point beyond continuity theory and deserve further study.

REFERENCES

Birch, D., Ison, J. R., & Sperling, S. E. Reversal learning under single stimulus presentation. JOURNAL OF EXPERIMENTAL PSYCHOLOGY, 1960, 60, 36-40.

Bitterman, M. E., & Elam, C. B. Discrimination following varying amounts of nondifferential reinforcement. AMERICAN JOURNAL OF PSYCHOLOGY, 1954, 67, 133-137.

Blough, D. S. Steady-state data and a quantitative model of operant generalization and discrimination. JOURNAL OF EXPERIMENTAL PSYCHOLOGY: ANIMAL BEHAVIOR PROCESSES, 1975, 1, 3-21.

Coate, W. B., & Gardner, R. A. Sources of transfer from original training to discrimination reversal. JOURNAL OF EXPERIMENTAL PSYCHOLOGY, 1965, 70, 94-97.

Couvillon, P. C., Tennant, W. A., & Bitterman, M. E. Interdimensional versus extradimensional transfer in the discriminative learning of goldfish and pigeons. ANIMAL LEARNING & BEHAVIOR, 1976, 4, 197-203.

D'Amato, M. R., & Jagoda, H. Analysis of the role of overlearning in discrimination reversal. JOURNAL OF EXPERIMENTAL PSYCHOLOGY, 1961, 61, 45-50.

Eninger, M. U. Habit summation in selective learning. JOURNAL OF COMPARATIVE AND PHYSIOLOGICAL PSYCHOLOGY, 1952, 45, 604-608.

Eninger, M. U. The role of generalized approach and avoidance tendencies in brightness discrimination. JOURNAL OF COMPARATIVE AND PHYSIOLOGICAL PSYCHOLOGY, 1953, 46, 398-402.

Gonzalez, R. C., & Bitterman, M. E. Two-dimensional discriminative learning in the pigeon. JOURNAL OF COMPARATIVE AND PHYSIOLOGICAL PSYCHOLOGY, 1968, 65, 427-432.

Goodwin, W. R., & Lawrence, D. H. The functional independence of two discrimination habits associated with a constant stimulus situation. JOURNAL OF COMPARATIVE AND PHYSIOLOGICAL PSYCHOLOGY, 1955, 48, 437-443.

Graf, V., & Tighe, T. J. Subproblem analysis of discrimination shift learning in the turtle (*Chrysemys picta picta*). PSYCHONOMIC SCIENCE, 1971, 25, 257-259.

Haberlandt, K. Transfer along a continuum in classical conditioning. LEARNING AND MOTIVATION, 1971, 2, 164-172.

Hall, G. Response strategies after overtraining in the jumping stand. ANIMAL LEARNING & BEHAVIOR, 1973, 1, 157-160.

Hull, C. L. A functional interpretation of the conditioned reflex. PSYCHOLOGICAL REVIEW, 1929, 36, 495-511.

Krechevsky, I. Hypotheses in rats. PSYCHOLOGICAL REVIEW, 1932, 39, 516-532.

Lashley, K. S. BRAIN MECHANISMS AND INTELLIGENCE. Chicago: University of Chicago Press, 1929.

Lawrence, D. H. Acquired distinctiveness of cues: I. Transfer between discriminations on the basis of familiarity with the stimulus. JOURNAL OF EXPERIMENTAL PSYCHOLOGY, 1949, 39, 770-784.

Lawrence, D. H. Acquired distinctiveness of cues: II. Selective association in a constant stimulus situation. JOURNAL OF EXPERIMENTAL PSYCHOLOGY, 1950, 40, 175-188.

Lawrence, D. H. The transfer of a discrimination along a continuum. JOURNAL OF COMPARATIVE AND PHYSIOLOGICAL PSYCHOLOGY, 1952, 45, 511-516.

Lawrence, D. H., & Mason, W. A. Systematic behavior during discrimination reversal and change of dimension. JOURNAL OF COMPARATIVE AND PHYSIOLOGICAL PSYCHOLOGY, 1955, 48, 1-7.

Lovejoy, E., & Russell, D. G. Suppression of learning about a hard cue by the presence of an easy cue. PSYCHONOMIC SCIENCE, 1967, 8, 365-366.

Lukaszewska, I. Some further failures to find the visual over-learning reversal effect in rats. JOURNAL OF COMPARATIVE AND PHYSIOLOGICAL PSYCHOLOGY, 1968, 65, 359-361.

Mackintosh, N. J. The effect of overtraining on a reversal and a nonreversal shift. JOURNAL OF COMPARATIVE AND PHYSIOLOGICAL PSYCHOLOGY, 1962, 55, 555-559.

Mackintosh, N. J. An analysis of overshadowing and blocking. QUARTERLY JOURNAL OF EXPERIMENTAL PSYCHOLOGY, 1971, 23, 118-125.

Mackintosh, N. J. A theory of attention: Variations in the associability of stimuli with reinforcement. PSYCHOLOGICAL REVIEW, 1975, 82, 276-298. (a)

Mackintosh, N. J. Blocking of conditioned suppression: Role of the first compound trial. JOURNAL OF EXPERIMENTAL PSYCHOLOGY: ANIMAL BEHAVIOR PROCESSES, 1975, 1, 335-345. (b)

Mackintosh, N. J., & Little, L. Intradimensional and extradimensional shift learning by pigeons. PSYCHONOMIC SCIENCE, 1969, 14, 5-6.

Mandler, J. M. The effect of overtraining on the use of positive and negative stimuli in reversal and transfer. JOURNAL OF COMPARATIVE AND PHYSIOLOGICAL PSYCHOLOGY, 1968, 66, 110-115.

Mandler, J. M., & Goldberg, J. Changes in response to S+ and S- during acquisition and overtraining of simultaneous discriminations in rats. ANIMAL LEARNING & BEHAVIOR, 1975, 3, 226-234.

Marsh, G. An evaluation of three explanations for the transfer of discrimination effect. JOURNAL OF COMPARATIVE AND PHYSIOLOGICAL PSYCHOLOGY, 1969, 68, 268-275.

Miles, R. C., & Jenkins, H. M. Overshadowing in operant conditioning as a function of discriminability. LEARNING AND MOTIVATION, 1973, 4, 11-27.

Mumma, R., & Warren, J. M. Two-cue discrimination learning by cats. JOURNAL OF COMPARATIVE AND PHYSIOLOGICAL PSYCHOLOGY, 1968, 66, 116-122.

Pavlov, I. P. CONDITIONED REFLEXES. Oxford: Oxford University Press, 1927.

Pullen, M. R., & Turney, T. H. Response modes in simultaneous and successive visual discriminations. ANIMAL LEARNING & BEHAVIOR, 1977, 5, 73-77.

Reid, L. S. The development of noncontinuity behavior through continuity learning. JOURNAL OF EXPERIMENTAL PSYCHOLOGY, 1953, 46, 107-112.

Rescorla, R. A.   Summation and retardation tests of latent inhibition. JOURNAL OF COMPARATIVE AND PHYSIOLOGICAL PSYCHOLOGY, 1971, 75, 77-81.

Rescorla, R. A.   Stimulus generalization: Some predictions from a model of Pavlovian conditioning. JOURNAL OF EXPERIMENTAL PSYCHOLOGY: ANIMAL BEHAVIOR PROCESSES, 1976, 2, 88-96.

Rescorla, R. A., & Wagner, A. R.   A theory of Pavlovian conditioning. In A. H. Black & W. F. Prokasy (Eds.), CLASSICAL CONDITIONING II: CURRENT THEORY AND RESEARCH. New York: Appleton-Century-Crofts, 1972.

Shepp, B. E., & Eimas, P. D.   Intradimensional and extradimensional shifts in the rat. JOURNAL OF COMPARATIVE AND PHYSIOLOGICAL PSYCHOLOGY, 1964, 57, 357-361.

Shepp, B. E., & Schrier, A. M.   Consecutive intradimensional and extradimensional shifts in monkeys. JOURNAL OF COMPARATIVE AND PHYSIOLOGICAL PSYCHOLOGY, 1969, 67, 199-203.

Siegel, S.   Overtraining and transfer processes. JOURNAL OF COMPARATIVE AND PHYSIOLOGICAL PSYCHOLOGY, 1967, 64, 471-477.

Spence, K. W.   The nature of discrimination learning in animals. PSYCHOLOGICAL REVIEW, 1936, 43, 427-449.

Sutherland, N. S., & Holgate, V.   Two-cue discrimination learning in rats. JOURNAL OF COMPARATIVE AND PHYSIOLOGICAL PSYCHOLOGY, 1966, 61, 198-207.

Sutherland, N. S., & Mackintosh, N. J.   MECHANISMS OF ANIMAL DISCRIMINATION LEARNING. New York: Academic Press, 1971.

Sweller, J.   A test between selective attention and stimulus generalization interpretations of the easy-to-hard effect. QUARTERLY JOURNAL OF EXPERIMENTAL PSYCHOLOGY, 1972, 24, 252-355.

Tennant, W. A., & Bitterman, M. E.   Some comparisons of intra- and extradimensional transfer in discriminative learning of goldfish. JOURNAL OF COMPARATIVE AND PHYSIOLOGICAL PSYCHOLOGY, 1973, 83, 134-139.

Tennant, W. A., & Bitterman, M. E.   Blocking and overshadowing in two species of fish. JOURNAL OF EXPERIMENTAL PSYCHOLOGY: ANIMAL BEHAVIOR PROCESSES, 1975, 1, 22-29.   (a)

Tennant, W. A., & Bitterman, M. E.   Extradimensional transfer in the discriminative learning of goldfish. ANIMAL LEARNING & BEHAVIOR, 1975, 3, 201-204.   (b)

Tighe, T. J., & Frey, K.   Subproblem analysis of discrimination shift learning in the rat. PSYCHONOMIC SCIENCE, 1972, 28, 129-133.

Tighe, T. J., & Graf, V. Subproblem analysis of discrimination shift learning in the pigeon. PSYCHONOMIC SCIENCE, 1972, 29, 139-141.

Turney, T. H. The easy-to-hard effect: Transfer along the dimension of orientation in the rat. ANIMAL LEARNING & BEHAVIOR, 1976, 4, 363-366.

Turrisi, F. D., Shepp, B. E., & Eimas, P. D. Intra- and extra-dimensional shifts with constant- and variable-irrelevant dimensions in the rat. PSYCHONOMIC SCIENCE, 1969, 14, 19-20.

Wagner, A. R. Incidental stimuli and discrimination learning. In R. M. Gilbert & N. S. Sutherland (Eds.), ANIMAL DISCRIMI-NATION LEARNING. London: Academic Press, 1969.

Waller, T. G. Effect of irrelevant cues on discrimination acquisition and transfer in rats. JOURNAL OF COMPARATIVE AND PHYSIOLOGICAL PSYCHOLOGY, 1971, 73, 477-480.

Warren, J. M. Additivity of cues in visual pattern discriminations by monkeys. JOURNAL OF COMPARATIVE AND PHYSIOLOGICAL PSY-CHOLOGY, 1953, 46, 484-486.

Warren, J. M., & Warren, H. B. Two-cue discrimination learning by rhesus monkeys. JOURNAL OF COMPARATIVE AND PHYSIOLOGICAL PSYCHOLOGY, 1969, 69, 688-691.

15. CONSTRAINTS ON LEARNING

15. CONSTRAINTS ON LEARNING

Vincent M. LoLordo

Dalhousie University
Halifax, Nova Scotia, Canada

Research on learning has concentrated on the search for general laws which transcend the specific choices of stimuli, responses, and reinforcers made by experimenters. Researchers have generally assumed that they could arbitrarily select the elements to be used in their classical conditioning and instrumental training experiments without markedly affecting the success of the experiments.

Although there have always been researchers who have taken exception to this view (e.g., Konorski, 1948; Mowrer, 1947; Turner & Solomon, 1962), only recently has a variety of data from diverse experimental settings forced most investigators to conclude that the choice of stimulus, response, and reinforcer is not a matter of indifference but may be one of the most critical determinants of the success of an experiment on learning (see reviews by Bolles, 1970, 1972; Garcia, McGowan, & Green, 1972; Hinde & Stevenson-Hinde, 1973; LoLordo, 1978; Rozin & Kalat, 1971; Segal, 1972; Seligman, 1970; Seligman & Hager, 1972; Shettleworth, 1972, 1978a).

Shettleworth (1972) characterized those experiments whose success was determined by the specific choice of stimulus, response, reinforcer, or combination of these elements as examples of constraints on learning or performance. She proposed a scheme for classifying such experiments in terms of the source of the constraint, in a strictly operational sense. According to this scheme, a constraint on learning or performance might result from the experimenter's choice of: (a) the stimulus, as in the case of an octopus that did not learn a "go-no go" discrimination based upon the weights of objects, although it did adjust its tentacles appropriately to

473

support weights lowered into its tank (Wells, 1964); (b) the response,
as in the case of Konorski's (1967) cats which could not be instrumentally trained to yawn or sneeze; (c) the reinforcer, e.g., in
the demonstration that two positive reinforcers, food and the opportunity for aggressive display, have very different effects upon instrumental responses of the male Siamese fighting fish (e.g., Hogan,
1967; Hogan, Kleist & Hutchings, 1970); (d) the combination of
stimulus and reinforcer, as in Garcia and Koelling's (1966) demonstration that rats are loathe to experience a taste that has been
paired with illness or an auditory-visual stimulus that has been
paired with painful shock but readily produce a taste that has been
paired with electric shock or an auditory-visual stimulus that has
been paired with illness; (e) the combination of stimulus and response, e.g., in the greater effectiveness of directional rather
than qualitative cues in left-right discriminations coupled with
the greater effectiveness of qualitative cues in "go-no go" discriminations (Dobrzecka & Konorski, 1967, 1968; Dobrzecka,
Szwejkowska, & Konorski, 1966; Lawicka, 1964, 1969; Szwejkowska,
1967a,b); or (f) the combination of response and reinforcer, e.g.,
in Sevenster's (1973) demonstration that male stickleback fish were
equally reinforced for swimming through a ring by a brief opportunity to court a female or a brief opportunity to fight another
male, whereas only the latter effectively reinforced the instrumental response of biting a rod.

Learning theorists have found the three classes of constraints
on the effectiveness of conditioning and training procedures that
depend on some combination of elements especially provocative because they suggest that associations between certain antecedent and
consequent events within a set may be formed very easily, whereas
associations between other antecedents and consequents within the
same set may not be formed at all, or only with great difficulty.
Perhaps this suggestion that there are selective associations (see
reviews cited earlier) will be clearer if it is contrasted with a
well-known model of associative learning, the Rescorla and Wagner
(1972) model. The latter asserts that two separate learning rate
parameters, α and β, characterize the contributions of the CS and
UCS, respectively, to the rate of growth of the association between
them. In contrast, a selective association would be said to occur
whenever the rate of growth of the association between a CS and a
UCS can only be characterized by a single learning rate parameter
which cannot be reduced to separate parameters for CS and UCS.
Several researchers who have asserted that some associations are
selective have further suggested that the laws of learning (i.e.,
of association formation) may be different for selective associations and other associations (e.g., Rozin & Kalat, 1971; Seligman,
1970). This assertion implies that variations in at least some of

the conditions known to affect associative learning will have very different effects on selective associations and non-selective ones.

An experimental analysis must be performed to determine whether each instance of a constraint on observed performance does reflect a selective association rather than the operation of some selective but non-associative factor. In this chapter, constraints arising from combinations of conditioned or discriminative stimuli and reinforcers and from combinations of responses and reinforcers will be analyzed. The distinction between selective, associative effects and selective but non-associative effects will be emphasized. Finally, the implications of selective associations for the generality of the laws of learning will be discussed.

EXPERIMENTAL DESIGNS WHICH DEMONSTRATE CONSTRAINTS

Constraints on performance which are attributable to combinations of stimuli and reinforcers will be considered first because these stimulus-reinforcer interactions have received the most experimental attention. Two basic experimental designs have been used to demonstrate such stimulus-reinforcer interactions. In the single-cue design, illustrated in Figure 1, groups of subjects receive the factorial combinations of the CSs and the UCSs. If, as in Figure 1, acquisition of a CR occurs more rapidly (or if performance reaches a higher asymptote) with CS 1 than with CS 2 when UCS 1 is the reinforcer, but the reverse is true when UCS 2 is the reinforcer, then a stimulus-reinforcer interaction has been demonstrated. A parallel design, using two instrumental responses and two positive reinforcers, that demonstrates a response-reinforcer interaction should be easy to construct. The occurrence of interactions like that depicted in Figure 1 cannot be attributed solely to a difference between the CSs, or solely to a difference between the UCSs, but must reflect some selective effect on learning or performance.

Stimulus-reinforcer interactions have also been demonstrated in a compound-cue design. Figure 2 illustrates this design, again applied to a classical conditioning procedure. Groups of subjects receive a simultaneous compound of the CSs during conditioning, but the compound CS precedes different UCSs in the two groups. Responses to the separate elements are assessed in a series of "attention" test trials following conditioning. Figure 2 illustrates an interactive effect of stimuli and reinforcers. It would be difficult to construct a parallel to the compound-cue design which would demonstrate a response-reinforcer interaction. However,

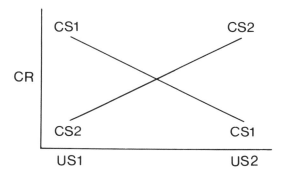

GROUP 1. CS1 – US1
2. CS2 – US1
3. CS1 – US2
4. CS2 – US2

Figure 15.1 Single-cue experimental design used to demonstrate
stimulus-reinforcer interactions. Groups of subjects
receive the factorial combinations of the two CSs and
the two UCSs. If acquisition of a CR occurs more
rapidly (or performance reaches a higher asymptote)
with CS 1 than with CS 2 when UCS 1 is the reinforcer,
but the reverse is true when UCS 2 is the reinforcer
(as in this hypothetical example), the outcome is
called a stimulus-reinforcer interaction.

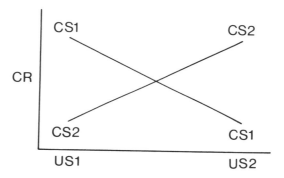

TRAIN   TEST

GROUP 1. $(CS1\ CS2)-US1$ CS1, CS2

2. $(CS1\ CS2)-US2$ CS1, CS2

Figure 15.2   Compound-cue experimental design used to demonstrate stimulus-reinforcer interactions. Two groups of subjects receive a simultaneous compound of the two CSs during conditioning, but the compound is paired with different UCSs for the two groups. The response to the separate stimulus elements is assessed in a series of test trials following conditioning. If there is more test responding to CS 1 than to CS 2 when UCS 1 is the reinforcer, but the reverse is true when UCS 2 is the reinforcer (as in this hypothetical example), the outcome is called a stimulus-reinforcer interaction.

a concurrent schedule procedure in which one reinforcer was contingent upon either of two topographically different responses in one
condition, but a different reinforcer was contingent upon either
response in another condition, would be parallel to the design illustrated in Figure 2 in some respects.

The compound-cue design is attractive because it allows
within-subject comparisons. On the other hand, because differences
in test responding to the two cues might result from some emergent
property of the compound conditioning procedure, it does not provide as direct a measure of the relative conditionability of the
cues as does the single-cue design.

Methodology for Stimulus-Reinforcer Interactions

Suppose that a stimulus-reinforcer interaction has been
demonstrated in either of the experimental designs described earlier.
Although such an outcome might reflect selective associations, there
are several reasonable alternatives to such an interpretation. To
simplify the argument, consider the case in which only CS 1 elicits
the measured response when UCS 1 is the reinforcer, whereas only
CS 2 elicits the response when UCS 2 is the reinforcer. Using a
label coined by Capretta (1961; see also Revusky & Garcia, 1970),
CS 1 is called "relevant" for UCS 1, and CS 2 is called "relevant"
for UCS 2.

Given the associationist bias which has permeated research
on learning, the first question that would be asked about the outcome which has just been described is whether the responding elicited
by the relevant stimuli could be attributed to factors other than
association by contiguity. Such factors include pseudoconditioning
and selective sensitization, which can be a viable account only if
the relevant stimuli initially elicit the to-be-conditioned responses in subthreshold form.

What control procedures would enable researchers to assess
the possibility that the responses evoked by relevant stimuli are
attributable to selective sensitization or pseudoconditioning? One
approach would be to include some of the standard controls for
these non-associative factors in the experimental designs described
earlier. For example, to the design shown in Figure 1 could be
added four groups, each receiving one CS-UCS combination, but with
CS and UCS unpaired. In some situations, the zero-contingency
procedure (Rescorla, 1967) would seem to be the best control for non-
associative effects, whereas in others, for example with response

systems for which there may be little conditioning to background cues and hence little inhibitory conditioning when the CS is negatively correlated with the UCS, a negative contingency between CS and UCS would be more appropriate (see Ch. 3). In cases where the relevant stimulus evokes a CR after as few as one or two trials, it may be more appropriate to demonstrate that the rate of acquisition of the CR decreases as the CS-UCS interval increases.

Recently, Rescorla and Holland (1976; see also Rescorla, 1978) have proposed a control procedure specifically designed to assess alternatives to selective association as an interpretation of stimulus-reinforcer interactions. In their design, all subjects receive both UCSs, but groups differ in the UCS that is paired with the CS. For example, in the compound-cue design, Group 1 would receive the CS 1—CS 2 compound paired with UCS 1, and UCS 2 would occur between trials. Group 2 would receive pairings of the compound CS with UCS 2, with UCS 1 occurring between trials. If the addition of the second UCS during the intertrial interval has no effect on subsequent responses to CS 1 and CS 2 during the test, i.e., the same interaction is observed, then selective, non-associative effects, which should be equivalent for the two groups, can be ruled out. Rescorla and his colleagues (Rescorla, 1978; Rescorla & Furrow, 1977) have used this design to demonstrate that excitatory associations are formed more rapidly between paired, physically similar stimuli than between physically dissimilar ones.

None of the aforementioned control procedures bears on another possibility: that an interaction like that shown in Figure 1 results from non-selective associations superimposed on different initial responses to the two CSs. Kandel (1976; cf. Konorski, 1948; LoLordo, 1978; Shettleworth, 1978a; Thompson, 1976) has recently suggested that a CS may elicit in subthreshold form all the responses that can subsequently be conditioned to it, and that various UCSs may facilitate some of these responses more than others, even leading to a motor CR. This view implies that constraints arising from combinations of CSs and UCSs should sometimes be observed, given variations in the strength of the neural responses elicited by diverse CSs, along with variations in the magnitude of the facilitative effects of various UCSs on those neural responses. If the facilitative effects of the UCSs on the initial responses to the CSs depend upon pairing CS and UCS, or on a positive contingency between the two events, then the observed interaction would be associative but not selective, because it would not require that the strength of the CS-UCS association grows at different rates in different groups. In principle, the initial responses to the CSs can be readily observed, and the hypothesis that the observed interaction results from differential initial responses to the CSs can be evaluated. However, for many response systems this experiment becomes difficult to perform once we concern ourselves with differential elicitation of some neural precursor of the to-be-conditioned effector responses by the two CSs (see Kandel, 1976; Thompson, 1976).

One more methodological point should be made before the discussion focuses on response-reinforcer interactions. Consider the case in which there is no evidence of conditioning to the less-relevant cue in a design like that in Figure 1. In such cases, the distinction between learning and performance must be made, and it must be recognized that animals might fail to make some response during a CS even though they had learned to associate the CS with a UCS (e.g., Holland, 1977; Riccio & Haroutunian, 1977; Weisman & Dodd, 1978). Holland has suggested several generally useful procedures for determining whether some less-relevant CS has been associated with a UCS. For example, the less-relevant CS could be used as the preconditioned "blocking" stimulus in Kamin's (1969) blocking procedure, or as the first-order CS in a second-order conditioning procedure. If the less-relevant CS promotes second-order conditioning, or if prior conditioning to the less relevant CS blocks conditioning to another added CS, then the organism must have associated the less-relevant CS with the UCS. The critical question remains, however, whether this association was formed more slowly than the association between the relevant CS and the UCS.

Methodology for Response-Reinforcer Interactions

A response-reinforcer interaction will be said to occur only when an experimental design like that in Figure 1 reveals a significant interactive effect of response and reinforcer upon the latency, rate, or pattern of instrumental responding. Outcomes of less complete designs do not reveal selective effects of any sort. For example, suppose our only observations were a high rate of response R 1 when a reinforcer was contingent upon R 1, coupled with a low rate of response R 2 in another group when the same reinforcer was contingent upon R 2. In this case, it may be that R 2 is not a response whose frequency can be increased instrumentally, regardless of the reinforcer used (see Dunham, 1977).

Once a response-reinforcer interaction has been demonstrated, we can ask whether it is the result of selective associations between responses and reinforcers, or of some other factor. Alternatives to selective association between responses and reinforcers are in many cases formally similar to the alternatives to selective associations between stimuli and reinforcers that were discussed earlier.

Suppose that four groups of animals receive the following treatments: (1) positive reinforcer S 1 contingent on instrumental response R 1; (2) S 1 contingent on R 2; (3) S 2 contingent on R 1; and (4) S 2 contingent on R 2. Suppose further that the to-be-reinforced responses have similar operant levels in the various groups, and that introduction of the response-reinforcer contingency results in a significant increase in response rate only in Groups 1 and 4. Although selective associations between responses and reinforcers could yield this result, it could also arise from some non-associative effect, i.e., one that does not require a contingency between the instrumental responses and reinforcers. For example, suppose that, in the absence of any contingencies between responses and reinforcers, the effect of presentations of S 1 is to increase the probability of R 1, and the effect of S 2 is to increase the probability of R 2. In such a case, the increase in response rates observed in Groups 1 and 4 could not be attributed to the acquisition of response-reinforcer associations in those groups.

Several investigators (e.g., Boakes, Poli, Lockwood, & Goodall, 1978; Jenkins & Moore, 1973; Moore, 1973; Sevenster, 1973) have asserted that Pavlovian conditioning of various responses to situational cues or features of the response manipulanda can result in a response-reinforcer interaction like the one in our example, if the form of the Pavlovian CRs interacts in appropriate ways with the form of the required instrumental responses. For example, if the CR 1 that becomes established when S 1 is repeatedly presented is compatible with the required R 1 for Group 1 but incompatible with the required R 2 for Group 2, and if the CR 2 that is acquired when S 2 is repeatedly presented is incompatible with the required R 1 for Group 3 but compatible with the required R 2 for Group 4, then the introduction of response-reinforcer contingencies would produce increases in instrumental response rates only in Groups 1 and 4, and this would occur even in the absence of acquired associations between responses and reinforcers.

Several sorts of control groups can be used to determine whether such direct, rather than response-contingent, effects of the reinforcers are responsible for an observed response-reinforcer interaction. These control procedures include systematic study of the effects of freely presented reinforcers upon various responses, as well as investigation of the delay-of-reinforcement gradient for various response-reinforcer pairs.

Another account of response-reinforcer interactions that does not depend upon selective associations maintains that the occurrence of a given reinforcer somehow causes the organism to attend to the internal stimuli correlated with, or to feedback from, a particular

response, making that response more easily associable with any reinforcer which is contingent upon it. This account is analogous
to an account of stimulus-reinforcer interactions described earlier
and can be evaluated by means of an analogous design, one in which
all subjects receive both reinforcers but differ in terms of the
reinforcer which is permitted to follow the instrumental response.

The final methodological question which must be asked about
response-reinforcer interactions like our example is: "Does the
failure of R 1 (R 2) to increase in rate when it is followed by
S 2 (S 1) imply that no R 1→S 2 (R 2→S 1) association has been
formed?" The answer is clearly "no." Suppose S 2 is contingent
upon occurrences of R 1, but a Pavlovian CR 2, which is physically
incompatible with R 1, is conditioned either to the experimental
situation or to the beginning of R 1. Then the rate of R 1 may
fail to increase even though R 1 has been associated with S 2. For
example, Shettleworth (1975) found that contingent food reinforcement failed to increase the rate of face washing in golden hamsters.
Furthermore, the average duration of bouts of face washing decreased
during this time, and increased again when food reinforcement was
omitted. Shettleworth argued that this decrease resulted from the
classical conditioning of the anticipation of food to the beginning
of face washing, indicating that the response and reinforcer had
been associated (cf. Shettleworth, 1978b).

As Shettleworth's experiment suggests, a study which fails to
demonstrate an increase in the frequency of a response R 1 upon
which a reinforcer S 2 is contingent may nonetheless provide other
evidence that the response and reinforcer have been associated. In
addition, there should be indirect methods of determining whether a
response-reinforcer association has been formed which are analogous
to those used to assess associations between stimuli and reinforcers.
For example, a blocking procedure could be used to determine whether
R 1 had been associated with S 2 in our example. Following a treatment in which S 2 was presented immediately after occurrences of
R 1, a brief exteroceptive stimulus would be presented during a
delay between R 1 and S 2. If the exteroceptive stimulus was a
less effective conditioned reinforcer after such a treatment than
after a control treatment, then it could be argued that a prior
association between R 1 and S 2 blocked conditioning to the added,
exteroceptive stimulus. Again, as in the case of stimulus-
reinforcer interactions, the critical question from an associative
viewpoint is whether associations between relevant responses and
reinforcers are formed more rapidly than associations between less-
relevant responses and reinforcers.

Several examples of stimulus-reinforcer and response-reinforcer

interactions will now be analyzed in light of the variety of
accounts which have just been offered.

STIMULUS-REINFORCER AND RESPONSE-REINFORCER INTERACTIONS

Illustrative Stimulus-Reinforcer Interactions

An experiment by Garcia and Koelling (1966) was largely responsible for arousing interest in stimulus-reinforcer interactions
and in the possibility of selective associations. Garcia and Koelling found that when a compound CS consisting of a gustatory stimulus and audio-visual stimulation was paired with visceral upset,
only the gustatory stimulus subsequently controlled passive avoidance of drinking. On the other hand, when the compound stimulus
was paired with electric shock, only the audio-visual stimulation
controlled avoidance of drinking (see Domjan & Wilson, 1972).

Garcia and Koelling offered two possible accounts of their
data. First, they suggested that common elements in the temporal-
intensity patterns of stimulation may have facilitated cross-modal
generalization from reinforcer to cue in two of the four groups,
but not in the other two. Audio-visual stimulation and pain presumably have relatively sudden onsets and terminations, whereas
taste and illness grow and diminish gradually. Thus, there should
be considerable stimulus generalization of the response to shock
to auditory and visual cues, and of illness to gustatory cues, but
little generalization in the other two combinations. This hypothesis, which implies that pairings of CSs and UCSs are not a necessary condition of the stimulus-reinforcer interaction, was not
favored by Garcia and Koelling (1966), who seemed to prefer a
selective, associative interpretation. They maintained that:

>   Natural selection may have favored mecha-
>   nisms which associate gustatory and ol-
>   factory cues with internal discomfort,
>   since the chemical receptors sample the
>   materials soon to be incorporated into the
>   internal environment (p. 124).

This view was based on the notion of adaptive specializations in
learning, namely that event-specific associative mechanisms evolved
to solve particular problems for a species (see Rozin & Kalat, 1971).

The question whether the stimulus-reinforcer interaction
observed by Garcia and Koelling (1966), Domjan and Wilson (1972),

and others is an associative effect has been hotly debated recently
(e.g., Bitterman, 1975, 1976; Garcia, 1978, Mitchell, 1977, 1978;
Mitchell, Scott, & Mitchell, 1977; Revusky, 1977b, 1978; Smith,
1978). LoLordo (1978) argued that illness-based taste aversions
and shock-based aversions to auditory and visual cues are associative effects, but that it has not been ascertained that these associative effects are selective in the sense that taste-illness and
tone-shock associations grow more rapidly than taste-shock and tone-
illness associations.

One alternative to a selective, associative account (Rescorla
& Holland, 1976) is that the first UCS presented determines which
stimulus modality will capture the rat's attention on subsequent
trials and so be most strongly associated with whatever UCS is presented on those trials. The experimental design proposed by
Rescorla and Holland, in which all groups receive both UCSs, would
permit evaluation of this possibility.

A second possibility is that a non-selective, associative
mechanism produces rapid acquisition of aversion to tastes paired
with illness and tones paired with electric shock (and not in the
other two pairs) because tastes (but not tones) initially elicit,
in subthreshold form, the same aversion reaction evoked by LiCl,
whereas tones (but not tastes) initially elicit the same aversion
reaction evoked by electric shock. Thus, even though associations
might grow at the same rate in all CS-UCS pairs, the CSs that began
the experiment "closer to threshold" for particular effector CRs
would, nonetheless, evoke those CRs first. The neural responses
to LiCl and electric shock that mediate their aversiveness are not
well understood. Consequently, we do not know where to look for the
different initial neural responses to the gustatory and audio-
visual CSs. Such responses should form the basis of stimulus-
reinforcer interactions like that observed by Garcia and Koelling,
according to the non-selective, associative account just described.
Thus, such an account remains a viable alternative to a selective,
associative interpretation.

The final question to be asked about the stimulus-reinforcer
interaction observed in the conditioned taste aversion paradigm
concerns the lack of responding to the less-relevant stimuli. Do
rats fail to suppress drinking in the presence of those stimuli
because they were not associated with the UCSs, or because the
associations were somehow not manifested in performance? Because
we already know that a variety of exteroceptive cues can be associated with illness (see Best, Best, & Henggeler, 1977; Riccio &
Haroutunian, 1977; Willner, 1978), and that taste cues can be associated with electric shock (Best et al., 1977; Krane & Wagner, 1975),

the question becomes "how rapidly are the associations between the less-relevant cues and the UCSs formed, relative to the associations between the relevant cues and the UCSs?"

To sum up the discussion of stimulus-reinforcer interactions in the taste aversion paradigm, it seems clear that the responses conditioned to the relevant stimuli do reflect the formation of associations. Whether the stimulus-reinforcer interaction reflects differences in the rates of growth of associations within relevant CS-UCS pairs and less-relevant pairs, or differences in the initial tendencies of various CSs to evoke the to-be-conditioned responses, or both of these phenomena, is unclear. A similar analysis will now be applied to a second area in which stimulus-reinforcer interactions have recently been demonstrated.

Recently, Testa (1974) maintained that the similarity of a paired CS and UCS can influence the rate at which an association between the two events is formed. Testa (1975) provided data compatible with this assertion. In a single-cue design, the similarity of the spatial loci and temporal-intensity patterns of visual CSs and air-blast UCSs significantly affected the rate of acquisition of conditioned suppression by rats when CSs and UCSs were paired. As Figure 3 illustrates, acquisition was faster when CS and UCS were similar than when they were dissimilar. As Testa noted, the lack of control groups in his experiment leaves open the possibility that the outcomes are attributable to pseudoconditioning or stimulus generalization rather than to modulation of the rate of growth of contiguity-based associations between CSs and UCSs by the similarity of those events. However, recent experiments by Rescorla and Furrow (1977) are not subject to such interpretations. These studies examined the effects of similarity in the context of second-order conditioning, thereby allowing the experimenters considerable flexibility to select events. In two experiments with rats, similar events were defined as those in the same sensory modality, either auditory or visual, whereas in a study with pigeons all stimuli were visual, and similar events were defined as those on the same dimension, either wavelength or line tilt.

All three experiments used variations of an experimental design in which all subjects received during the second-order conditioning phase two UCSs that were, in this case, both first-order CSs. The various groups thus differed in terms of which first-order CS was paired with a second-order CS, and which first-order CS was presented alone. Figure 4 provides a representative experimental design. Some animals received pairings of similar events during the second-order conditioning phase; others received pairings of dissimilar events. Conditioned responses, whether conditioned pecking by pigeons, conditioned approach to a food hopper by rats, or

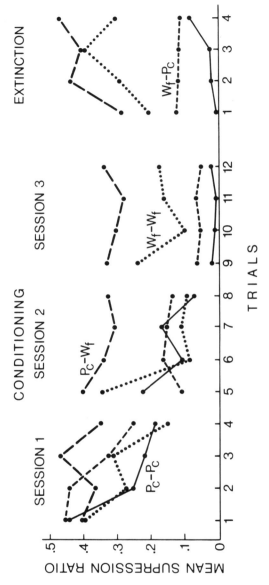

Figure 15.3   Mean suppression ratios [= number of lever press responses during
CS/(number of responses during CS + number of responses during a
comparable pre-CS period)] for four groups of rats in an experiment
by Testa (1975). The groups received the four possible combinations
of a square wave visual CS located at the floor (Wf) or a pulsed
visual CS located at the ceiling (Pc) and a square wave air blast
UCS originating from the floor (Wf) or a pulsed air blast UCS
originating at the ceiling (Pc).

PHASE

	FIRST-ORDER	SECOND-ORDER

GROUP 1. $T_1 \rightarrow US,\ L_1 \rightarrow US,\ T_2 -$ $T_2 \rightarrow T_1,\ L_1 -$

2. $T_1 \rightarrow US,\ L_1 \rightarrow US,\ T_2 -$ $T_2 \rightarrow L_1,\ T_1 -$

3. $T_1 \rightarrow US,\ L_1 \rightarrow US,\ L_2 -$ $L_2 \rightarrow T_1,\ L_1 -$

4. $T_1 \rightarrow US,\ L_1 \rightarrow US,\ L_2 -$ $L_2 \rightarrow L_1,\ T_1 -$

Figure 15.4   Design of an experiment to demonstrate the effects of stimulus similarity on the acquisition of a second-order conditioned response, redrawn from Rescorla and Furrow (1977). During first-order conditioning, tone and light CSs (T_1, L_1) are separately paired with the UCS, but the stimulus that is to be used as the second-order CS (either T_2 or L_2) is not reinforced. In the second-order conditioning phase, Groups 1 and 4 receive pairings of similar second- and first-order stimuli, i.e., those from the same modality (T_2-T_1 or L_2-L_1). Groups 2 and 3 receive pairings of dissimilar events (T_2-L_1 or L_2-T_1). All groups receive the other first-order CS between trials, so that all receive equal exposure to both first-order CSs during the second-order conditioning phase.

conditioned suppression of food-reinforced behavior in rats, were acquired more rapidly when the first- and second-order CSs were similar than when they were dissimilar.

The design of these experiments rules out a non-associative account of the observed interaction as well as the possibility that presentation of a particular first-order CS simply increased attention to similar events, thus leading to rapid acquisition of CRs to those events regardless of which reinforcer followed them. Such an hypothesis would predict equal conditioning in groups given similar and dissimilar CSs and UCSs. Moreover, because the response to the two first-order CSs was the same at the start of second-order conditioning, the faster conditioning with similar first- and second-order CSs than with dissimilar ones cannot be attributed to differences in the initial probability of elicitation of a subthreshold form of the to-be-conditioned response by the two CSs. In the absence of differential rates of growth of associations in the various groups, such a mechanism would result in better second-order conditioning with one second-order CS than with the other, regardless of which first-order CS was paired with the second-order CS.

AVOIDANCE

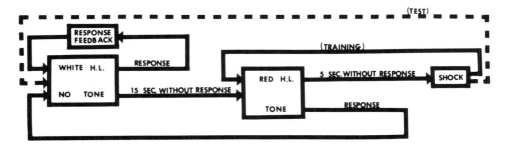

APPETITIVE

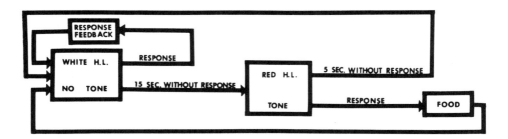

Figure 15.5   Diagramatic representations of the avoidance (top)
              and appetitive training procedures used by Foree and
              LoLordo (1973). Note the similar consequences of
              responding in the presence of "white houselight-no
              tone" in the two procedures.

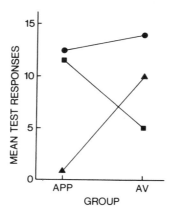

Figure 15.6   Mean test responses to the compound (circles), the red
               houselight (squares), and the tone (triangles) for
               groups of pigeons given appetitive (App) or avoidance
               (Av) training in a study by Foree and LoLordo (1973).

    These experiments demonstrate that selective associations can
occur, i.e., that the rate of formation of a contiguity-based association between CS and US can be affected by the similarity of those
two events.   Rescorla and Furrow (1977) have noted that little is
understood about the mechanism(s) of such selective associations.

    Another stimulus-reinforcer interaction was discovered by
Foree and LoLordo (1973).   In a design like that in Figure 2, groups
of food-deprived pigeons were trained to depress a foot treadle in
the presence of a compound stimulus consisting of tone and red houselights to avoid electric shock or to obtain grain.   Responding in
the absence of the compound stimulus postponed its next appearance.
The final training procedure is illustrated in Figure 5.

    When the pigeons were responding on more than 75% of the com-
pound stimulus trials but responding infrequently between trials,
the degree to which the compound and each element controlled treadle
pressing was determined.   Twenty test trials each of the compound,
the tone, and the red houselight were presented in random order
with responses during all trials reinforced.   Figure 6 shows that
the compound and the red light exerted strong control over treadle-
pressing in the appetitive test, but the tone exerted very little.
On the other hand, during the test which followed avoidance training,
the compound and the tone controlled much more treadle-pressing than
the light.   This outcome has been replicated by LoLordo and Furrow
(1976), who more nearly equated the feedback for treadle-pressing in

the appetitive and aversive conditions, and by LoLordo, Foree, and Jacobs (see LoLordo, 1978), who used the single-cue design.

None of these outcomes included controls for non-associative factors, largely because non-associative accounts of the discriminated, instrumental behavior of the pigeons seemed implausible. Shapiro, Jacobs, and LoLordo (in LoLordo, 1978) extended the research with pigeons to the classical conditioning paradigm in an attempt to control for sensitization and other non-associative effects. In a compound-cue design, two groups of pigeons received repeated pairings of the auditory-visual compound CS with a UCS. For one group, the UCS was a 3-sec access to mixed grain; for the other, it was a brief electric shock. Two additional groups of birds received uncorrelated presentations of the CS and either food or shock. Three independent observers described any changes in the birds' behavior in response to the CS.

All birds that received pairings of the CS and food came to peck either around the food hopper or on the wall above the hopper during the compound CS. Birds that received pairings of the compound CS and shock typically pranced vigorously and made elaborate head-bobbing movements in response to the compound CS. During a subsequent test session in which red light, tone, and no signal (Blank) trials were interspersed among presentations of the compound CS, the three birds in the appetitive, experimental group responded (pecked) on nearly all presentations of the compound and the red light, but responded infrequently to the tone or on blank trials. The three birds in the aversive, experimental group responded (pranced) most frequently during presentations of the compound, somewhat less frequently during presentations of the tone, and even less frequently on presentations of the red light or on blank trials. Thus the interactive effect of stimuli and reinforcers observed by Foree and LoLordo (1973) was also observed in the absence of any explicit contingencies between responses and reinforcers.

None of the birds in the control groups exhibited any differential responding to the CSs, suggesting that the responding of the experimental groups has an associative basis. However, it is unclear whether selective association had occurred. Two alternative accounts which have been presented earlier are also plausible explanations of these results. The first account maintains that the presentation of food (shock) increases attention to visual (auditory) cues, so that those cues are more strongly associated with whatever UCS follows them. This account could be tested if the experimental design in which each animal receives both UCSs during a session were applied to the classical conditioning procedure used by Shapiro et al. This has not been done, but Foree and

LoLordo (1975) have obtained some data that bear on the account.

Six pigeons were trained to depress a treadle in the presence of a compound discriminative stimulus to obtain food. Once they were responding on 75% of the trials, a punishment contingency was added. Each intertrial response produced a brief electric shock. Shock intensity increased gradually until the shocks evoked consistent reactions. In a subsequent test, all birds pressed the treadle on most presentations of the compound and the red light but rarely pressed it during the tone. The presence of shock neither attenuated visual control nor increased auditory control of the treadle press response relative to the outcome of the appetitive procedure used by Foree and LoLordo (1973). This result suggests that the auditory dominance observed when the compound CS was paired with shock did not simply result from a shock-induced increase in attention to auditory cues, which should have resulted in auditory control of food-reinforced treadle-pressing.

On the other hand, Foree (1974) obtained data which are compatible with the assertion that presentation of food simply increases attention to visual cues. In a concurrent schedule, pigeons were required to peck a key to receive intermittent presentations of food and also to depress a treadle in order to avoid shock whenever the compound discriminative stimulus was presented. In a test that followed the acquisition of avoidance responding, most of the pigeons depressed the treadle on an equal, and large, proportion of auditory and visual trials. This outcome is compatible with the assertion that the presence of food in the situation increases the bird's attention to red light, yielding a stronger association between red light and shock than would have occurred otherwise. Perhaps this same mechanism produced the visual control observed in Foree and LoLordo's (1973) appetitive condition. In any case, further tests must be made of the hypothesis that attentional biases induced by UCS presentation account for the results observed by Foree and LoLordo (1973).

Finally, it is not clear whether the stimulus-reinforcer interaction under discussion is the result of a greater initial tendency of tone than red light to evoke some subthreshold aversive reaction, and/or a greater initial tendency of red light than tone to elicit some subthreshold appetitive reaction. If such initial reactions did occur, the observed stimulus-reinforcer interaction would follow, even if the acquired associative strength superimposed on this base grew non-selectively. The unconditioned effects of the red houselight and tone have not been thoroughly assessed, thus the possibility that the observed effects depend upon differential initial reactions to tone and red light cannot be ruled out.

     To summarize, the stimulus-reinforcer interaction observed by
Foree and LoLordo (1973) is an associative effect. Whether this
associative effect is selective, i.e., reflects faster growth of
a tone-shock than of a red light-shock association and faster
growth of a red light-food than of a tone-food association, is yet
to be determined.

Illustrative Response-Reinforcer Interactions

     There have been only two published demonstrations of response-
reinforcer interactions (Sevenster, 1968, 1973; Shettleworth, 1973,
1975, 1978b), and only these demonstrations will be analyzed here.
There have been other provocative studies that suggest the possibility of additional response-reinforcer interactions, but these
studies do not include data from all four groups in a design like
that in Figure 1 (Bolles, 1970, 1972). Additional experiments
have demonstrated that some instrumental responses are much more
affected by motivational variables than others, with the reinforcer
or punisher held constant (Black & Young, 1972; Walters & Herring,
1978; Young & Black, 1977).

     Sevenster (1968, 1973) demonstrated that for male stickleback
fish either a brief opportunity to court a female or a brief opportunity to fight another male effectively reinforced swimming through
a ring, whereas only the opportunity to fight reinforced the instrumental response of biting a transparent rod with a green tip.
Figure 7 illustrates, for each response-reinforcer combination, the
distribution of intervals between reinforcement and the next instrumental response when a continuous reinforcement schedule was
in effect. The distribution of rod-bites when courtship was the
reinforcer contains very few short intervals and is markedly different from the other distributions. Moreover, the mean of the
former distribution did not decrease over repeated sessions.

     Additional research by Sevenster revealed that: (1) When
only one instance of rod-biting in three produced the opportunity
to court, the response latency following a reinforced response was
much longer than the latency following a non-reinforced response,
and (2) a brief, response-independent exposure to a female resulted in a substantial increase in the latency of the next response.
These outcomes suggest that courtship had a suppressive effect on
rod biting. Observation of the stickleback during the long intervals between reinforcement and rod biting cast some light on the
nature of the suppressive effect. The fish tended to approach the
rod soon after courting but then circled "around it with zig-zag

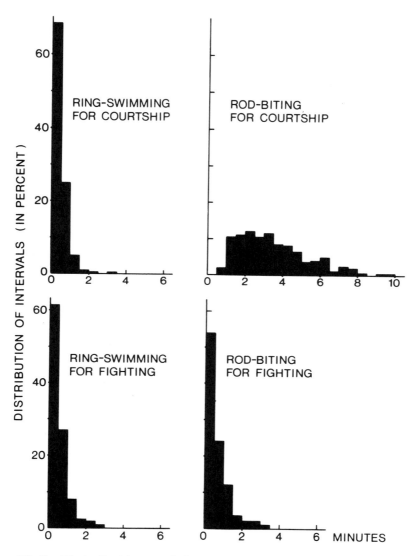

Figure 15.7   Distributions of intervals between the end of the
              reinforcing activity and the next instrumental re-
              sponse for the four combinations of instrumental
              responses and reinforcers studied by Sevenster (1973).

like jumps and often with open mouth, sometimes making snapping movements at the tip or softly touching it" (Sevenster, 1973, p. 277). Then the male might swim off, or with a probability much less than 0.5, he might bite the rod.

Sevenster maintained that the tendency to bite the rod had been increased by the contingency between rod biting and courtship, using the relatively high rate of rod biting early in extinction following continuous reinforcement as evidence for this assertion. He further argued that conditioned rod biting was suppressed because the rod had become associated with the female and thus was being treated like a "dummy" female, i.e., approached with zig-zag jumps. Sevenster obtained one additional bit of evidence that a classically conditioned, directed response was competing with the instrumental response. When care was taken to prevent association of rod biting with presentation of the female, the effect of brief presentations of the female on rod biting (that was being reinforced by the opportunity to fight) was much less disruptive, and the probability that approach to the rod would culminate in a bite was 0.75.

To summarize, there is no strong support for the assertion that the response-reinforcer interaction observed by Sevenster (1968, 1973) reflects the selective formation of associations. Sevenster's account of the low rate of rod biting when that response resulted in the opportunity to court a female is compatible with the assertion that response-reinforcer associations formed equally rapidly in the four conditions. On the other hand, Sevenster's account is also compatible with the possibility of selective associations, and he has presented no unequivocal evidence against the contention that the association between biting the rod and the opportunity to court a female was learned especially slowly.

Shettleworth (1973, 1975, 1978b) examined the effects of response-contingent food and response-contingent shock upon various activities of golden hamsters. This account will focus on three of those activities, face washing, open rearing or standing on hind legs with both forepaws off floor and walls, and scrabbling or scraping with forepaws against a wall while standing erect. The reinforcer was presented whenever the hamster accumulated 2 sec of the instrumental response.

Food reinforcement of scrabbling or open rearing resulted in marked increases in the amount of time the hamster spent on those activities, but food reinforcement of face washing resulted in only a very small and delayed increase in that response. Moreover, unlike scrabbling or open rearing, the mean length of bouts of face washing declined markedly when face washing produced food, and increased again in extinction.

When 2 sec of scrabbling resulted in shock, the time spent scrabbling declined precipitously, i.e., scrabbling was affected by contingent shock as well as by contingent food. If the differential effects of food reinforcement upon open rearing and face washing simply reflect the relative modifiability of the two responses by their consequences, then response-contingent shock should have suppressed open rearing more than face washing. However, the reverse was true. The total time spent face washing, as well as the mean bout length of that response, decreased when shock was contingent upon face washing. Comparable measures of open rearing were less affected when that response was punished. Thus a 2 X 2 design including face washing vs. open rearing combined factorially with food vs. shock revealed a response-reinforcer interaction.

Does the response-reinforcer interaction studied by Shettleworth reflect the selective formation of associations? Since response-independent food depressed both scrabbling and open rearing (Shettleworth, 1975), the increases in food-reinforced scrabbling and open rearing do seem to depend on the response-reinforcer contingencies. That is, these increases are associative effects. Despite the failure of response-contingent food to increase the amount of time spent face washing, it is clear that this response also was associated with food. First, hamsters which received food pellets when they washed their faces picked up the pellets sooner than hamsters which received response-independent pellets, suggesting that the former were able to anticipate pellet delivery. Second, the reduction in the bout length of face washing during conditioning suggested to Shettleworth that anticipation of food was classically conditioned to the initiation of face washing and, thus, competed with its continuation. Shettleworth did not identify the classically conditioned response with approach to the feeder, which also would have depressed the bout lengths of food-reinforced scrabbling or open rearing, but rather with a conditioned state of anticipation of food. Her suggestion that only responses which the hamster makes when anticipating food can be performed at a high rate for food reinforcement is roughly analogous to Bolles' claim that only species-specific defense reactions can be rapidly acquired as avoidance responses. Thus face washing was associated with food in Shettleworth's experiment, and there is no reason to believe that this association grew more slowly than the associations between open rearing or scrabbling and food.

Next, consider the three groups that received electric shock. The suppressive effects of punishment on scrabbling and face washing seem to be associative effects since response-independent shock did not suppress either response (Shettleworth, 1978b). Shettleworth maintained that open rearing was also associated with shock because in additional experiments, punishment of every bout of open rearing resulted in marked suppression of that response. Moreover, in one experiment, the mean bout length of open rearing remained below its pre-punishment value after punishment had been removed. Since open rearing was facilitated during a signal for response-independent shock, Shettleworth speculated that the ineffectiveness of response-contingent shock in suppressing open rearing, as compared with face washing and scrabbling, may be the result of an interaction between a suppressive effect arising from the response-reinforcer association and a facilitative effect arising from the association of the situation with shock. Again, there seems to be no reason to postulate differential rates of growth of the associations between the various responses and shock.

In summary, the response-reinforcer interaction observed by Shettleworth occurred despite the formation of associations between the instrumental response and the reinforcer in all groups. As in Sevenster's demonstration, responses arising from stimulus-reinforcer associations played an important role in preventing the response-reinforcer associations from being manifested in performance in certain cases. However, Shettleworth (1978b) notes that the nature of the interaction between behaviors arising from stimulus-reinforcer and response-reinforcer associations in her experimental situation is not yet well understood (see Boakes, 1978).

IMPLICATIONS OF STIMULUS-REINFORCER AND RESPONSE-REINFORCER INTERACTIONS

The study of constraints on learning and performance, including stimulus-reinforcer and response-reinforcer interactions, is still in its early stages. In addition to a selective, associative account, analysis of these interactions has suggested a variety of other associative interpretations as well as some non-associative ones. The three stimulus-reinforcer interactions described earlier are all associative effects, although only Rescorla and Furrow's (1977) experiments on similarity conclusively demonstrate selective association, i.e., that the rate at which the association between a CS and a UCS grows depends upon a single parameter which is characteristic for that combination of events. Neither of the response-reinforcer interactions described earlier clearly reflects the selective formation of associations. Instead, they seem to be the result of interactions between the behavior generated by the instrumental response-reinforcer contingency and behaviors engendered by associations between various other stimuli and the reinforcer.

Interactions between response-reinforcer and stimulus-reinforcer
associations will certainly be diverse, with the behavioral outcomes depending upon a number of contextual variables that are just
beginning to be studied (see Boakes, 1978).

     Suppose that future research reveals a number of situations in
which selective associations do occur. What implications would such
a state of affairs have for the laws of learning? Does behavior
in response to so-called relevant stimuli, or those which enter into
selective associations, obey special laws?

     Revusky (1977a) has reviewed many taste aversion experiments
that bear on this question. He concluded that nothing in the literature on conditioned taste aversions demands the postulation of more
than parametric differences between taste-illness associations and
those formed in other experiments on classical conditioning. Even
long-delay learning poses no special problems because the form of
the delay-of-reinforcement gradient is the same in taste aversion
learning and in other paradigms (e.g., Revusky, 1968).

     LoLordo, Jacobs, and Foree (1975, see LoLordo, 1978) have
suggested that responses to the relevant stimuli from the experiment
by Foree and LoLordo (1973), red light when food is the reinforcer
and tone when shock avoidance is the reinforcer, are not affected
like responses to arbitrarily selected stimuli in a standard paradigm for the study of stimulus control, the blocking paradigm.
In two experiments prior conditioning to another discriminative
stimulus failed to block conditioning to the relevant stimulus when
the latter was added to the former, and reinforcement continued as
before. Since blocking has been widely observed in classical conditioning and instrumental learning, LoLordo et al. argued that
control by relevant stimuli may obey special laws, as only such
control is impervious to blocking. However, a recent experiment
by Hall, Mackintosh, Goodall, and dal Martello (1977) suggests that
it will be difficult to block control by any very salient stimulus.
Thus, there is no firm evidence that the tone-shock or red light-
food associations studied by Foree and LoLordo (1973) obey special
laws.

     Researchers seeking implications of selective vs. other asso-
ciations might look for different products of learning in the two
cases. Along these lines, Garcia, Hankins, and Rusiniak (1974)
have suggested that the product of at least one putatively selective
association, that between tastes and illness, is different from the
products of associations between other sorts of exteroceptive
stimuli and other aversive events. They argued that whereas
pairings of a tone or light CS with shock cause the rats to treat
the CS as a signal for shock, taste-illness pairings render the

taste unpleasant. However, this difference may not be essential because rejection of an unpleasant taste is a form of withdrawal from a localized cue, and rats and pigeons have been shown to withdraw from localized visual signals for shock (e.g. Biederman, D'Amato, & Keller, 1964; Karpicke, Christoph, Peterson, & Hearst, 1977). Thus, a localized visual cue which has been paired with shock should be called unpleasant as well. In any case, the implications of selective associations for the product of learning are unclear (see Rescorla & Furrow, 1977).

The study of constraints on learning and performance is still at the stage of accumulation of isolated entries in a catalogue, but signs of hope that it will advance beyond this stage can be found in the increased scope of learning theorists' vision that has accompanied the study of the examples described earlier in this chapter and many others. Several potentially general accounts of stimulus-reinforcer and response-reinforcer interactions have been proposed, and our understanding of the mechanisms of association and performance will almost certainly increase as we evaluate these proposals. It may be well, in closing, to take note, at least, of another set of constraints on learning not dealt with in this chapter but alluded to in previous ones. These are the constraints of taxonomic status or phylogeny. Research on learning has been concentrated on the search for general laws that transcend specific choices, not only of stimuli, responses, and reinforcers, but of species as well. The methodological and conceptual problems presented by the questions of generality within and between species are similar in many respects, as are our reasons for wanting to solve them. They have been reviewed recently by Bitterman (1975).

REFERENCES

Best, P. J., Best, M. R., & Henggeler, S. The contribution of environmental noningestive cues in conditioning with aversive internal consequences. In L. M. Barker, M. R. Best, & M. Domjan (Eds.), LEARNING MECHANISMS IN FOOD SELECTION. Waco, Texas: Baylor University Press, 1977.

Biederman, G. B., D'Amato, M. R., & Keller, D. M. Facilitation of discriminated avoidance learning by dissociation of CS and manipulandum. PSYCHONOMIC SCIENCE, 1964, 1, 229-230.

Bitterman, M. E. The comparative analysis of learning. SCIENCE, 1975, 188, 699-709.

Bitterman, M. E. Flavor aversion studies. SCIENCE, 1976, 192, 266-267.

Black, A. H., & Young, G. A. Constraints on the operant conditioning of drinking. In R. M. Gilbert & J. R. Millenson (Eds.), REIN- FORCEMENT: BEHAVIORAL ANALYSES. New York: Academic Press, 1972.

Boakes, R. A. Interactions between type 1 and type 2 processes involving positive reinforcement. In A. Dickinson & R. A. Boakes (Eds.), MECHANISMS OF LEARNING AND MOTIVATION: A MEMORIAL TO JERZY KONORSKI. Hillsdale, N. J.: Lawrence Erlbaum Associates, 1978.

Boakes, R. A., Poli, M., Lockwood, M. J., & Goodall, G. A study of misbehavior: token reinforcement in the rat. JOURNAL OF THE EXPERIMENTAL ANALYSIS OF BEHAVIOR, 1978, 29, 115-134.

Bolles, R. C. Species-specific defense reactions and avoidance learning. PSYCHOLOGICAL REVIEW, 1970, 77, 32-48.

Bolles, R. C. Reinforcement, expectancy, and learning. PSYCHOLOGI- CAL REVIEW, 1972, 79, 394-409.

Capretta, P. An experimental modification of food preferences in chickens. JOURNAL OF COMPARATIVE AND PHYSIOLOGICAL PSYCHOLOGY, 1961, 54, 238-242.

Dobrzecka, C., & Konorski, J. Qualitative versus directional cues in differential conditioning. I. Left leg-right leg differ- entiation to cues of a mixed character. ACTA BIOLOGIAE EXPERIMENTALIS (Warsaw), 1967, 27, 63-68.

Dobrzecka, C., & Konorski, J. Qualitative versus directional cues in differential conditioning. IV. Right leg-left leg differentiation to non-directional cues. ACTA BIOLOGIAE EXPERIMENTALIS (Warsaw), 1968, 28, 61-69.

Dobrzecka, C., Szwejkowska, G., & Konorski, J. Qualitative versus directional cues in two forms of differentiation. SCIENCE, 1966, 153, 87-89.

Domjan, M., & Wilson, N. E. Specificity of cue to consequence in aversion learning in rats. PSYCHONOMIC SCIENCE, 1972, 26, 143-145.

Dunham, P. The nature of reinforcing stimuli. In W. K. Honig & J. E. R. Staddon (Eds.), HANDBOOK OF OPERANT BEHAVIOR. Englewood Cliffs, N. J.: Prentice-Hall, Inc., 1977.

Foree, D. D. Stimulus-reinforcer interactions in the pigeon. Unpublished Doctoral dissertation, University of North Carolina, Chapel Hill, 1974.

Foree, D. D., & LoLordo, V. M. Attention in the pigeon: The dif- ferential effects of food-getting vs. shock-avoidance pro- cedures. JOURNAL OF COMPARATIVE AND PHYSIOLOGICAL PSYCHOLOGY, 1973, 85, 551-558.

Foree, D. D., & LoLordo, V. M. Stimulus-reinforcer interactions in
    the pigeon: The role of electric shock and the avoidance
    contingency. JOURNAL OF EXPERIMENTAL PSYCHOLOGY: ANIMAL
    BEHAVIOR PROCESSES, 1975, 104, 39-46.

Garcia, J. Mitchell, Scott, and Mitchell are not supported by
    their own data. ANIMAL LEARNING & BEHAVIOR, 1978, 6, 116.

Garcia, J., Hankins, W. G., & Rusiniak, K. W. Behavioural regula-
    tion of the *milieu interne* in man and rat. SCIENCE, 1974,
    185, 823-831.

Garcia, J., & Koelling, R. A. Relation of cue to consequence in
    avoidance learning. PSYCHONOMIC SCIENCE, 1966, 4, 123-124.

Garcia, J., McGowan, B. K., & Green, K. F. Biological constraints
    on conditioning. In A. H. Black & W. F. Prokasy (Eds.),
    CLASSICAL CONDITIONING II: CURRENT RESEARCH AND THEORY. New
    York: Appleton-Century-Crofts, 1972.

Hall, G., Mackintosh, N. J., Goodall, G., & dal Martello, M. Loss
    of control by a less valid or less salient stimulus compounded
    with a better predictor of reinforcement. LEARNING AND
    MOTIVATION, 1977, 8, 145-158.

Hinde, R. A., & Stevenson-Hinde, J. CONSTRAINTS ON LEARNING.
    London: Academic Press, 1973.

Hogan, J. Fighting and reinforcement in the Siamese fighting fish
    (*Betta splendens*). JOURNAL OF COMPARATIVE AND PHYSIOLOGICAL
    PSYCHOLOGY, 1967, 64, 356-359.

Hogan, J. A., Kleist, S., & Hutchings, C. S. L. Display and food as
    reinforcers in the Siamese fighting fish (*Betta splendens*).
    JOURNAL OF COMPARATIVE AND PHYSIOLOGICAL PSYCHOLOGY, 1970, 70,
    351-357.

Holland, P. C. Conditioned stimulus as a determinant of the form
    of the Pavlovian conditioned response. JOURNAL OF EXPERIMENTAL
    PSYCHOLOGY: ANIMAL BEHAVIOR PROCESSES, 1977, 3, 77-104.

Jenkins, H. M., & Moore, B. R. The form of the auto-shaped response
    with food or water reinforcers. JOURNAL OF THE EXPERIMENTAL
    ANALYSIS OF BEHAVIOR, 1973, 20, 163-181.

Kamin, L. J. Predictability, surprise, attention, and conditioning.
    In B. A. Campbell & R. M. Church (Eds.), PUNISHMENT AND
    AVERSIVE BEHAVIOR. New York: Appleton-Century-Crofts, 1969.

Kandel, E. R. CELLULAR BASIS OF BEHAVIOR. San Francisco: W. H.
    Freeman, 1976.

Karpicke, J., Christoph, G., Peterson, G., & Hearst, E. Signal
    location and positive and negative conditioned suppression
    in the rat. JOURNAL OF EXPERIMENTAL PSYCHOLOGY: ANIMAL
    BEHAVIOR PROCESSES, 1977, 3, 105-118.

Konorski, J. CONDITIONED REFLEXES AND NEURON ORGANIZATION. Cam-
    bridge: Cambridge University Press, 1948.

Konorski, J. INTEGRATIVE ACTIVITY OF THE BRAIN. Chicago: Univer-
    sity of Chicago Press, 1967.

Krane, R. V., & Wagner, A. R. Taste aversion learning with delayed
    shock US: Implications for the "generality of the laws of
    learning" JOURNAL OF COMPARATIVE AND PHYSIOLOGICAL PSYCHOLOGY,
    1975, 88, 882-889.

Lawicka, W. The role of the stimulus modality in successive discri-
    mination and differentiation learning. BULLETIN OF THE POLISH
    ACADEMY OF SCIENCES, 1964, 12, 35-38.

Lawicka, W. Differing effectiveness of auditory quality and location
    cues in two forms of differentiation learning. ACTA BIOLOGIAE
    EXPERIMENTALIS (Warsaw), 1969, 29, 83-92.

LoLordo, V. M. Selective associations. In A. Dickinson & R. A.
    Boakes (Eds.), MECHANISMS OF LEARNING AND MOTIVATION: A
    MEMORIAL TO JERZY KONORSKI. Hillsdale, N. J.: Lawrence
    Erlbaum Associates, 1978.

LoLordo, V. M., & Furrow, D. R. Control by the auditory or the
    visual element of a compound discriminative stimulus: Effects
    of feedback. JOURNAL OF THE EXPERIMENTAL ANALYSIS OF BEHAVIOR,
    1976, 25, 251-256.

LoLordo, V. M., Jacobs, W. J., & Foree, D. D. Failure to block
    control by a relevant stimulus. Paper presented at the 1975
    meeting of the Psychonomic Society, Denver, Colorado.

Mitchell, D. Reply to Revusky. ANIMAL LEARNING & BEHAVIOR, 1977,
    5, 321-322.

Mitchell, D. The psychological vs. the ethological rat: Two views
    of the poison avoidance behavior of the rat compared. ANIMAL
    LEARNING & BEHAVIOR, 1978, 6, 121-124.

Mitchell, D., Scott, D. W., & Mitchell, L. K. Attentuated and en-
    hanced neophobia in the taste-aversion "delay of reinforce-
    ment" effect. ANIMAL LEARNING & BEHAVIOR, 1977, 5, 99-102.

Moore, B. R. The role of directed Pavlovian reactions in simple
    instrumental learning in the pigeon. In R. A. Hinde & J.
    Stevenson-Hinde (Eds.), CONSTRAINTS ON LEARNING. New York:
    Academic Press, 1973.

Mowrer, O. H. On the dual nature of learning--a reinterpretation
        of "conditioning" and "problem-solving." HARVARD EDUCATIONAL
        REVIEW, 1947, 17, 102-148.

Rescorla, R. A. Pavlovian conditioning and its proper control pro-
        cedures. PSYCHOLOGICAL REVIEW, 1967, 74, 71-80.

Rescorla, R. A. Some implications of a cognitive perspective on
        Pavlovian conditioning. In S. H. Hulse, H. Fowler, & W. K.
        Honig (Eds.), COGNITIVE PROCESSES IN ANIMAL BEHAVIOR. Hills-
        dale, N. J.: Lawrence Erlbaum Associates, 1978.

Rescorla, R. A., & Furrow, D. R. Stimulus similarity as a determi-
        nant of Pavlovian conditioning. JOURNAL OF EXPERIMENTAL
        PSYCHOLOGY: ANIMAL BEHAVIOR PROCESSES, 1977, 3, 203-215.

Rescorla, R. A., & Holland, P. C. Some behavioral approaches to
        the study of learning. In M. R. Rosenzweig & E. L. Bennett
        (Eds.), NEURAL MECHANISMS OF LEARNING AND MEMORY. Cambridge,
        Massachusetts: The MIT Press, 1976.

Rescorla, R. A., & Wagner, A. R. A theory of Pavlovian conditioning:
        Variations in the effectiveness of reinforcement and non-
        reinforcement. In A. H. Black & W. F. Prokasy (Eds.), CLAS-
        SICAL CONDITIONING II: CURRENT RESEARCH AND THEORY. New
        York: Appleton-Century-Crofts, 1972.

Revusky, S. Aversion to sucrose produced by contingent X-irradia-
        tion: Temporal and dosage parameters. JOURNAL OF COMPARA-
        TIVE AND PHYSIOLOGICAL PSYCHOLOGY, 1968, 65, 17-22.

Revusky, S. Learning as a general process with an emphasis on data
        from feeding experiments. In N. W. Milgram, L. Krames, &
        T. M. Alloway (Eds.), FOOD AVERSION LEARNING. New York:
        Plenum Press, 1977. (a)

Revusky, S. Correction of a paper by Mitchell, Scott, and Mitchell.
        ANIMAL LEARNING & BEHAVIOR, 1977, 5, 320. (b)

Revusky, S. Reply to Mitchell. ANIMAL LEARNING & BEHAVIOR, 1978,
        6, 119-120.

Revusky, S., & Garcia, J. Learned associations over long delays.
        In G. H. Bower (Ed.), THE PSYCHOLOGY OF LEARNING AND MOTIVA-
        TION. Vol. 4. New York: Academic Press, 1970.

Riccio, D. C., & Haroutunian, V. Failure to learn in a taste
        aversion paradigm: Associative or performance deficit?
        BULLETIN OF THE PSYCHONOMIC SOCIETY, 1977, 10, 219-222.

Rozin, P., & Kalat, J. W. Specific hungers and poison avoidance
        as adaptive specializations of learning. PSYCHOLOGICAL RE-
        VIEW, 1971, 78, 459-486.

Segal, E. F. Induction and the provenance of operants. In R. M.
    Gilbert & J. R. Millenson (Eds.), REINFORCEMENT: BEHAVIORAL
    ANALYSES. New York: Academic Press, 1972.

Seligman, M. E. P. On the generality of the laws of learning.
    PSYCHOLOGICAL REVIEW, 1970, 77, 406-418.

Seligman, M. E. P., & Hager, J. L. BIOLOGICAL BOUNDARIES OF LEARN-
    ING. New York: Appleton-Century-Crofts, 1972.

Sevenster, P. Motivation and learning in sticklebacks. In D.
    Ingle (Ed.), THE CENTRAL NERVOUS SYSTEM AND FISH BEHAVIOR.
    Chicago: University of Chicago Press, 1968.

Sevenster, P. Incompatability of response and rewards. In R. A.
    Hinde & J. Stevenson-Hinde (Eds.), CONSTRAINTS ON LEARNING.
    London: Academic Press, 1973.

Shapiro, K., Jacobs, W. J., & LoLordo, V. M. In V. M. LoLordo,
    Selective associations. In A. Dickinson & R. A. Boakes (Eds.),
    MECHANISMS OF LEARNING AND MOTIVATION: A MEMORIAL TO JERZY
    KONORSKI. Hillsdale, N. J.: Lawrence Erlbaum Associates,
    1978.

Shettleworth, S. J. Constraints on learning. ADVANCES IN THE STUDY
    OF BEHAVIOR, 1972, 4, 1-68.

Shettleworth, S. J. Food reinforcement and the organization of
    behavior in golden hamsters. In R. A. Hinde & J. Stevenson-
    Hinde (Eds.), CONSTRAINTS ON LEARNING. London: Academic
    Press, 1973.

Shettleworth, S. J. Reinforcement and the organization of behavior
    in golden hamsters: Hunger, environment, and food reinforce-
    ment. JOURNAL OF EXPERIMENTAL PSYCHOLOGY: ANIMAL BEHAVIOR
    PROCESSES, 1975, 1, 56-87.

Shettleworth, S. J. "Constraints on conditioning" in the writings
    of Konorski. In A. Dickinson & R. A. Boakes (Eds.), MECHA-
    NISMS OF LEARNING AND MOTIVATION: A MEMORIAL TO JERZY
    KONORSKI. Hillsdale, N. J.: Lawrence Erlbaum Associates,
    1978. (a)

Shettleworth, S. J. Reinforcement and the organization of behavior
    in golden hamsters: Punishment of three action patterns.
    LEARNING AND MOTIVATION, 1978, 9, 99-123. (b)

Smith, J. C. Comments on paper by Mitchell, Scott, and Mitchell.
    ANIMAL LEARNING & BEHAVIOR, 1978, 6, 117-118.

Szwejkowska, G. Qualitative versus directional cues in differential
    conditioning. II. Go--no go differentiation to cues of a
    mixed character. ACTA BIOLOGIAE EXPERIMENTALIS (Warsaw),
    1967, 27, 169-175. (a)

Szwejkowska, G. Qualitative versus directional cues in differential
    conditioning. III. The role of qualitative and directional
    cues in differentiation of salivary reflexes. ACTA BIOLOGIAE
    EXPERIMENTALIS (Warsaw), 1967, 27, 413-420. (b)

Testa, T. J. Causal relationships and the acquisition of avoidance
    responses. PSYCHOLOGICAL REVIEW, 1974, 81, 491-505.

Testa, T. J. Effects of similarity of location and temporal inten-
    sity pattern of conditioned and unconditioned stimuli on the
    acquisition of conditioned suppression in rats. JOURNAL OF
    EXPERIMENTAL PSYCHOLOGY: ANIMAL BEHAVIOR PROCESSES, 1975,
    104, 114-121.

Thompson, R. F. The search for the engram. AMERICAN PSYCHOLOGIST,
    1976, 31, 209-227.

Turner, L. H., & Solomon, R. L. Human traumatic avoidance learning:
    Theory and experiments on the operant-respondent distinction
    and failures to learn. PSYCHOLOGICAL MONOGRAPHS, 1962, 76,
    Whole No. 559.

Walters, G. C., & Herring, B. Differential suppression by punish-
    ment of non-consummatory licking and lever-pressing. JOURNAL
    OF EXPERIMENTAL PSYCHOLOGY: ANIMAL BEHAVIOR PROCESSES, 1978,
    4, 170-187.

Weisman, R. G., & Dodd, P. W. D. The study of association: Metho-
    dology and basic phenomena. In A. Dickinson & R. A. Boakes
    (Eds.), MECHANISMS OF LEARNING AND MOTIVATION: A MEMORIAL
    TO JERZY KONORSKI. Hillsdale, N. J.: Lawrence Erlbaum
    Associates, 1978.

Wells, M. J. Learning and movement in octopuses. ANIMAL BEHAVIOR
    SUPPLEMENT, 1964, 1, 115-128.

Willner, J. A. Blocking of a taste aversion by prior pairings of
    exteroceptive stimuli with illness. LEARNING AND MOTIVATION,
    1978, 9, 125-140.

Young, G. A., & Black, A. H. A comparison of operant licking and
    lever pressing in the rat. LEARNING AND MOTIVATION, 1977,
    8, 387-403.